Nonparametric Function Estimation, Modeling, and Simulation

Nonparametric Function Estimation, Modeling, and Simulation

James R. Thompson

Richard A. Tapia

siam®

Society for Industrial and Applied Mathematics • Philadelphia

Library of Congress Cataloging-in-Publication Data

Tapia, Richard A.
Nonparametric function estimation, modeling, and simulation / Richard A. Tapia, James R. Thompson.
p. cm.
Includes bibliographical references and index.
ISBN 0-89871-261-0 : $32.50
1. Estimation theory. 2. Nonparametric statistics. I. Thompson, James R., 1938– . II. Title
QA276.8.T37 1990
519.5'44–dc20 90-10222
CIP

To my mother, Mary Haskins Thompson,
my aunt, Willie T. Jackson,
and to the memory of my aunt, Maycil Haskins Morton
—James R. Thompson

To my mother, Magda Tapia,
and to my aunt, Gloria Casillas
—Richard A. Tapia

Contents

Preface

Every book should have a niche, a special constituency for whom it is written. This book is written for those who wish to use exploratory devices, such as nonparametric density estimation, as a step toward better understanding of a real world process. An emphasis is given to processes which require several characterizing parameters and which may have multidimensional data outputs. Case studies are given which go from the exploration stage to a model to the practical consequences of that model.

Those whose predominant interest is nonparametric density estimation *sui generis*, particularly in one dimension, are referred to the excellent books by Devroye and Gyorfi, Prakasa-Rao, and Silverman. Those who have particular interest in a graphical approach to nonparametric density estimation are referred to the forthcoming monograph of David Scott.

The nonparametric density estimation project at Rice was begun in 1971 and was motivated by an interest at NASA for the identification of crops using satellite data. After NASA successfully landed two people on the Moon in 1969, Congress, which leaves no act of excellence unpunished, cut back drastically on its funding. Scrambling around for ways to apply the enormous technology developed in the years of NASA's existence, someone conceived the idea of using remote sensing as a means of forecasting Soviet grain production. Using pilot studies on American crop lands obtained by aircraft overflights, the project promised to be feasible and relatively inexpensive. The hardware device used (the multispectral scanner) measured intensities of reflectivity in 12 different channels of the visible and near visible spectrum. Detection was done by a likelihood ratio algorithm which assumed Gaussianity. Misclassification probabilities were high, around 15%. The hardware fix proposed by some was to increase the number of channels to 24. Our pilot work indicated that misclassification was due, in large measure, to multimodal signature densities. Naturally, this pathology is intensified, not lessened by increasing the dimensionality of the data.

At Rice, we started looking at one and two dimensional detection algorithms based on nonparametric density estimators including kernel and penalized likelihood techniques. Richard Tapia and I had come to Rice in 1970, and we decided to see what might be done pooling the resources of statistics and constrained optimization theory. Quickly it became possible to lower misclassification probabilities to 5%. This was only the beginning, and Richard Heydorn's group at NASA developed algorithms which worked nicely in classification using one, two, three and four channels from satellite sensors (this group, having achieved spectacular success was, of course, ultimately abolished). It was possible not only to identify crops, but also their condition and maturity. The very difficult boundary closure problem, never completely solved, was noted to be marginal due to the enormous size of *sovkhozes* and *kolkhozes* devoted to the growth of cereals, and also due to the fact that what was really relevant was the total number of pixels of healthy wheat. Boundaries of the fields really were not that important.

In the early 1970's I was giving a talk at a national meeting of statisticians and meteorologists. I noted how, using remote sensing, we had passed the Soviets in the ability to forecast Soviet grain production. At the end of my talk, a very red-faced *refusenik* immigrant to the United States introduced himself as a former bureau chief of Soviet agricultural statistics. He railed against the temerity of the Americans spying on the Soviet farms and my own bad taste in belittling Soviet agricultural statistics, which was well known to be the glory of modern science.

In 1978, I was giving a lecture to around 50 graduate students at the Polish Academy of Sciences dealing with remote sensing algorithms for the purpose of forecasting agricultural production. The slides I showed were all from Kansas (one did not remove Soviet data from NASA and take it anywhere, certainly not behind the Iron Curtain), but when one of the audience breathlessly asked from where the data had come, I could not resist answering, in my best fractured Polish, *"z Ukrainy."* The students cheered.

Both the *refusenik* scientist and the Polish graduate students perceived that the NASA remote sensing project represented something really basic about the balance of power between the Soviets and the Americans. If American technology was such that not only was per person productivity in the agricultural sector an order of magnitude better than that in the Soviet Union, but American scientists could take satellite data and forecast Soviet grain production better than the Soviet bureaucrats on the ground could, then something was pretty much different about the quality of technology in the two countries.

It is conjectured by some that the Soviet system is coming to an end, largely due to a thousand technological cuts inflicted on it by American science. If this is so, then workers in the Rice nonparametric density estimation project can take some satisfaction in its part in inflicting one of those cuts.

The Tapia-Thompson collaboration came essentially to a close after the manuscript of *Nonparametric Probability Density Estimation* was finally completed for the Johns Hopkins University Press in January of 1978. Our NASA remote sensing work was at a natural termination point; Richard Tapia was returning to his first love, constrained optimization theory; and I was looking forward to extending what we had learned to cancer modeling and defense related models. I recall, when finishing the last uncompleted task in the book (the proof of Lemma 1, on page 126) while waiting for a plane to Warsaw to work at the Curie-Sklodowska Tumor Institute, having a real sense that one project had been completed and another was about to begin.

The current book is, in a sense, a partial record of that second project. The emphasis of the book is modeling and simulation, which are obvious next steps beyond such exploratory devices as nonparametric probability density estimation. Since the discussion of one dimensional nonparametric density estimation is a natural entrée to the new material and since the out-of-print *Nonparametric Probability Density Estimation* dealt with that topic rather handily, I proposed to SIAM and Richard Tapia that we use the material in *Nonparametric Probability Density Estimation* as an introduction to the current book. This book is almost twice as long as *Nonparametric Probability Density Estimation.* The new material is my sole responsibility. The division of work in this book is roughly: Thompson (Chapters 1, 2, 6, 7, 8, Appendices III and IV). Tapia (Chapters 3, 4, Appendices I and II), both authors (Chapter 5).

Over the years, I have been discouraged by how much emphasis is still being given to the done-to-death problem of one dimensional nonparametric density estimation. When I recently raised the question as to why we are still stuck in one dimension to a master of the 1-d NPDE craft, the reply was "because there is so much there we still do not understand." Well, human beings will never in this life understand anything fully. But the fact is that we passed diminishing returns in 1-d NDE around 1978, or, some would argue, much earlier than that.

Investigations of higher dimensional problems are rather sparse in the statistical literature. Such attention as is given to higher dimensional problems is mostly graphical in nature. Such analyses may be quite valuable at the exploratory level for dimensions up to, say, four. However, I feel that the graphics oriented density estimation enthusiasts fall fairly clearly into the exploratory data analysis camp, which tends to replace statistical theory by the empiricism of a human observer. Exploratory data analysis, including nonparametric density estimation, should be a first step down a road to understanding the mechanism of data generating systems. The computer is a mighty tool in assisting us with graphical displays, but it can help us even more fundamentally in revising the way we seek out the basic underlying mechanisms of real world phenomena via stochastic modeling.

The current book attempts to be a kind of road map to investigators who try to make sense out of multiparametric models of, frequently, multidimensional data. I have been down many a false trail in data analysis, and perhaps this book will spare some from needlessly repeating my mistakes.

The first five chapters give an historical background of nonparametric density estimation and a review of much of the seminal work in the estimation of densities in one and two dimensions.

Chapter 6 addresses the problems of multidimensional data and some means of coping with them. A discussion is given of multivariate nonparametric density estimation, including graphical representation. In this chapter, we also address the nonparametric density based random number generator SIMDAT developed by Malcolm Taylor and myself for coping with resampling in those situations where the Dirac-comb resampling algorithm (the bootstrap) will not work. It is argued that global graphical display of a nonparametric density estimator is seldom the answer for the analyst of high dimensional data. It is further argued that, in many situations, the way to go is to find the centers of high density of the data and use these as base points for further investigation. Some parallels between the mode finding algorithms of Steven Boswell and the *Stein Paradox* are drawn.

Chapter 7 notes the relationship between "nonparametric regression" and the EDA curve fitting techniques of John Tukey. The quest for a good nonparametric intensity function estimator is shown to lead, very naturally, to the penalized likelihood approach.

Chapter 8 spends a great deal of time tracing an exploratory data analysis via the nonparametric estimation of the intensity of metastatic display through an attempt to infer an aspect of the spread of cancer through a mathematical model. The "closed form" likelihood approach of estimating the parameters of the model is developed and then superceded by the SIMEST algorithm developed at Rice and the M.D. Anderson Cancer Center by myself and my colleagues, Neely Atkinson, Robert Bartoszyński and Barry Brown. SIMEST goes directly from the microaxioms to the estimation of the underlying parameters via a forward simulation. A model is developed showing the likely effectiveness of adjuvant chemotherapy using both a direct mathematical solution and a simulation based approach. A speculative model for the extreme virulence of the AIDS epidemic in the United States is developed showing what could have been done to avoid it and what might yet be done to minimize its death toll. Finally, the currently popular science of chaos theory is examined in the light of stochasticity.

Appendix I gives a brief survey of the mathematical theory of optimization. Appendix II looks at some of the classical algorithms for the numerical solution of constrained optimization problems. Appendix III looks at two algorithms for dealing with optimization in the presence of noise. Appendix IV gives a brief review of the subject of simulation.

The support by the Army Research Office (Durham) under DAAL-03-88-K0131 for the major part of this book is gratefully acknowledged. Also, the work in chaos theory and simulation based estimation has been greatly facilitated by the purchase of a desktop Levco Translink computer made available through DAAL-03-88-G-0074. I would also like to acknowledge the support of work in the first five chapters by the Office of Naval Research under NR-042-283, the Air Force under AFOSR 76-2711, the National Science Foundation under ENG 74-17955, and ERDA under E-(40-1)-5046. The NASA remote sensing data in Figures 5.8 through 5.12 were supplied by David van Rooy and Ken Baker of NASA. Jean-Pierre Carmichael supplied the snowfall data in Figure 5.7. The graphics in Chapters 2 and 5 were supplied by David Scott. Figures 8.9 and 8.10 were supplied by Kerry Go. Figures 8.11 through 8.16 as well as the cover figure were supplied by David Stivers.

Over the years, a number of Rice graduate students have received their doctorates in topics related to nonparametric function estimation and modeling based issues. These include Neely Atkinson, John Bennett, Steven Boswell, Richard Hathaway, Joyce Husemann, Rodney Jee, Gilbert de Montricher, Donna Nezames, Roland Sanchez, David Scott, Melvyn Smith, and Ferdie Wang. Thanks are due to all these former students for their willingness to sail in generally uncharted waters. Also, I would like to thank Gerald Andersen, Joe Austin, Robert Bartoszyński, Eileen Bridges, Barry Brown, Diane Brown, John Chandler, Jagdish Chandra, Andrew Coldman, John Dennis, Tim Dunne, Joe Ensor, Katherine Ensor, Randall Eubank, James Gentle, Rui de Figueiredo, Danuta Gadomska, Kerry Go, James Goldie, I.J. Good, Corey Gray, Peter Hall, Tom Kauffman, Vickie Kearn, Marek Kimmel, Jacek Koronacki, Tadeusz Koszarowski, Martin Lawera, Ray McBride, Ennis McCune, Emanuel Parzen, Michael Pearlman, Paul Pfeiffer, William Schucany, Robin Sickles, David Stivers, Malcolm Taylor, George Terrell, Ewa Thompson, Patrick Tibbits, Virginia Torczon, Chris Tsokos, John Tukey, Matt Wand, and C.C. Wang.

James R. Thompson
Houston, Texas
July, 1990

CHAPTER

1 Historical Background

1.1. *Moses as Statistician*

A major difference in the orientation of probabilists and statisticians is the fact that the latter must deal with data from the real world. Realizing the basic nonstationarity of most interesting data-generating systems, the statistician is frequently as much concerned with the parsimonious representation of a particular data set as he is with the inferences which can be made from that data set about the larger population from which the data was drawn. (In many cases, it makes little sense to talk about a larger population than the observed data set unless one believes in parallel worlds.)

Let us consider an early statistical investigation—namely, Moses's census of 1490 B.C. (in Numbers 1 and 2) given below in Table 1.1.

Table 1.1

Tribe	Number of militarily fit males
Judah	74,600
Issachar	54,400
Zebulun	57,400
Reuben	46,500
Simeon	59,300
Gad	45,650
Ephraim } Joseph	40,500
Manasseh } Joseph	32,200
Benjamin	35,400
Dan	62,700
Asher	41,500
Naphtali	53,400
Levi	—
TOTAL	603,550

The tribe of Levi was not included in this or any subsequent census, owing to special priestly exemptions from the misfortunes which inevitably accrue to anyone who gets his name on a governmental roll. A moment's reflection on the table quickly dispels any notions as to a consistent David versus Goliath pattern in Israeli military history. No power in the ancient world could field an army in a Palestinian campaign which could overwhelm a united Israeli home guard by sheer weight of numbers.

Table 1.1 does not appear to have many stochastic attributes. Of course, it could have been used to answer questions, such as "What is the probability that if 20 militiamen are selected at random from the Hebrew host, none will be from the tribe of Judah?" The table is supposedly an exhaustive numbering of eleven tribes; it is not meant to be a random sample from a larger population. And yet, in a sense, the data in Table 1.1 contain much of the material necessary for the construction of a histogram. Two essential characteristics for such a construction are lacking in the data. First, and most important, a proper histogram should have as its domain intervals of the reals (or of R^n). As long as indexing is by qualitative attributes, we are stuck with simply a multinomial distribution. This prevents a natural inferential pooling of adjacent intervals. For example, there is likely to be more relationship between the number of fighters between the heights of 60 and 61 inches and the number of fighters between the heights of 61 and 62 inches than there is between the number of fighters in the two southern tribes of Judah and Benjamin.

Second, no attempt has been made to normalize the table. The ancient statistician failed to note, for example, that the tribe of Issachar contributed 9 percent of the total army. We can only conjecture as to Moses's "feel" about the relative numbers of the host via such things as subsequent battle dispositions.

1.2. The Haberdasher and the "Histogram"

Perhaps one can gain some insight as to the trauma of the perception of the real number system when it is realized that the world had to wait another 3,150 years after the census of Moses for the construction of something approaching a proper histogram. In 1538 A.D., owing to a concern with decreases in the English population caused by the plague, Henry VIII had ordered parish priests of the Church of England to keep a record of christenings, marriages, and deaths [2]. By 1625, monthly and even weekly

Table 1.2

Age interval	Probability of death in interval	Probability of death after interval
0–6	.36	.64
6–16	.24	.40
16–26	.15	.25
26–36	.09	.16
36–46	.06	.10
46–56	.04	.06
56–66	.03	.03
66–76	.02	.01
76–86	.01	0

birth and death lists for the city of London were printed and widely disseminated.

The need for summarizing these mountains of data led John Graunt, a London haberdasher, to present a paper to the Royal Society, in 1661, on the "bills of mortality." This work, *Natural and Political Observations on the Bills of Mortality*, was published as a book in 1662. In that year, on the basis of Graunt's study and on the recommendation of Charles II, the members of the Royal Society enrolled Graunt as a member. (Graunt was dropped from the rostrum five years later, perhaps because he exhibited the anti-intellectual attributes of being (1) a small businessman, (2) a Roman Catholic, and (3) a statistician.) Graunt's entire study is too lengthy to dwell on here. We simply show in Table 1.2 [16, p. 22] his near histogram, which attempts to give probabilities that an individual in London will die in a particular age interval.

Much valid criticism may be made of Graunt's work (see, eg., [2] and [16]). For example, he avoided getting into the much more complicated realm of stochastic processes by tacitly assuming first order stationarity. Then, he gave his readers no clear idea what the cut-off age is when going from one age interval to the next. Finally, he did not actually graph his results or normalize the cell probabilities by the interval widths.

The point to be stressed is that Graunt had given the world perhaps its first empirical cumulative distribution function and was but a normalizing step away from exhibiting an empirical probability density function. To have grasped the desirability of these representations, which had passed uncreated for millenia, is surely a contribution of Newtonian proportions

Again, the statistical (as opposed to the probabilistic) nature of Graunt's work is clear. Unlike his well-connected contemporary, Pascal, Graunt was not concerned with safe and tightly reasoned deductive arguments on idealized games of chance. Instead, he was involved with making sense out of a mass of practical and dirty data. The summarizing and inferential aspects of his work are equally important and inextricably connected.

It is interesting to speculate whether Graunt visualized the effect of letting his age intervals go to zero; i.e., did he have an empirical feel for modeling the London mortality data with a continuous density function? Of course, derivatives had not yet been discovered, but he would have started down this road empirically if he had constructed a second histogram from the same data, but using a different set of age intervals. It appears likely that Graunt's extreme practicality and his lack of mathematical training did not permit him to get this far. Probably, in Graunt's mind his table was an entity *sui generis* and only vaguely a reflection of some underlying continuous model. In a sense, the empirical cumulative distribution function and empirical probability density function appear to antedate the theoretical entities for which they are estimators.

In the years immediately following publication of Graunt's work, a number of investigators employed histogram and histogram-like means of tabulation. These include Petty, Huygens, van Dael, and Halley [16]. However, it seems clear that each of these built on a knowledge of Graunt's work. We are unaware of any discovery of a near histogram independent of Graunt's treatise.

It is not surprising that a number of years passed before successful attempts were made to place Graunt's work on a solid mathematical footing. His approach in its natural generality is an attempt to estimate a continuous probability density function (of unknown functional form) with a function characterized by a finite number of parameters, where the number of the parameters nevertheless increases without limit as the sample size goes to infinity. The full mathematical arsenal required to undertake the rigorization (and improvement) of Graunt's approach would not be available for two hundred years. We shall not dwell here on the several important examples of continuous probability density models proposed by early investigators for specific situations; e.g., DeMoivre's discovery of the normal density function [15]. Our study is concerned with the estimation of probability densities of unknown functional form. There was very little progress in this task between the work of John Graunt and that of Karl Pearson.

A major difficulty of the histogram (developed in the next chapter) is, of course, the large number of characterizing parameters. The last column of Table 1.2 is unity minus a form of the empirical cumulative distribution function. It requires eight parameters for its description. Should we decide to decrease the interval width, the number of parameters will increase accordingly. In the absence of a computer, it is desirable to attempt to characterize a distribution by a small number of parameters. A popular solution for over one hundred years has been the assumption that every continuous distribution is either normal or nearly normal. Such an optimistic assumption of normality is still common. However, already in the late 1800s, Karl Pearson, that great enemy of "myth" and coiner of the term "histogram" [12], had by extensive examination of data sets perceived that a global assumption of normality was absurd. What was required was the creation of a class of probability densities which possess the properties of fitting a large number of data sets, while having a small number of characterizing parameters.

1.3. The Pearson Distributions

In 1895 Karl Pearson proposed a family of distributions which (as it turned out) included many of the currently more common univariate probability densities as members. "We have, then, reached this point: that to deal effectively with statistics we require generalized probability curves which include the factors of skewness and range" [13, p. 58]. For a thorough discussion of the subject, we refer the reader to Pearson's original work or Johnson's updated and extensively revised version of Elderton's *Frequency Curves and Correlation* [3]. More concise treatments are given by Kendall and Stuart [8], Fisher [4], and Ord [11]. This family of densities is developed, according to the tastes of the time, with a differential equation model.

The hypergeometric distribution is one of the most important in the work of turn-of-the-century physicists and probabilists. We recall that this "sampling without replacement" distribution is the true underlying distribution in a number of situations, where for purposes of mathematical tractability we assume an asymptotically valid approximation such as the binomial or its limit the Gaussian.

Let X be a random variable which the number of black balls drawn in a sample of size n from an urn with Np black balls and $N(1-p)$ white balls. Then, following the argument of Kendall and Stuart [8],

$$f_x = P[X = x] = \frac{\binom{Np}{x}\binom{N(1-p)}{n-x}}{\binom{N}{n}}$$

and

$$\frac{\Delta f_x}{f_x} = \frac{P(X = x) - P(X = x - 1)}{P(X = x)} = 1 - \frac{P(X = x - 1)}{P(X = x)}$$

$$= 1 - \frac{x(x-1)!(Np-x)!(N(1-p)-n+x)(N(1-p)-n+x-1)!(n-x)!}{(x-1)!(Np-x+1)(Np-x)!(N(1-p)-n+x-1)!(n-x+1)(n-x)!}$$

$$= 1 - \frac{x(N(1-p)-n+x)}{(Np-x+1)(n-x+1)}$$

$$= \frac{x - (n+1)(Np+1)/(N+2)}{-(n+1)(Np+1)/(N+2) + (Np+n+2)/(N+2)x - (N+2)^{-1}x^2}$$

$$= \frac{(x-a)}{b_0 + b_1 x + b_2 x^2}.$$

Let us consider the associated differential equation

$$\frac{d(\log f(x))}{dx} = \frac{x-a}{b_0 + b_1 x + b_2 x^2}, \tag{1}$$

where $f(x)$ is the probability density function of X. Rewriting (1), we have

$$(b_0 + b_1 x + b_2 x^2)\, df = (x-a) f\, dx. \tag{2}$$

Multiplying both sides of (2) by x^n and integrating the left-hand side by parts over the range of x, we have

$$(b_0 + b_1 x + b_2 x^2) x^n f \Big|_{x=\ell_1}^{x=\ell_2} - \int_{\ell_1}^{\ell_2} [n b_0 x^{n-1} + (n+1) b_1 x^n + (n+2) b_2 x^{n+1}] f\, dx = \int_{\ell_1}^{\ell_2} x^{n+1} f\, dx - a \int_{\ell_1}^{\ell_2} x^n f\, dx. \tag{3}$$

Assuming that the first term vanishes at the extremities, ℓ_1 and ℓ_2, we have the recurrence relationship

$$-n b_0 \mu'_{n-1} - (n+1) b_1 \mu'_n - (n+2) b_2 \mu'_{n+1} = \mu'_{n+1} - a\mu'_n \tag{4}$$

or

$$\mu'_{n+1} = [(a - (n+1)b_1)\mu'_n - n b_0 \mu'_{n-1}]/[(n+2)b_2 + 1], \tag{5}$$

where $\mu'_n = E[x^n] = \int x^n f\, dx$ is the nth moment of x about the origin. Let us assume that the origin of x has been placed such that $\mu'_1 = 0$, so that $\mu_n = E[(x - \mu'_1)^n] = \mu'_n$. Letting $n = 0$, 1, 2, 3, and solving for a, b_0, b_1, and b_2 in terms of μ_2, μ_3, and μ_4, we have

$$\begin{aligned}(6)\qquad a &= -\mu_3(\mu_4 + 3\mu_2^2)/Q = -\sqrt{\mu_2}\sqrt{\beta_1}(\beta_2 + 3)/Q' \\ b_0 &= -\mu_2(4\mu_2\mu_4 - 3\mu_3^2)/Q = -\mu_2(4\beta_2 - 3\beta_1)/Q' \\ b_1 &= -u_3(\mu_4 + 3\mu_2^2)/Q = -\sqrt{\mu_2}\sqrt{\beta_1}(\beta_2 + 3)/Q' \\ b_2 &= [-2\mu_2\mu_4 + 3\mu_3^2 + 6\mu_2^3]/Q = -(2\beta_2 - 3\beta_1 - 6)/Q',\end{aligned}$$

where $Q = 10\mu_2\mu_4 - 18\mu_2^3 - 12\mu_3^2$, $Q' = 10\beta_2 - 18 - 12\beta_1$, $\beta_1 = \mu_3^2/\mu_2^3$ and $\beta_2 = \mu_4/\mu_2^2$. We note that, since $a = b_1$, the Pearson System after translation of the data to make $\mu_1' = 0$ is actually a three-parameter family.

The relations in (6) give us a ready means of estimating a, b_0, b_1, and b_2 via the method of moments. Given a random sample $\{x_1, x_2, \ldots, x_n\}$, we might use as approximation to μ_n', $\frac{1}{n}\sum x_j^n$, the nth sample moment.

Clearly, the solution of (1) depends on the roots of the denominator $b_0 + b_1x + b_2x^2$. Letting $\varkappa = b_1^2/4b_0b_2$, we can easily characterize whether we have two real roots with opposite signs (Pearson's Type I), two complex roots (Pearson's Type IV), or two real roots with the same sign (Pearson's Type VI), according as $\varkappa < 0$, $0 < \varkappa < 1$ or $\varkappa > 1$ respectively. The special cases when $\varkappa = 0$ or $\varkappa = 1$ will be discussed later. The development below follows the arguments of Elderton and Johnson [3].

Type I ($\varkappa < 0$)

$$(7)\qquad \frac{d\log f}{dx} = \frac{(x-a)}{b_2(x+A_1)(x-A_2)} \qquad \text{where } A_1, A_2 > 0$$

$$= \frac{1}{b_2}\frac{A_1 + a}{A_1 + A_2}\frac{1}{x + A_1} + \frac{1}{b_2}\frac{A_2 - a}{A_1 + A_2}\frac{1}{x - A_2},$$

giving

$$\log f = \log c + \frac{A_1 + a}{b_2(A_1 + A_2)}\log(x + A_1) + \frac{1}{b_2}\frac{A_2 - a}{A_1 + A_2}\log(A_2 - x)$$

or

$$(8)\qquad f(x) = c(x + A_1)^{(A_1 + a)/b_2(A_1 + A_2)}(A_2 - x)^{(A_2 - a)/b_2(A_1 + A_2)},$$

where c is chosen so that $\int f(x)\,dx = 1$ and $-A_1 < x < A_2$.

This is a bcta distribution (of thc first kind). Wc arc morc familiar with thc standard form

$$(9)\quad f(x) = \frac{1}{B(\alpha,\beta)}x^{\alpha-1}(1-x)^{\beta-1};\ \alpha,\beta > 0, 0 < x < 1 \text{ and } B(\alpha,\beta) = \frac{\Gamma(\alpha)\Gamma(\beta)}{\Gamma(\alpha+\beta)}.$$

The applications of this distribution include its use as the natural conjugate prior for the binomial distribution. We shall shortly note the relation between Pearson's Type I and his Type VI distributions.

Type IV $(0 < \varkappa < 1)$

$$\frac{d \log f}{dx} = \frac{(x-a)}{b_2[(x+c)^2 + d^2]} \tag{10}$$

$$= \frac{2(x+c)}{2b_2[(x+c)^2+d^2]} - \frac{(c+a)}{b_2}\frac{1}{(x+c)^2+d^2},$$

giving

$$\log f = \log k + \frac{1}{2b_2}\log[(x+c)^2+d^2] - \frac{c+a}{db_2}\arctan\left(\frac{x+c}{d}\right)$$

or

$$f(x) = k[(x+c)^2+d^2]^{1/(2b_2)}\exp\left[-\frac{c+a}{db_2}\arctan\left(\frac{x+c}{d}\right)\right] \tag{11}$$

where $-\infty < x < \infty$ and k is chosen so that $\int f(x)\,dx = 1$.

This distribution is mainly of historical interest, as it is seldom used as a model in data analysis.

Type VI $(\varkappa > 1)$

$$\frac{d \log f(x)}{dx} = \frac{(x-a)}{b_2(x-A_1)(x-A_2)} \tag{12}$$

(where the real valued A_1 and A_2 have the same sign)

$$= \frac{1}{b_2}\frac{A_1-a}{A_1-A_2}\frac{1}{x-A_1} - \frac{1}{b_2}\frac{A_2-a}{A_1-A_2}\frac{1}{x-A_2},$$

giving

$$\log f = \log c + \frac{A_1-a}{b_2(A_1-A_2)}\log(x-A_1) - \frac{1}{b_2}\frac{A_2-a}{A_1-A_2}\log(x-A_2)$$

or

$$f = c(x-A_1)^{(A_1-a)/b_2(A_1-A_2)}(x-A_2)^{-(A_2-a)/b_2(A_1-A_2)}, \tag{13}$$

where $\operatorname{sgn}(a)\,x > \min(|A_1|, |A_2|)$ and c is chosen so that $\int f(x)\,dx = 1$.

This is a beta-distribution of the second kind and is more commonly seen in the standard form

$$f(x) = \frac{1}{B(\alpha,\beta)}\frac{x^{\alpha-1}}{(x+1)^{\alpha+\beta}}, \qquad \text{where } \alpha, \beta > 0 \quad \text{and} \quad x \geq 0. \tag{14}$$

Perhaps the most important use of this distribution is in analyses of variance, because the F distribution is a beta distribution of the second kind. Because of the easy transformation of a beta variable of the first or second kind into a beta variable of the other kind, the existence of both Type I and Type VI distributions in the Pearsonian system provides for a kind of redundancy.

If we are to use sample moments in (6) to estimate a, b_0, b_1, and b_2, then we really need only the three members already considered. $\varkappa = \pm\infty$, 0 or 1, the cases not yet examined, will occur very rarely and only as a result of truncation of the decimal representations of the data values. This is reasonably unsettling, since it shows that a direct hammer-and-tongs method of moments on the data will give us only one of two forms of the beta distribution or the rather unwieldy density in (11). For the Pearson family to be worthwhile, it is clear that there must be other distributions of interest for some of the values $\varkappa = \pm\infty$, 0, 1, and that we will not be able to rely simply on data-oriented estimation of a, b_0, b_1, and b_2. Let us consider briefly these important "transition" cases.

Type III ($\varkappa = \infty$)

Let us consider the case where $\varkappa = \infty$. This can occur only if $b_2 = 0$. (Note that $b_0 = 0$ is an impossibility, for then we must have $4\mu_2\mu_4 - 3\mu_3^2 = 0$. But by the Cauchy–Schwarz Inequality, $\mu_2\mu_4 \geq \mu_3^2$. Hence the above equality is impossible). We have, then,

$$\text{(15)} \qquad \frac{d \log f}{dx} = \frac{x-a}{b_0 + b_1 x} = \frac{1}{b_1}\frac{x-a}{x + b_0/b_1},$$

giving

$$\log f(x) = \log c + \frac{1}{b_1}x - \frac{ab_1 + b_0}{b_1^2}\log(x + b_0/b_1)$$

or

$$\text{(16)} \qquad f(x) = ce^{x/b_1}(x + b_0/b_1)^{-(ab_1+b_0)/b_1^2}.$$

Changing the origin, we have more concisely

$$\text{(17)} \qquad f(x) = c_0 e^{-\gamma x}\left(1 + \frac{x}{c_1}\right)^{\gamma c_1},$$

where $x > -c_1$, $\gamma = 2\mu_2/\mu_3$, $c_1 = \dfrac{2\mu_2^2}{\mu_3} - \dfrac{\mu_3}{2\mu_2}$ and c_0 is chosen so that $\int f(x)\,dx = 1$.

Of course, by shifting the origin to 0 and changing scale via the transformation $x = zc_1 - c_1$, we can obtain the standard gamma distribution

$$\text{(18)} \qquad f(z) = \frac{1}{\Gamma(\alpha)\beta^\alpha} z^{\alpha-1}e^{-z/\beta}, \qquad \text{where } \alpha = \gamma c_1 + 1 > 0, \quad \beta = 1/(\gamma c_1) > 0.$$

Let us consider the case where $\varkappa = 0$

$$\frac{d \log f}{dx} = \frac{x}{b_0 + b_2 x^2}. \tag{19}$$

Then,

$$d(\log f) = \frac{1}{2b_2} \frac{d(b_0 + b_2 x^2)}{b_0 + b_2 x^2}, \tag{20}$$

giving

$$\begin{aligned} f &= c[b_0 + b_2 x^2]^{1/(2b_2)} \\ &= c'[1 + (b_2/b_0)x^2]^{1/(2b_2)}. \end{aligned} \tag{21}$$

Now, from (6) we recall that

$$b_2/b_0 = (2\beta_2 - 6)/(4\mu_2\beta_2).$$

Thus the coefficient of the x^2 term in (21) is negative or positive, depending on whether $\beta_2 < 3$ or $\beta_2 > 3$.

Type II ($\varkappa = 0, \beta_2 < 3$)

$$f = c'[1 - |b_2/b_0|x^2]^{1/(2b_2)}, \tag{22}$$

where $-\sqrt{|b_0/b_2|} < x < \sqrt{|b_0/b_2|}$ and c' is chosen so that $\int f(x)\,dx = 1$.

Type VII ($\varkappa = 0, \beta_2 > 3$)

$$f = c'[1 + (b_2/b_0)x^2]^{1/(2b_2)}, \tag{23}$$

where (since the exponent $1/(2b_2) < -1$) $-\infty < x < \infty$ and c' is chosen so that $\int f(x)\,dx = 1$.

Gaussian Distribution ($\varkappa = 0, b_1 = b_2 = 0$)

One further case should be considered when $\varkappa = 0$, namely, that in which $b_1 = b_2 = 0$

$$\frac{d \log f}{dx} = \frac{x}{b_0}. \tag{24}$$

From (6) it can be seen that under these conditions b_0 is negative. Thus,

$$\log f = \log c - \frac{x^2}{2|b_0|}.$$

$$f(x) = c \exp\left[-\frac{x^2}{2|b_0|}\right] \tag{25}$$

where $-\infty < x < \infty$ and c is chosen so that $\int f(x)\,dx = 1$.

Type V ($\varkappa = 1$)

When $\varkappa = 1$, $b_0 + b_1 x + b_2 x^2$ has a repeated real root at $-\dfrac{b_1}{2b_2}$

(26) $$\frac{d \log f}{dx} = \frac{1}{b_2} \frac{x - a}{(x + b_1/2b_2)^2},$$

giving

$$\log f = \log c + \frac{1}{b_2} \log\left(x + \frac{b_1}{2b_2}\right) + \frac{2ab_2 + b_1}{2b_2^2} \frac{1}{(x + b_1/2b_2)}$$

$$= \log c + \log(x + b_1/2b_2)^{1/b_2} + \log\left[\exp\left\{\frac{a + b_1/2b_2}{b_2(x + b_1/2b_2)}\right\}\right],$$

so

(27) $$f(x) = c(x + b_1/2b_2)^{1/b_2} \exp\left[\frac{a + b_1/2b_2}{b_2(x + b_1/2b_2)}\right]$$

or, letting $x = z - \dfrac{b_1}{b_2}$,

(28) $$f(z) = cz^{1/b_2} \exp \frac{a + b_1/2b_2}{b_2 z}$$

$$= cz^{-\alpha} e^{-\beta/z},$$

where $\alpha = -\dfrac{1}{b_2}$, $\beta = -\dfrac{a + b_1/2b_2}{b_2}$, $z > 0$ and c is chosen so that $\int f(z)\, dz = 1$.

"Student's" Discovery of the t Distribution

The most famous result in statistics which came about by using the method of moments with the Pearson family appears in a 1908 paper of W. S. Gosset (pseudonym "Student") [6]. Although before Gosset it was well known that the distribution of the sample mean $\bar{x}$ of independent observations from a Gaussian distribution with mean μ and variance σ^2 is Gaussian with mean μ and variance $\dfrac{\sigma^2}{n}$ where n is the sample size, it was unclear what sort of statement could be made when σ^2 was unknown. The usual procedure was to substitute the sample variance s^2 for σ^2; i.e., it was assumed that $\bar{x}$ was Gaussian with mean μ and variance $\dfrac{s^2}{n}$. Gosset has empirically observed that the distribution of $t = \sqrt{n}(\bar{x} - \mu)/s$ is close to the Gaussian for n large, but that for n small the tails of the t distribution were heavier than those of the Gaussian. We sketch below "Student's" derivation of the t distribution.

First, Gosset determined the first four population moments of

$$s^2 = \frac{1}{n-1} \sum_{j=1}^{n} (x_j - \bar{x})^2$$

where the $\{x_j\}$ were assumed to be Gaussian with mean μ and variance σ^2. He found that he obtained a, b_0, b_1 and b_2 values which corresponded in the Pearson family to a Type III curve with origin at 0, namely,

$$f(s^2) = c(s^2)^{(n-3)/2} e^{-ns^2/2\sigma^2}. \tag{29}$$

Next, Gosset found that s and $(\bar{x} - \mu)$ were uncorrelated. Then, assuming that this implied independence of s and $(\bar{x} - \mu)$ (as it turns out is true in this case), he proceeded to derive the distribution of $\sqrt{n}(x - \mu)/s = t$, obtaining

$$f(t) = \frac{\Gamma\left(\frac{n}{2}\right)}{\sqrt{(n-1)\pi}\,\Gamma\left(\frac{n-1}{2}\right)} \frac{1}{\left(1 + \frac{t^2}{n-1}\right)^{n/2}}. \tag{30}$$

It is interesting to note that Gossett actually used the method of moments to obtain the distribution of s^2—not to obtain the distribution of t. The last step he carried out in a manner similar to that usually employed in a contemporary elementary statistics course, where one starts out with the joint density of a x^2 variate and an independent Gaussian variate, introduces t via a transformation, and integrates out the dummy variable.

We note that the t distribution is of Type VII. Had "Student" employed the method of moments on data to infer the distributions of either s^2 or t, he would not have landed in the Type III or Type VII sets—these being of measure zero in the parameter space of (a, b_0, b_1, b_2) as estimated by the method of moments.

This example shows one reason the Pearsonian approach is seldom used at present. If an exact distribution of a measurable function of random variables is to be found, we must know the underlying distribution of the original random variables. We then compute the first four moments of the function and search for the distribution type in the Pearsonian system. (This may be more difficult than finding the distribution directly.) When we finish our work, we have no assurance that the true distribution really is a member of the Pearsonian system. Note that multimodal distributions, for example, are excluded from the system.

Finally, if we seek only an approximation to the underlying distribution for a particular data set, we realize we will land in the parameter sets corresponding to Types I, IV, and VI—a system neither particularly tractable nor flexible. Having made the above comments, we must remember that the most common distributions dealt with in contemporary statistics are members of the Pearsonian family. It is simply that Pearson's approach is rather too ad hoc for theoretical derivation, rather too impractical for ad hoc density estimation.

If the development of statistics proceeded by a sequence of steady Teutonic increments, one might suppose that following Pearson's breakthrough, the main channel of statistics would have proceeded from his work. Accordingly, we might expect to have seen a succession of important generalizations of Pearson's family. Indeed there were a number of such papers, e.g., [7], [10], [19]. However, these studies did not actually push toward the natural goal, namely, the creation of practical algorithms which enable the stable and consistent estimation of probability densities in very general classes. (In a real sense, the subsequent studies of the foremost investigator of probability distributions, Norman Lloyd Johnson, may be said to be motivated in part by Pearson's work.)

Instead, R. A. Fisher appeared on the scene with the concept of maximum likelihood estimation [4] and deflected the thrust and direction of the Pearsonian methodology. The really important concept in Pearson's approach is the desirability in many situations of leaving the a priori functional form of the unknown probability density as unspecified as possible. Unfortunately, the Pearson–Fisher controversy was fought out on the line of the method of moments versus maximum likelihood. The Fisherian victory was nearly complete.

1.4. Density Estimation by Classical Maximum Likelihood

Although over half a century old, maximum likelihood is still the most used of any estimation technique. To motivate this procedure, let us consider a probability density of a random variable x characterized by a real parameter θ:

$$f(x;\theta)=f(x|\theta), \tag{31}$$

where

$$-\infty < x < +\infty$$

and

$$a \le \theta \le b.$$

We have a random sample (collection of n identically and independently distributed random variables) of x's $\{x_1, x_2, \ldots, x_n\}$. Let us suppose we have some knowledge of the true value of θ, which can be characterized by a *prior probability density* $p(\theta)$. The joint density of $\mathbf{x} = \{x_1, x_2, \ldots, x_n\}$ and θ is given by

$$h(\mathbf{x}, \theta) = \prod_{j=1}^{n} f(x_j|\theta)p(\theta) = f_n(\mathbf{x}|\theta)p(\theta), \tag{32}$$

where $f_n(\mathbf{x}|\theta)$ is the *likelihood function.* (It was pointed out to the authors by Salomon Bochner that Fisher was apparently the first researcher to exploit the fact that under the conditions given above, $f_n(\mathbf{x}|\theta) = \prod f(x_j|\theta)$.) The marginal density of $\mathbf{x}$ is given by

$$\ell(\mathbf{x}) = \int_a^b f_n(\mathbf{x}|\theta)p(\theta)\, d\theta. \tag{33}$$

The conditional density of θ given $\mathbf{x}$ (i.e., the *posterior density* of θ) is then

$$g(\theta|\mathbf{x}) = \frac{h(\mathbf{x}, \theta)}{\ell(\mathbf{x})} = \frac{f_n(\mathbf{x}|\theta)p(\theta)}{\int f_n(\mathbf{x}|\theta')p(\theta')\, d\theta'}. \tag{34}$$

This is simply a version of Bayes's Theorem. To obtain a good estimate of θ we might use some measure of centrality of the posterior distribution of θ. For example, if we attempt to minimize

$$E[(\tilde{\theta}(\mathbf{x}) - \theta)^2] = \int_x \int_\theta [\tilde{\theta}(\mathbf{x}) - \theta]^2 p(\theta) \prod_{j=1}^{n} f(x_j|\theta)\, dx_j\, d\theta \tag{35}$$

we may do so by selecting for each $\mathbf{x}$ that value which minimizes

$$\int_\theta [\tilde{\theta}(\mathbf{x}) - \theta]^2 \prod_{j=1}^{n} f(x_j|\theta)p(\theta)\, d\theta. \tag{36}$$

Now, this may be obtained by differentiating with respect to the real valued $\tilde{\theta}(\mathbf{x})$ and setting the derivative equal to zero to give

$$\tilde{\theta}(\mathbf{x}) = \int \theta g(\theta|\mathbf{x})\, d\theta \tag{37}$$

or, simply the mean of the posterior distribution $g(\theta|\mathbf{x})$.

Alternatively, we might seek that value of θ such that

$$\int_a^{\theta_{me}} g(\theta|\mathbf{x})\, d\theta = \int_{\theta_{me}}^b g(\theta|\mathbf{x})\, d\theta = .5. \tag{38}$$

That is, we could use the posterior median as an estimate for θ.

As a third possibility we might seek a value of θ which maximizes the posterior density function of θ. If we are fortunate, the posterior mode will be unique.

Let us consider the case where we have only very vague notions as to the true value of θ. We only know that θ cannot possibly be smaller than some a or larger than some b. Then we might decide to use as the prior distribution for θ the uniform distribution on the interval $[a, b]$, namely,

$$p(\theta) = \frac{1}{b-a} \quad \text{if} \quad a \le \theta \le b \tag{39}$$

$$= 0, \qquad \text{otherwise.}$$

Then,

$$g(\theta|\boldsymbol{x}) = \frac{f_n(\boldsymbol{x}|\theta)\dfrac{1}{(b-a)}}{\displaystyle\int_a^b f_n(\boldsymbol{x}|\theta')\frac{1}{b-a}\,d\theta'}, \tag{40}$$

$$= \frac{f_n(\boldsymbol{x}|\theta)}{\displaystyle\int_a^b f_n(\boldsymbol{x}|\theta')\,d\theta'} \qquad \text{if } a \le \theta \le b$$

$$= 0, \qquad \text{otherwise.}$$

Thus, to obtain a value of θ which maximizes $g(\theta|\boldsymbol{x})$ we need only find a value of θ which maximizes the likelihood

$$f_n(\boldsymbol{x}|\theta) = \prod_{j=1}^{n} f(\boldsymbol{x}_j|\theta). \tag{41}$$

Such an estimator is called a *maximum likelihood estimator for θ*. We note that in comparison to the other procedures mentioned for obtaining an estimator for θ, the maximum likelihood approach has the advantage of (relative) computational simplicity. However, we note that its use is equivalent to going through a Bayesian argument using a rather noninformative prior. In fact, we have in effect used Bayes's Axiom; i.e., in the absence of information to the contrary, we have assumed all values of θ in $[a, b]$ to be equally likely. Such an assumption is certainly open to question.

Moreover, we have, in choosing maximum likelihood, opted for using the mode of the posterior distribution $g(\theta|\boldsymbol{x})$ as an estimate for θ based on the data. The location of the global maximum of a stochastic curve is generally much more unstable than either the mean or the median of the curve. Consequently, we might expect maximum likelihood estimators frequently to behave poorly for small samples. Nonetheless, under very general conditions they have excellent large sample properties, as we shall show below.

Following Wilks [18], let us consider the case where the continuous probability density function $f(.|\theta)$ is regular with respect to its first θ derivative in the parameter space Θ, i.e.,

$$E\left[\frac{\partial \log f(x|\theta)}{\partial\theta}\right] = \frac{\partial}{\partial\theta}\int f(x|\theta)\,dx = 0. \tag{42}$$

Let us assume that for any given sample $\boldsymbol{x} = (x_1, x_2, \ldots, x_n)$ equation (43) below has a unique solution.

$$\left[\frac{\partial}{\partial\theta}\log f_n(\boldsymbol{x}|\theta)\right]_{\hat{\theta}_n} = \sum_{j=1}^{n}\left[\frac{\partial}{\partial\theta}\log f(x_j|\theta)\right]_{\hat{\theta}_n} = 0. \tag{43}$$

Now, let us suppose our sample $(x_1, x_2, \ldots, x_n)$ has come from $f(.|\theta_0)$, i.e., the actual value of θ is the point $\theta_0 \in \Theta$. We wish to examine the stochastic convergence of $\hat{\theta}_n$ to θ_0. We shall assume that

$$B^2(\theta,\theta) = \operatorname{var}\left[\frac{\partial \log f(x|\theta)}{\partial\theta}\right] = \int\left(\frac{\partial \log f(x|\theta)}{\partial\theta}\right)^2 f(x|\theta)\,dx < \infty.$$

Let us examine

$$H(\theta_0, \theta) = \int[\log f(x|\theta)]f(x|\theta_0)\,dx. \tag{44}$$

Considering the second difference of $H(\theta_0, .)$ about θ_0, we have (taking care that $[\theta_0 - h, \theta_0 + h]$ is in Θ)

$$\begin{aligned}
(45)\quad \Delta^2_{\theta_0,h}H(\theta_0, \theta) \\
&= \int[\log f(x|\theta_0 + h)]f(x|\theta_0)\,dx \\
&\quad + \int[\log f(x|\theta_0 - h)]f(x|\theta_0)\,dx - 2\int[\log f(x|\theta_0)]f(x|\theta_0)\,dx \\
&= \int\log\left[\frac{f(x|\theta_0 + h)}{f(x|\theta_0)}\right]f(x|\theta_0)\,dx + \int\log\left[\frac{f(x|\theta_0 - h)}{f(x|\theta_0)}\right]f(x|\theta_0)\,dx \\
&< \log\int\frac{f(x|\theta_0 + h)}{f(x|\theta_0)}f(x|\theta_0)\,dx + \log\int\frac{f(x|\theta_0 - h)}{f(x|\theta_0)}f(x|\theta_0)\,dx
\end{aligned}$$

(by the strict concavity of the logarithm and Jensen's Inequality)

$$= \log 1 + \log 1.$$

Thus,

$$\Delta^2_{\theta_0,h}H(\theta_0, \theta) < 0.$$

But then

$$A(\theta_0, \theta) = \int \left[\frac{\partial \log f(x|\theta)}{\partial \theta} \right] f(x|\theta_0)\, dx$$

is strictly decreasing in a neighborhood $(\theta_0 - \delta, \theta_0 + \delta) \in \Theta$ about θ_0.

Now, $\frac{1}{n}\frac{\partial}{\partial \theta} \log f_n(\mathbf{x}|\theta) = \frac{1}{n} \sum \frac{\partial}{\partial \theta} \log f(x_j|\theta)$ is the sample mean of a sample of size n of the random variable $Y = \frac{\partial}{\partial \theta} \log f(x|\theta)$. Since

$$E[Y] = \int \frac{\partial}{\partial \theta} [\log f(x|\theta)]\, f(x|\theta_0) = A(\theta_0, \theta),$$

by the Strong Law of Large Numbers, we know that $\frac{1}{n}\frac{\partial}{\partial \theta} \log f_n(\mathbf{x}|\theta)$ converges almost surely to $A(\theta_0, \theta)$. Moreover, by (42) we have $A(\theta_0, \theta_0) = 0$. Thus, for any $\epsilon > 0$, $0 < \delta' < \delta$, there may be found an $n(\delta', \epsilon)$, such that probability exceeds $1 - \epsilon$ that the following inequalities hold for all $n > n(\delta', \epsilon)$:

$$\frac{1}{n}\frac{\partial}{\partial \theta} \log f_n(\mathbf{x}|\theta) > 0 \quad \text{if} \quad \theta = \theta_0 - \delta'$$

$$\frac{1}{n}\frac{\partial}{\partial \theta} \log f_n(\mathbf{x}|\theta) < 0 \quad \text{if} \quad \theta = \theta_0 + \delta'.$$

Consequently, if $\frac{\partial}{\partial \theta} \log f_n(\mathbf{x}|\theta)$ is continuous in θ over $(\theta_0 - \delta', \theta_0 + \delta')$

$$P\left[\frac{1}{n}\frac{\partial}{\partial \theta} \log f_n(\mathbf{x}|\theta) = 0, \right. \tag{46}$$

for some θ in $(\theta_0 \pm \delta')$ for all $n > n(\delta', \epsilon)|\theta_0] > 1 - \epsilon$. But since we have assumed that $\hat{\theta}_n$ uniquely solves (43), we have that the maximum likelihood estimator converges almost surely to θ_0; i.e.,

$$\lim_{n \to \infty} P[\hat{\theta}_n = \theta_0] = 1. \tag{47}$$

Thus, we have

Theorem 1. Let $(x_1, x_2, \ldots, x_n)$ be a sample from the distribution with probability density function $f(.|\theta)$. Let $f(.|\theta)$ be regular with respect to its first θ derivative in $\Theta \subset R_1$. Let $\frac{\partial}{\partial \theta} \log f(.|\theta)$ be continuous in θ for all values of $x \in R$, except possibly for a set of probability zero. Suppose (43) has a unique solution, say, $\hat{\theta}_n(x_1, x_2, \ldots, x_n)$, for any n and almost all $(x_1, x_2, \ldots, x_n) \in R_n$. Then the sequence $\{\hat{\theta}_n(x_1, x_2, \ldots, x_n)\}$ converges almost surely to θ_0.

More generally, Wald [17] has considered the case where there may be a multitude of relative maxima of the log likelihood. He has shown that, assuming a number of side conditions, if we take for our estimator a value of $\hat{\theta}_n \in \Theta$, which gives an absolute maximum of the likelihood, then $\hat{\theta}_n$ will converge almost surely to θ_0. We have presented the less general result of Wilks because of its relative brevity. (Simplified and abbreviated versions of Wald's proof of which we are aware tend to be wrong, e.g., [9, pp. 40–41].)

Now, an estimator $\hat{\theta}_n$ for a parameter θ_0 is said to be consistent if $\{\hat{\theta}_n\}$ converges to θ_0 in probability. Since almost sure convergence implies convergence in probability, we note that maximum likelihood estimates are consistent. That consistency is actually a fairly weak property is seen by noting that if $\hat{\theta}_n(\mathbf{x})$ is consistent for θ_0, so also is $\hat{\theta}_n(\mathbf{y})$ where $\mathbf{y}$ is the censored sample obtained by throwing away all but every billionth observation. Clearly, consistency is not a sufficient condition for an estimator to be satisfactory. From a theoretical standpoint consistency is not a necessary condition either. Since we will never expect to know precisely the functional form of the probability density, and since the true probability density is probably not precisely stationary during the period (in time, space, etc.) over which the observations were collected, it is unlikely that any estimator, say $\check{\theta}$, we might select will converge in probability to θ_0. In fact, it is unlikely that θ_0 can itself be well defined. Of course, this sort of argument can be used to invalidate all human activity whatever. As a practical matter formal consistency is a useful condition for an estimator to possess.

Had Fisher only been able to establish the consistency of maximum likelihood estimators, it is unlikely he would have won his contest with Pearson. But he was able to show [4] a much stronger asymptotic property.

To obtain an upper bound on the rate of convergence of an estimator for θ, let us, following Kendall and Stuart [9], consider

$$\int \cdots \int f_n(\mathbf{x}|\theta) \prod dx_j = 1. \tag{48}$$

Assuming (48) is twice differentiable under the integral sign, we have differentiating once

$$\begin{aligned} 0 = \int \cdots \int \frac{\partial f_n}{\partial \theta} \prod dx_j &= \int \cdots \int \left(\frac{1}{f_n}\frac{\partial f_n}{\partial \theta}\right) f_n \prod dx_j \\ &= \int \cdots \int \frac{\partial \log f_n}{\partial \theta} f_n \prod dx_j. \end{aligned} \tag{49}$$

Differentiating again

$$\int \cdots \int \left[f_n \frac{\partial}{\partial \theta} \left(\frac{1}{f_n} \frac{\partial f_n}{\partial \theta} \right) + \left(\frac{\partial f}{\partial \theta} \right)^2 \frac{1}{f_n} \right] \prod dx_j$$

$$= \int \cdots \int \left[\frac{\partial^2 \log f_n}{\partial \theta^2} + \left(\frac{1}{f_n} \frac{\partial f_n}{\partial \theta} \right)^2 \right] f_n \prod dx_j = 0,$$

which gives

$$E\left[\frac{\partial^2 \log f_n}{\partial \theta^2} \right] = -E\left[\left(\frac{\partial \log f_n}{\partial \theta} \right)^2 \right]. \tag{50}$$

Now, if we have an unbiased estimator for θ, say $\breve{\theta}$, then,

$$\int \cdots \int (\breve{\theta}(\mathbf{x}_n) - \theta) f_n(\mathbf{x}|\theta) \prod dx_j = 0. \tag{51}$$

Differentiating with respect to θ

$$\int \cdots \int (\breve{\theta} - \theta) \frac{\partial \log f_n}{\partial \theta} f_n \prod dx_j = 1 \tag{52}$$

or

$$E\left[(\breve{\theta} - \theta) \left(\frac{\partial \log f_n}{\partial \theta} \right) \right] = 1,$$

giving by the Cauchy–Schwarz Inequality:

$$E[(\breve{\theta} - \theta)^2] \geq \frac{1}{E\left[\left(\frac{\partial \log f_n}{\partial \theta} \right)^2 \right]} = \frac{-1}{E\left[\frac{\partial^2 \log f_n}{\partial \theta^2} \right]}. \tag{53}$$

The Cramér-Rao Inequality in (53), shows that no unbiased estimator can have smaller mean square error than $\dfrac{-1}{E\left[\frac{\partial^2 \log f_n}{\partial \theta^2} \right]}$. Since the observations are independent, this can easily be shown to be, simply $\dfrac{-1}{nE\left[\left(\frac{\partial^2 \log f(x|\theta)}{\partial \theta^2} \right) \right]}$.

Theorem 2. Let the first two derivatives of $f_n(\mathbf{x}_n|\theta)$ with respect to θ exist in an interval about the true value θ_0. Furthermore, let

$$E\left[\frac{\partial \log f_n}{\partial \theta} \right] = 0 \quad \text{and}$$

$$B_n^2(\theta) = E\left[\left(\frac{\partial \log f_n}{\partial \theta} \right)^2 \right] = -E\left[\frac{\partial^2 \log f_n}{\partial \theta^2} \right] = -nE\left[\frac{\partial^2 \log f}{\partial \theta^2} \right]$$

be nonzero for all θ in the interval. Then the maximum likelihood estimator $\hat{\theta}_n$ is asymptotically Gaussian with mean θ_0 and variance equal to the Cramer-Rao Lower Bound $1/B_n^2(\theta_0)$.

Proof

Expanding $\left(\frac{\partial \log f_n}{\partial \theta}\right)_{\hat{\theta}_n}$ about θ_0, we have

$$\left(\frac{\partial \log f_n}{\partial \theta}\right)_{\hat{\theta}_n} = \left(\frac{\partial \log f_n}{\partial \theta}\right)_{\theta_0} + (\hat{\theta}_n - \theta_0)\left(\frac{\partial^2 \log f_n}{\partial \theta^2}\right)_{\theta_n'}, \tag{54}$$

where θ_n' is between θ_0 and $\hat{\theta}_n$. Since the left-hand side of (54) is zero,

$$(\hat{\theta}_n - \theta_0) = -\frac{\left(\frac{\partial \log f_n}{\partial \theta}\right)_{\theta_0}}{\left(\frac{\partial^2 \log f_n}{\partial \theta^2}\right)_{\theta_n'}}, \quad \text{where } \theta_n' \text{ is between } \theta_0 \text{ and } \hat{\theta}_n, \tag{55}$$

or, rewriting

$$(\hat{\theta}_n - \theta_0)B_n(\theta_0) = \frac{\left(\frac{\partial \log f_n}{\partial \theta}\right)_{\theta_0} \Big/ B_n(\theta_0)}{\left(\frac{\partial^2 \log f_n}{\partial \theta^2}\right)_{\theta_n'} \Big/ -B_n^2(\theta_0)} \tag{56}$$

$$= \frac{\sum_{j=1}^{n} \left(\frac{\partial \log f(x_j)}{\partial \theta}\right)_{\theta_0} \Big/ B_n(\theta_0)}{\frac{1}{n}\sum_{j=1}^{n} \left(\frac{\partial^2 \log f(x_j)}{\partial \theta^2}\right)_{\theta_n'} \Big/ E\left[\left(\frac{\partial^2 \log f}{\partial \theta^2}\right)_{\theta_0}\right]}.$$

But we have seen that $\lim_{n\to\infty} P(\hat{\theta}_n = \theta_0) = 1$. Hence, in the limit the denominator on the right-hand side of (56) becomes unity.

Moreover, we note that

$$\frac{\partial \log f_n}{\partial \theta} = \sum_{j=1}^{n} \frac{\partial \log f(x_j)}{\partial \theta} = \sum_{j=1}^{n} y_j, \tag{57}$$

where the $\{y_j\}$ are independent random variables with mean zero and variance

$$E\left[\left(\frac{\partial \log f(x|\theta)}{\partial \theta}\right)^2_{\theta_0}\right] = \frac{1}{n} B_n^2(\theta_0).$$

Thus, the numerator of the right-hand side of (56) converges in distribution to a Gaussian variate with mean zero and variance 1. Finally, then, $\hat{\theta}_n$

converges in distribution to the Gaussian distribution with mean θ_0 and variance $1/B_n^2(\theta_0)$. ■

Now, the efficiency of an estimator $\theta_n^{(1)}$ for the scalar parameter θ_0 relative to a second estimator $\theta_n^{(2)}$ may be defined by

$$\text{Eff}[\theta_n^{(1)}|\theta_n(2)] = \frac{E[(\theta_n^{(2)}(x_1, x_2, \ldots, x_n) - \theta_0)^2]}{E[(\theta_n^{(1)}(x_1, x_2, \ldots, x_n) - \theta_0)^2]}. \tag{58}$$

But if a maximum likelihood estimator for θ_0 exists, it asymptotically achieves the Cramér-Rao Lower Bound $1/B_n^2(\theta_0)$ for variance. Thus, if we use as the standard $\theta_n^{(2)}$, a hypothetical estimator which achieves the Cramér-Rao lower variance bound, we have as the asymptotic efficiency of the maximum likelihood estimator 100%. Although he left the rigorous proofs for later researchers (e.g., Cramér [1]), Fisher, in 1922, stated this key result [4] while pointing out in great detail the generally poor efficiencies obtained in the Pearson family if the method of moments is employed. Fisher neatly side-stepped the question of what to do in case one did not know the functional form of the unknown density. He did this by separating the problem of determining the form of the unknown density (in Fisher's terminology, the problem of "specification") from the problem of determining the parameters which characterize a specified density (in Fisher's terminology, the problem of "estimation"). The specification problem was to be solved by extramathematical means: "As regards problems of specification, these are entirely a matter for the practical statistician, for those cases where the qualitative nature of the hypothetical population is known do not involve any problems of this type. In other cases we may know by experience what forms are likely to be suitable, and the adequacy of our choice may be tested *a posteriori*: We must confine ourselves to those forms which we know how to handle, or for which any tables which may be necessary have been constructed."

Months before his death [14] Pearson *seemingly* pointed out the specification-estimation difficulty of classical maximum likelihood. "To my astonishment that method depends on first working out the constants of the frequency curve by the Method of Moments, and then superposing on it, by what Fisher terms the "Method of Maximum Likelihood," a further approximation to obtain, what he holds he will thus get, "more efficient values" of the curve constants." Actually, to take the statement in context, Pearson was objecting to the use of moments estimators as "starters" for an iterative algorithm to obtain maximum likelihood estimators. Apparently, Pearson failed to perceive the major difficulty with parametric maximum likelihood estimation. His arguments represent a losing rear-guard action. Nowhere does he make the point that since maximum likelihood assumes a great deal to be "given," it should surprise no one that it is more efficient than more general procedures *when the actual functional form of the density is that assumed.* But what about the fragility of maximum likelihood when the wrong functional form is assumed?

Fisher's answering diatribe [5] shortly after Pearson's death was unnecessary. By tacitly conceding Fisher's point that the underlying density's functional form is known prior to the analysis of a data set. Pearson had lost the controversy.

Neither in the aforementioned harsh "obituary" of one of Britain's foremost scientific intellects, nor subsequently, did Fisher resolve the specification-estimation difficulty of classical maximum likelihood. The victory of Fisher forced upon statistics a parametric straightjacket which it wears to this day. Although Fisher was fond of pointing out the difficulties of assuming the correct prior distribution, $p(\theta)$, in (32), he did not disdain to make a prodigious leap of faith in his selection of $f(x|\theta)$. In the subsequent chapters of this book we wish to examine techniques whereby we may attack the problem of estimating the density function of x without prior assumptions (other than smoothness) as to its functional form.

References

[1] Cramér, Harald (1946). *Mathematical Methods of Statistics*. Princeton: Princeton University Press.

[2] David, F. N. (1962). *Games, Gods and Gambling*. New York: Hafner.

[3] Elderton, William Palen, and Johnson, Norman Lloyd (1969). *Systems of Frequency Curves*. Cambridge: Cambridge University Press.

[4] Fisher, R. A. (1922). "On the mathematical foundations of theoretical statistics." *Philosophical Transactions of the Royal Society of London, Series A* 222: 309–68.

[5] ——— (1937). "Professor Karl Pearson and the method of moments." *Annals of Eugenics* 7: 303–18.

[6] Gosset, William Sealy (1908). "The probable error of a mean." *Biometrika* 6: 1–25.

[7] Hansmann, G. H. (1934). "On certain non-normal symmetrical frequency distributions." *Biometrika* 26: 129–95.

[8] Kendall, Maurice G., and Stuart, Alan (1958). *The Advanced Theory of Statistics*, Vol. I. New York: Hafner.

[9] ——— (1961). *The Advanced Theory of Statistics*, Vol. II. New York: Hafner.

[10] Mouzon, E. D. (1930). "Equimodal frequency distributions." *Annals of Mathematical Statistics* 1: 137–58.

[11] Ord, J. K. (1972). *Families of Frequency Distributions*. New York: Hafner.

[12] Pearson, E. S. (1965). "Studies in the history of probability and statistics. XIV. Some incidents in the early history of biometry and statistics, 1890–94." *Biometrika* 52: 3–18.

[13] Pearson, Karl (1895). "Contributions to the mathematical theory of evolution. II. Skew variation in homogeneous material." *Philosophical Transactions of the Royal Society of London, Series A* 186; 343–414.

[14] ——— (1936). "The method of moments and the method of maximum likelihood." *Biometrika* 28: 34–59.

[15] Walker, Helen M. (1929). *Studies in the History of Statistical Method.* Baltimore: Williams and Wilkins.

[16] Westergaard, Harald (1968). *Contributions to the History of Statistics.* New York: Agathon.

[17] Wald, Abraham (1949). "Note on the consistency of the maximum likelihood estimate." *Annals of Mathematical Statistics* 20: 595–601.

[18] Wilks, Samuel S. (1962). *Mathematical Statistics.* New York: John Wiley and Sons.

[19] Zoch, R. T. (1934). "Invariants and covariants of certain frequency curves." *Annals of Mathematical Statistics* 5: 124–35.

CHAPTER

2 Some Approaches to Nonparametric Density Estimation

2.1 The Normal Distribution as Universal Density

Francis Galton, perhaps more than any other researcher, deserves the title of Prophet of the Normal Distribution: "I know of scarcely anything so apt to impress the imagination as the wonderful form of cosmic order expressed by the 'Law of Frequency of Error.' The law would have been personified by the Greeks and deified, if they had known of it. It reigns with serenity and in complete self-effacement amidst the wildest confusion. The huger the mob and the greater the apparent anarchy, the more perfect is its sway. It is the supreme law of Unreason" [20, p. 66]. Moreover, Galton perceived the heuristics of the Central Limit Theorem: "The (normal) Law of Error finds a footing whenever the individual peculiarities are wholly due to the combined influence of a multitude of 'accidents'" [20, p. 55].

As one form of this theorem, we give Rényi's version [37, p. 223] of Lindeberg's 1922 result [31].

Theorem 1. (Central Limit Theorem). Let $x_1, x_2, \ldots, x_n, \ldots,$ be a sequence of independent random variables. Let $E(x_n) = 0$ and $E(x_n^2) < \infty$ for all n. Let $F_n(x) = P(x_n \le x)$.
Let

$$z_n = \sum_{j=1}^{n} x_j \tag{1}$$

and

$$s_n^2 = E(z_n^2) = \sum_{j=1}^{n} E(x_j^2). \tag{2}$$

For $\epsilon > 0$, let

$$B_n^*(\epsilon) = \frac{1}{s_n^2} \sum_{j=1}^{n} \int_{|x| > \epsilon s_n} x^2 \, dF_j(x). \tag{3}$$

Suppose that for every $\epsilon > 0$,

$$\lim_{n \to \infty} B_n^*(\epsilon) = 0. \tag{4}$$

Then one has uniformly (in x):

$$\lim_{n \to +\infty} P\left(\frac{z_n}{s_n} < x\right) = N(x), \tag{5}$$

where $N(x) = \dfrac{1}{\sqrt{2\pi}} \displaystyle\int_{-\infty}^{x} e^{-u^2/2} \, du.$

Proof.

Let $\quad \|f\| = \sup_x |f(x)| \quad \text{for} \quad f \in C(-\infty, \infty),$

where $C(-\infty, \infty)$ is the set of all bounded continuous functions defined on the real line.

For $F(\cdot)$ any cumulative probability distribution function, let us define the operator A_F on $C(-\infty, \infty)$ by

$$A_F f(x) = \int_{-\infty}^{\infty} f(x + y) \, dF(y). \tag{6}$$

Clearly, A_F is linear. Also, since no shifted average of f can be greater than the supremum of f,

$$\|A_F f(x)\| \leq \|f(x)\|, \tag{7}$$

i.e., A_F is a contraction operator. And if $G(\cdot)$ is another *cdf*

$$\begin{aligned} A_F A_G f(x) = A_G A_F f(x) &= \int_{-\infty}^{\infty} \int_{-\infty}^{\infty} f(x + y + z) \, dF(y) \, dG(z) \\ &= \int_{-\infty}^{\infty} f(x + u) \, dH(u) = A_H f(x), \quad \text{where } H = F^*G. \end{aligned} \tag{8}$$

Now if $A_1, A_2, \ldots, A_n, B_1, B_2, \ldots, B_n$ are linear contraction operators, since

$$\prod_{j=1}^{n} A_j - \prod_{\ell=1}^{n} B_\ell = \sum_{k=1}^{n} \left[\left(\prod_{i<k} A_i \right) (A_k - B_k) \prod_{j>k} B_j \right]. \tag{9}$$

We have

$$\|A_1 A_2 \cdots A_n f - B_1 B_2 \cdots B_n f\| \leq \sum_{k=1}^{n} \|A_k f - B_k f\|. \tag{10}$$

Let $G_n(x) = P[z_n/s_n \leq x]$.

Then,

$$G_n(x) = G_{n1}(x)*G_{n2}(x)* \cdots *G_{nn}(x), \tag{11}$$

where

$$\begin{aligned} G_{nk}(x) &= F_k(xs_n) \quad \text{for} \quad k = 1, 2, \ldots, n \\ &= P[x_k \le xs_n] = P[x_k/s_n \le x]. \end{aligned} \tag{12}$$

So

$$\begin{aligned} \int_{-\infty}^{\infty} x \, dG_{nk}(x) &= \int_{-\infty}^{\infty} x \, dF_k(xs_n) \\ &= \frac{1}{s_n} \int \zeta \, dF_k(\zeta) \\ &= \frac{1}{s_n} E(x_k) = 0 \end{aligned} \tag{13}$$

and

$$\int_{-\infty}^{\infty} x^2 \, dG_{nk}(x) = \frac{1}{s_n^2} E(x_k^2). \tag{14}$$

Now,

$$N(x) = N_{n1}(x)*N_{n2}(x)* \cdots *N_{nn}(x), \tag{15}$$

where

$$N_{nk}(x) = N\left(\frac{xs_n}{\sqrt{E(x_k^2)}}\right). \tag{16}$$

But

$$\int_{-\infty}^{\infty} x \, dN_{nk}(x) = 0 \tag{17}$$

and

$$\int x^2 \, dN_{nk}(x) = \frac{E(x_k^2)}{s_n^2}. \tag{18}$$

From (10), we have for every $f \in C(-\infty, \infty)$

$$\|A_{G_n} f - A_N f\| \le \sum_{k=1}^{n} \|A_{G_{nk}} f - A_{N_{nk}} f\|. \tag{19}$$

Now, denoting by $C^3(-\infty, \infty)$ the set of functions having the first three derivatives continuous and bounded, then if $f \in C^3(-\infty, \infty)$,

$$(20) \qquad A_{G_{nk}} f = f(x) + \frac{f''(x)E(x_k^2)}{2s_n^2} + \frac{1}{6}\int_{|y| \le \epsilon} y^3 f'''(x + \zeta_1 y)\, dG_{nk}(y) + \frac{1}{2}\int_{|y| > \epsilon} y^2[f''(x + \zeta_2 y) - f''(x)]\, dG_{nk}(y),$$

where $0 \le \zeta_1 \le 1$ and $0 \le \zeta_2 \le 1$.

Similarly,

$$(21) \qquad A_{N_{nk}} f = f(x) + \frac{f''(x)E(x_k^2)}{2s_n^2} + \frac{1}{6}\int_{|y| \le \epsilon} y^3 f'''(x + \zeta_1 y)\, dN_{nk}(y) + \frac{1}{2}\int_{|y| > \epsilon} y^2[f''(x + \zeta_2 y) - f''(x)]\, dN_{nk}(y).$$

Thus,

$$(22) \quad \|A_{G_{nk}} f - A_{N_{nk}} f\| \le k_1\left[\int_{|y| > \epsilon} y^2\, dG_{nk}(y) + \int_{|y| > \epsilon} y^2\, dN_{nk}(y)\right] + \frac{k_2 \epsilon}{6}\left[\int_{|y| \le \epsilon} y^2\, dG_{nk}(y) + \int_{|y| \le \epsilon} y^2\, dN_{nk}(y)\right],$$

where $\sup|f''(x)| = k_1$ and $\sup|f'''(x)| = k_2$. Now,

$$\frac{k_2 \epsilon}{6}\left[\int_{|y| \le \epsilon} y^2\, dG_{nk}(y) + \int_{|y| \le} y^2\, dN_{nk}(y)\right] \le \frac{k_2 \epsilon E(x_k^2)}{3s_n^2};$$

$$\int_{|y| > \epsilon} y^2\, dG_{nk}(y) = \frac{1}{s_n^2}\int_{|y| > \epsilon s_n} y^2\, dF_k(y);$$

$$\int_{|y| > \epsilon} y^2\, dN_{nk}(y) = \frac{E(x_k^2)}{s_n^2}\int_{|u| > \epsilon s_n/D_k} \frac{u^2 e^{-u^2/2}\, du}{\sqrt{2\pi}}.$$

So

$$(23) \quad \|A_{G_n} f - A_N f\| \le \frac{k_2 \epsilon}{3} + \frac{k_1}{2}\left(B_n^*(\epsilon) + \frac{1}{\sqrt{2\pi}}\int_{|u| > \epsilon s_n/\Delta_n} u^2 e^{-u^2/2}\, du\right),$$

where $\Delta_n = \max\limits_{1 \le k \le n} \sqrt{E(x_k^2)}$.

But for each $k \leq n$ and every $\epsilon' > 0$

$$\frac{E(x_k^2)}{s_n^2} = \frac{1}{s_n^2}\int_{-\infty}^{\infty} x^2\, dF_k(x) \tag{24}$$

$$= \frac{1}{s_n^2}\int_{|x|>\epsilon' s_n} x^2\, dF_k(x) + \frac{1}{s_n^2}\int_{|x|\leq \epsilon' s_n} x^2\, dF_k(x)$$

$$\leq B_n^*(\epsilon') + \epsilon'^2.$$

Hence,

$$\frac{\Delta_n^2}{s_n^2} \leq \epsilon'^2 + B_n^*(\epsilon'). \tag{25}$$

So, by (4),

$$\lim_{n\to\infty} \frac{\Delta_n}{s_n} \leq \epsilon'.$$

Since ϵ' can be chosen arbitrarily small,

$$\lim_{n\to\infty} \frac{\Delta_n}{s_n} = 0.$$

Thus, we have uniformly in x

$$\lim_{n\to\infty} A_{G_n} f = A_N f. \tag{26}$$

Therefore, we have for each $f \in C^3(-\infty, \infty)$

$$\lim_{n\to\infty} \int_{-\infty}^{\infty} f(y)\, dG_n(y) = \int_{-\infty}^{\infty} f(y)\, dN(y). \tag{27}$$

Now (27) holds for all f in $C^3(-\infty, \infty)$. Consider the following C^3 function:

$$f(\epsilon, x, y) = \begin{cases} 1 & \text{for } -\infty < y < x \\ \left[1 - \left(\dfrac{x-y}{\epsilon}\right)^4\right]^4 & \text{for } x \leq y \leq x + \epsilon. \\ 0 & \text{for } x + \epsilon < y. \end{cases}$$

Further, by (27) we know that

$$\lim_{n\to\infty} \int_{-\infty}^{\infty} f(\epsilon, x, y)\, dG_n(y) = \int_{-\infty}^{\infty} f(\epsilon, x, y)\, dN(y).$$

But

$$\int_{-\infty}^{\infty} f(\epsilon, x, y)\, dN(y) \leq N(x+\epsilon) \quad \text{for all } \epsilon > 0.$$

Hence

$$\limsup_{n\to\infty} G_n(x) \leq N(x+\epsilon).$$

Similarly by looking at $f(\epsilon, x-\epsilon, y)$ we find that

$$\liminf_{n\to\infty} G_n(x) \geq N(x-\epsilon).$$

Thus

$$\lim_{n\to\infty} G_n(x) = N(x). \quad \blacksquare$$

The Central Limit Theorem does not state that a random sample from a population satisfying the Lindeberg conditions will have an approximately normal distribution if the sample size is sufficiently large. It is rather the sample sum, or equivalently the sample mean, which will approach normality as the sample size becomes large. Thus, if one were to obtain average IQ's from say, 1,000 twelfth-grade students chosen at random from all twelfth-grade classes in the United States, it might be reasonable to apply normal theory to obtain a confidence interval for the average IQ of American twelfth-grade students. It would not be reasonable to infer that the population of IQ's of twelfth-grade students is normal. In fact, such IQ populations are typically multimodal.

Yet Galton treated almost any data set as though it had come from a normal distribution. Such practice is common to this day among many applied workers—particularly in the social sciences. Frequently, such an assumption of normality does not lead to an erroneous analysis. There are many reasons why this may be so in any particular data analysis.

For example, the Central Limit Theorem may indeed be driving the data to normality. Let us suppose that each of the observations $\{x_1, x_2, \ldots, x_n\}$ in a data set is the average of N independent and identically distributed random variables $\{y_{i1}, y_{i2}, \ldots, y_{iN}\}$, $(i = 1, 2, \ldots, n)$, having finite mean μ and variance σ^2. Then, for N very large each x will be approximately normal with mean μ and variance σ^2/N. We observe that the size of n has nothing to do in bringing about the approximate normality of the data. It need not be the case that the $\{y_{ij}\}$ be observables—in fact, it is unlikely that one would even know what y is.

Generalizations on the above conditions can be made which will still leave the $\{x_i\}$ normal. For example [10, p. 218], suppose

$$x = g(y_1, y_2, \ldots, y_N), \tag{28}$$

where g has continuous derivatives of the first and second orders in the neighborhood of $\mu = (\mu_1, \mu_2, \ldots, \mu_N)$, where $\mu_j = E(y_j)$. Then we may write

$$g(y_1, y_2, \ldots, y_N) = g(\mu_1, \mu_2, \ldots, \mu_N) + \sum_{\upsilon=1}^{N} \frac{\partial g}{\partial \mu_\upsilon}\bigg|_{\mu} (y_\upsilon - \mu_\upsilon) + R, \tag{29}$$

where R contains derivatives of the second order. The first term is a constant. The second term has zero average. Under certain conditions, the effect of the R term is negligible for N large. Hence, if the Lindeberg condition (4) is satisfied for the random variables $\left\{\left.\frac{\partial g}{\partial \mu_\upsilon}\right|_\mu (y_\upsilon - \mu_\upsilon)\right\}$, we have that x is asymptotically (in N) distributed as a normal variate with mean $g(\mu_1, \mu_2, \ldots, \mu_N)$.

Many other generalizations are possible. For example, we may make N a random variable and introduce dependence between the $\{y_i\}$. Still, to assume that a random data set is normal due to some mystical ubiquity of the Central Limit Theorem would seem to be an exercise in overoptimism. Yet Galton was anything but naive. His use of the normal distribution is laced with caveats, e.g., "I am satisfied to claim that the Normal Curve is a fair average representation of the Observed Curves during nine-tenths of their course; that is, for so much of them as lies between the grades of 5% and 95%." [20, p. 57], and "It has been shown that the distribution of very different human qualities and faculties is approximately Normal, and it is inferred that with reasonable precautions we may treat them as if they were wholly so, in order to obtain approximate results" [20, p. 59].

2.2. The Johnson Family of Distributions

There is no doubting that the examination of a vast number of disparate data sets reveals that the normal distribution, for whatever reasons, is a good approximation. This being so, it is fair to hope that a nonnormal data set might be the result of passing a normal predata set through some filter. Galton suggested that one might seek to find a transformation which transformed a nonnormal data set $(x_1, x_2, \ldots, x_n)$ into a normal data set $\{z_1, z_2, \ldots, z_n)$. One such transformation, proposed by Galton [19], was

$$z = \gamma + \delta \log\left(\frac{x - \xi}{\lambda}\right). \tag{30}$$

Thus, if z is $N(0, 1)$, then x is said to be log normal. Moreover, since z may be written

$$\begin{aligned} z &= (\gamma - \delta \log \lambda) + \delta \log(x - \xi) \\ &= \gamma' + \delta \log(x - \xi), \end{aligned}$$

we may estimate the three characterizing parameters from the first three sample moments. The density of $w = \frac{x - \xi}{\lambda}$ is given by

$$g(w) = \frac{\delta}{\sqrt{2\pi}\, w} \exp\left[-\frac{1}{2}(\gamma + \delta \log w)^2 \right], \qquad w \geq 0. \tag{31}$$

The density is unimodal and positively skewed.

N. L. Johnson [25] (see also [27, p. 167]) has referred to data sets which may be transformed to approximate normality by the transformation in (30) as members of the "S_L system." He has suggested that one should consider a more general transformation to normality, namely,

$$z = \gamma + \delta f\left(\frac{x - \xi}{\lambda}\right), \tag{32}$$

where f is nondecreasing in $\left(\frac{x - \xi}{\lambda}\right)$ and is reasonably easy to calculate. One such f is

$$f(y) = \log\left(\frac{y}{1 - y}\right). \tag{33}$$

(Bounded) data sets which may be transformed to approximate normality by (33) are members of the "S_B system." Various methods of estimating γ and δ are given, e.g., when both endpoints are known, we have

$$\hat{\gamma} = -\bar{f}/s_f$$
$$\hat{\delta} = 1/s_f,$$

where

$$\bar{f} = \frac{1}{n} \sum_{i=1}^{n} f_i$$

$$s_f^2 = \frac{1}{n} \sum_{i=1}^{n} (f_i - \bar{f})^2$$

and

$$f_i = \log\left(\frac{x_i - \xi}{\xi + \lambda - x_i}\right), \qquad i = 1, 2, \ldots, n.$$

Densities of the S_B class are unimodal unless

$$\delta < 1/\sqrt{2}, \qquad |\gamma| < \delta^{-1}\sqrt{1 - 2\delta^2} - 2\delta \tanh^{-1}(\sqrt{1 - 2\delta^2}).$$

In the latter case, the density is bimodal. Letting $w = \frac{x - \xi}{\lambda}$, the characterizing density of the S_B class is

$$(34)\quad g(w) = \frac{\delta}{\sqrt{2\pi}\, w(1-w)} \exp\left[-\frac{1}{2}\left(\gamma + \delta \log \frac{w}{1-w}\right)^2\right], \qquad 0 \le w \le 1.$$

A third "transformation to normality" proposed by Johnson is

$$(35)\qquad f = \sinh^{-1}(y).$$

Since

$$\mu_1' = E(x) = -\sqrt{\omega}\,\sinh(\Omega)$$

$$\mu_2 = E(x - \mu_1')^2 = \frac{1}{2}(\omega - 1)(\omega \cosh 2\Omega + 1)$$

$$\mu_3 = E(x - \mu_1')^3 = -\frac{1}{4}(\omega - 1)^2\sqrt{\omega}\{\omega(\omega + 2)\sinh 3\Omega + 3 \sinh \Omega\}$$

$$\mu_4 = E(x - \mu_1')^4 = \frac{1}{8}(\omega - 1)^2\{\omega^2(\omega^4 + 2\omega^3 + 3\omega^2 - 3)\cosh 4\Omega + 4\omega^2(\omega + 2)\cosh 2\Omega + 3(2\omega + 1)\}$$

where $\omega = e^{\delta^{-2}}$ and $\Omega = \gamma/\delta$,
the method of moments might be used to compute the parameters of the transformation. Data sets transformable to normality by the use of (35) are said to be members of the S_u (unbounded) system. S_u densities are unimodal with mode lying between the median and zero. They are positively or negatively skewed, according to whether γ is negative or positive [27, p. 172]. Letting $w = \frac{x - \xi}{\lambda}$, we have as the characterizing S_u density:

$$(36)\qquad g(w) = \frac{\delta}{\sqrt{2\pi}} \frac{1}{\sqrt{w^2 + 1}} \exp\left[-\frac{1}{2}[\gamma + \delta \log\{w + \sqrt{w^2 + 1}\}]^2\right], \qquad -\infty \le w \le \infty.$$

The approach of Johnson is clearly philosophically related to that employed by Pearson with his differential equation family of densities. In both situations four parameters are to be estimated. The selection of a transformation by Johnson may be viewed as having a similar theoretical-heuristic mix as the selection by Pearson of his differential equation model. This selection is a solution of Fisher's problem of "specification." It is to be noted that Fisher's specification-estimation dichotomy is present in the Johnson approach. The data are used in estimating the four parameters but not in the determination of the transformation to normality.

There exists strong evidence that a number of data sets may be transformed to approximate normality by the use of one of Johnson's transformations. Moreover, we might extend the Johnson class of transformations to take care of additional sets. However, an infinite number of transformations would be required to take care of random samples from all possible continuous probability distributions. Any attempt to develop an exhaustive class of data-oriented transformations to normality would be, naturally, of the same order of complexity as the task of general probability density estimation itself.

2.3. The Symmetric Stable Distributions

We have seen that the normal distribution arises quite naturally, because a random observation may itself be the sum of random variables satisfying the conditions of Theorem 1. In situations where a set of observations is not normal, it might still be the case that each observation arises by the additivity of random variables. For example, let us suppose that an observation $x = s_n = \sum_{j=1}^{n} z_j$ where the $\{z_j\}$ are independently identically distributed with *cdf* F. If there exist normalizing constants, $a_n > 0$, b_n, such that the distribution of $(s_n - b_n)/a_n$ tends to some distribution G, then F is said to belong to the *domain of attraction* of G. It is shown by Feller [16, p. 168] that G possesses a domain of attraction if and only if G is *stable.* G is stable if and only if a set of i.i.d. random variables $\{w_1, w_2, \ldots, w_n\}$ with *cdf* G has the property that for some $0 < \alpha < 2$,

$$n^{-1/\alpha}[w_1 + w_2 + \cdots + w_n]$$

has itself *cdf* G.

Fama and Roll [13, 14] have considered a subset of the class of stable distributions, namely, those having characteristic function

$$\varphi_x(t) = E(e^{itx}) = \exp[i\,\delta t - |ct|^{\alpha}], \tag{37}$$

where $1 \leq \alpha \leq 2$.

It is interesting to note that this family of symmetric stable distributions runs the gamut of extremely heavy tails (Cauchy, $\alpha = 1$) to sparse tails (normal, $\alpha = 2$). It requires only three parameters to characterize a member of this family. Hence, it would appear that a methodology might be readily developed to handle the specification and estimation problem for many nonnormal data sets. There is one apparent drawback, namely, whereas the characteristic function has a very simple form, there is no known simple form for the corresponding density. Bergstrom [3] gives the series representation

$$g(u) = \frac{1}{\pi\alpha} \sum_{k=0}^{\infty} (-1)^k \frac{\Gamma\left(\frac{2k+1}{\alpha}\right)}{(2k)!} u^{2k}, \tag{38}$$

where $u = \dfrac{x - \delta}{c}$

and $1 < \alpha \leq 2$.

Thus, Fama and Roll resorted to an empirical Monte Carlo approach for the estimation of the characterizing parameters. They suggest as an estimator for the location parameter δ the 25% trimmed mean, i.e.,

$$\hat{\delta} = \frac{2}{n}\left[x_{([.25n])} + x_{([.25n])+1)}, \ldots, + x_{([.75n])}\right]. \tag{39}$$

For an estimate of the scale parameter c, they suggest

$$\hat{c} = \frac{1}{(.827)2}\left[x_{([.72n])} - x_{([.28n])}\right]. \tag{40}$$

As one would expect, the most difficult parameter to be estimated is α. They suggest the estimation of α by the computation

$$\hat{v} = \frac{x_{([.97n])} - x_{([.03n])}}{2\hat{c}},$$

followed by a table look-up using a chart given in [14]. Only time will tell whether, in practice, a substantial number of nonnormal data sets can be approximated by members of the symmetric stable family. Clearly, only unimodal symmetric densities can be so approximated. Moreover, a very large data set would be required to obtain a reliable estimator for α.

In many situations most of the practically relevant information in a random sample would be obtained if we could only estimate the center of the underlying density function. One might naively assume this could easily be accomplished by using the sample mean $\bar{x} = \frac{1}{n}\sum_{j=1}^{n} x_j$. However, if the observations come from the Cauchy distribution with density

$$f(x) = \frac{\sigma}{\pi} \frac{1}{(x-\mu)^2 + \sigma^2}$$

and characteristic function

$$\varphi_x(t) = e^{it\mu - |t|\sigma}.$$

Then the characteristic function of $\bar{x}$ is given by

$$\varphi_{\bar{x}}(t) = \left[\varphi_x\left(\frac{t}{n}\right)\right]^n = e^{it\mu - |t|\sigma}.$$

Thus, using the sample mean from a sample of size 10^{10} would be no more effective in estimating the location parameter μ than using a single observation.

One might then resort to the use of the sample median

$$\hat{x}_{me} = x_{[n/2]}.$$

However, if the data came from a normal distribution with mean μ and variance σ^2, then

$$\frac{E[(\hat{x}_{me} - \mu)^2]}{E[(\bar{x} - \mu)^2]} \sim 1.57.$$

Thus, the efficiency of the sample median relative to that of the sample mean is only 64%.

The task of obtaining location estimators which may be used for the entire gamut of symmetrical unimodal densities and yet have high efficiency when compared to the best that could be obtained if one knew a priori the functional form of the unknown density is one of extreme complexity. We shall not attempt to cover the topic here, since the monumental work of Tukey and his colleagues on the subject has been extensively explicated and catalogued elsewhere [1] and since our study is concerned with those cases where it is important to estimate the unknown density.

However, one recent proposal [49] for estimating the location parameter μ is given by the iterative algorithm

$$\hat{\mu}_{m+1} = \frac{\sum_{i=1}^{n} w_{mi} x_i}{\sum_{i=1}^{n} w_{mi}}, \tag{41}$$

where the *biweight function*

$$w_{mi} = (1 - u_{mi}^2)^2 \quad \text{if} \quad u_{mi}^2 < 1$$
$$= 0 \quad \text{otherwise}$$

$$u_{mi} = \frac{x_i - \hat{\mu}_m}{cs}$$

s is a scale estimator (e.g., the interquartile range or median absolute deviation)

c is an appropriately chosen constant,

and

$\hat{\mu}_0$ is the sample median.

2.4. Series Estimators

One natural approach in the estimation of a probability density function is first to devise a scheme for approximating the density, using a simple basis assuming perfect information about the true density. Then the data can be used to estimate the parameters of the approximation. Such a two-step approach is subject to some danger. We might do a fine job of devising an approximation in the presence of perfect information, which would be a catastrophe as the front end of an estimation procedure based on data.

For example, suppose we know that a density exists and is infinitely differentiable from $-\infty$ to $+\infty$. Then if we actually knew the functional form of the unknown density, we could use a Maclaurin's expansion representation

$$f(x) = \sum_{j=0}^{\infty} x^j f^{(j)}(0)/j! = \sum_{j=0}^{\infty} a_j x^j. \tag{42}$$

But to estimate the $\{a_j\}$ from a finite data set $\{x_1, x_2, \ldots, x_n\}$ to obtain

$$\hat{f}(x) = \sum_{j=0}^{\infty} \hat{a}_j x^j \tag{43}$$

is impossible.

If we decide to use

$$\hat{f} = \sum_{j=0}^{m} \hat{a}_j x^j \quad \text{with} \quad m \ll n \tag{44}$$

and are satisfied with a decent estimate over some finite interval $[a, b]$, we can come up with some sort of estimate—probably of little utility.

We might first preprocess the data via the transformation

(45) $$y_j = \frac{x_j - \text{sample median}}{\text{sample interquartile range}}.$$

This will help if the unknown density is symmetrical and unimodal. But if the density is multimodal, we are still in big trouble.

If we assume that the unknown density is close to normal, we might first transform the data via

$$y_j = \frac{x_j - \bar{x}}{s}$$

and then use an approximation which would be particularly good if y is $N(0, 1)$. Such an approximation is the Gram–Charlier series

(46) $$f(y) = \alpha(y) \sum_{j=0}^{s} a_j H_j(y)/j!$$

where $\alpha(y) = \frac{1}{\sqrt{2\pi}} e^{-y^2/2}$

and the Hermite polynomials $\{H_j(y)\}$ are given by

(47) $$H_j(y) = y^j - \frac{j^{[2]}y^{j-2}}{2.1!} + \frac{j^{[4]}y^{j-4}}{2^2.2!} - \frac{j^{[6]}y^{j-6}}{2^3.3!} + \cdots.$$

We recall that the Hermite polynomials are orthonormal with respect to the weight function α, i.e.,

(48) $$\int_{-\infty}^{\infty} H_i(y)H_j(y)\alpha(y)\,dy = 1 \quad \text{if} \quad i = j$$
$$= 0 \quad \text{otherwise.}$$

If we select the a_j so as to minimize

(49) $$S(a) = \int (f_0(y) - \alpha(y) \sum_{j=0}^{s} a_j H_j(y))^2 \alpha^{-1}(y)\,dy,$$

then the $\{a_j\}$ are given for all s [34, p. 26] by

(50) $$\begin{aligned} a_0 &= 1 \\ a_1 &= 0 \\ a_2 &= (\mu_2 - 1)/2! \\ a_3 &= \mu_3/3! \\ a_4 &= (\mu_4 - 6\mu_2 + 3)/4! \\ a_5 &= (\mu_5 - 10\mu_3)/5! \\ a_6 &= (\mu_6 - 15\mu_4 + 45\mu_2 - 15)/6! \quad \text{etc.} \end{aligned}$$

and

$$\mu_j = \int_{-\infty}^{\infty} y^j f_0(y)\, dy.$$

If y is $N(0, 1)$, then

$$\mu_{2k} = \frac{(2k)!}{2^k k!}$$

$$\mu_{2k+1} = 0.$$

So

$$a_j = 0 \qquad \text{for} \qquad j \neq 1.$$

Thus, for y $N(0, 1)$, we have simply

$$f(y) = \frac{1}{\sqrt{2\pi}} \exp\left[-\frac{y^2}{2}\right].$$

Naturally, any other density will require a more complicated Gram–Charlier representation. Nevertheless, to the extent that y is nearly $N(0, 1)$, it is reasonable to hope that the $\{a_j\}$ will quickly go to zero as j increases (in fact, this is one measure of closeness of a density to $N(0, 1)$). Cramér [9, see also 27, p. 161] has shown that if f_0 is of bounded variation in every interval, and if $\int |f_0(y)|\alpha^{-1}(y)\, dy < \infty$, then the Gram–Charlier representation (46) exists and is uniformly convergent in every finite interval.

The situation is rather grim for Gram–Charlier representations when f_0 is significantly different from $N(0, 1)$ and the coefficients $\{a_j\}$ must be estimated from data. The method of moments appears a natural means for obtaining $\{\hat{a}_j\}$. However, as a practical matter, sample moments past the fourth generally have extremely high variances. Shenton [46] has examined the quality of fitting $\hat{f}$, using the method of moments for the simple case where

$$f_0(y) = \sum_{j=0}^{4} a_j H_j(y)/j!. \tag{51}$$

He found that even in this idealized case, the procedure is not particularly promising. It appears clear that as a practical matter the naive Gram–Charlier estimation procedure is unsatisfactory unless f_0 is very close to α (in which case we would probably do as well to assume y is normal).

Kronmal and Tarter [29, 47, 48] have considered the representation for the density f defined on [0, 1]

$$f(x) = \sum_{j=-\infty}^{\infty} B_k \psi_k(x), \tag{52}$$

where

$$\psi_k(x) = e^{2\pi i k x}$$

and

$$B_k = \int_0^1 \psi_k(x) f(x)\, dx.$$

For the practical case, where f is unknown, they propose as an estimate of f based on the data set $\{x_1, x_2, \ldots, x_n\}$

(53) $$\hat{f}(x) = \sum_{k=-\infty}^{\infty} \hat{B}_k \psi_k(x),$$

where

$$\hat{\hat{B}}_k = \frac{1}{n} \sum_{j=1}^{n} e^{-2\pi i k x_j}$$

and

$$\hat{B}_k = \hat{\hat{B}}_k \quad \text{if} \quad \hat{\hat{B}}_k \hat{\hat{B}}_{-k} > \frac{2}{n+1}, \qquad k \leq K(n)$$

$$= 0 \quad \text{otherwise.}$$

The fact that the Kronmal–Tarter approach requires that the unknown density have finite support is small liability. It would be unreasonable to expect any "nonparametric" data-based approach to give an estimate of the density outside some interval $[a, b]$ within the range of the data. A transformation from $[a, b]$ to $[0, 1]$ is easily obtained via

$$y = (x - a)/(b - a).$$

Of greater difficulty is the possibility of high frequency wiggles in the estimated density. Naturally, in order to have any hope of consistency, we should have $\lim_{n \to \infty} K(n) = \infty$. However, as a practical matter, since n is never infinite, consistency is not a particularly important criterion. Accordingly, Kronmal and Tarter suggest using $k = 10$. Of course, with such a truncation, we would not expect $\hat{f}(x)$ to be nonnegative for all x or that $\int \hat{f}(x)\, dx \equiv 1$.

More generally, let us approximate a known $f \in L^2[0, 1]$ defined on $[0, 1]$ by

(54) $$f_N(x) = f_0(x) \sum_{k=0}^{N} a_k \varphi_k(x),$$

where

$$\int_0^1 \varphi_j(x) \varphi_k(x) f_0(x)\, dx = 1 \quad \text{if} \quad j = k$$

$$= 0 \quad \text{if} \quad j \neq k$$

$$\varphi_0(x) = 1$$

(55) $$a_k = \int_0^1 \varphi_k(x) f(x)\, dx = E_f(\varphi_k(x))$$

$f_0(x)$ is a density defined on $[0, 1]$.
Now, defining

(56) $$\|g(x)\|^2 = \int_0^1 g^2(x) f_0^{-1}(x)\, dx,$$

we have

(57) $$S(a) = \|f - f_0(x) \sum_{k=0}^{N} a_k \varphi_k(x)\|^2$$

$$= \|f\|^2 - \sum_{k=0}^{N} \left(\int_0^1 f(x)\varphi_k(x)\, dx \right)^2 + \sum_{k=0}^{N} \left(\int_0^1 f(x)\varphi_k(x)\, dx - a_k \right)^2.$$

Thus, for any N, the $\{a_k\}$ given in (55) will minimize $S(a)$.

Now, in the practical case, where f is known only through a random sample $\{x_1, x_2, \ldots, x_n\}$, a naive estimate for f is given by

(58) $$\hat{f}_n(x) = \frac{1}{n} \sum_{j=1}^{n} \delta(x - x_j),$$

where the Dirac delta function $\delta(.)$ is defined via

$$\int_{-\infty}^{\infty} \delta(y)\, dy = 1$$

$$\delta(y) = 0 \quad \text{if} \quad y \neq 0.$$

Using $\hat{f}_n$ in (55), we have as an estimate for a_k

(59) $$\hat{a}_k = \int_0^1 \left(\frac{1}{n} \sum_{j=1}^{n} \delta(x - x_j) \right) \varphi_k(x)\, dx$$

$$= \frac{1}{n} \sum_{j=1}^{n} \varphi_k(x_j),$$

giving

(60) $$\hat{f}(x) = f_0(x) \sum_{k=0}^{N} \hat{a}_k \varphi_k(x).$$

If $\{\varphi_k\}_{k=1}^{\infty}$ is a basis for $L^2[0, 1]$ then (54) with $N = \infty$ gives f. Consider the integrated mean square error of the estimator in (60)

(61) $$\text{MISE} = E_f \int_0^1 (f(x) - \hat{f}(x))^2 f_0(x)\, dx$$

$$= \sum_{k=N+1}^{\infty} a_k^2 + \frac{1}{n} \left[\sum_{k=0}^{N} \int_0^1 \varphi_k^2(x) f(x)\, dx - \sum_{k=0}^{N} a_k^2 \right].$$

We note by the Cauchy–Schwarz Inequality that the second term in (61) is nonnegative. The first term in (61) shows that for the MISE to go to zero, it is necessary that N go to infinity, as n goes to infinity. However, the second term shows that N must not go to infinity too rapidly relative to the sample size n.

Watson [54] considers the estimate

$$\hat{f}_w(x) = \sum_{k=0}^{\infty} \lambda_k(n)\hat{a}_k\varphi_k(x),$$

where the $\{\lambda_k(n)\}$ are chosen so as to minimize the mean integrated square error

$$J(\boldsymbol{\lambda}(n)) = E_f \int (f(x) - \hat{f}_w(x))^2\,dx = \sum_{k=0}^{\infty}\left\{a_k^2(1-\lambda_k(n))^2 + n^{-1}\lambda_k^2(n)\operatorname*{var}_f(\varphi_k(x))\right\}. \tag{62}$$

Hence, $\lambda_k(n) = a_k^2 \Big/ \left[a_k^2 + n^{-1}\operatorname*{var}_f(\varphi_k(x))\right]$.

Unfortunately, we know neither the $\{a_k\}$ nor $\left\{\operatorname*{var}_f(\varphi_k(x))\right\}$.

One ad hoc approach, not suggested by Watson, would be to replace a_k and $(\mathrm{var}(\varphi_k))$ in the expression for $\lambda_k(n)$ by their representations, using a prior guess f_0 for the unknown density:

$$\lambda_k(n) = \begin{cases} \left[1 + n^{-1}\dfrac{1 + \int \varphi_k^2(x) f_0(x)\,dx}{\left(\int \varphi_k(x) f_0(x)\,dx\right)^2}\right]^{-1} & \text{for } k \le \sqrt{n} \\ 0 & \text{for } k > \sqrt{n} \end{cases}$$

Brunk [7] uses a Bayesian approach to resolve this difficulty. He notes that it is quite natural to use as f_0 our best prior guess for f. Then natural prior guesses for the a_k are

$$\begin{aligned} E[a_k] &= 1 \quad \text{for} \quad k = 0 \\ &= 0 \quad \text{for} \quad k > 0. \end{aligned} \tag{63}$$

The second prior moments are more difficult to postulate. Brunk suggests

$$\begin{aligned} \mathrm{Cov}(a_j, a_k) &= \sigma_j^2 \quad \text{if} \quad j = k \\ &= 0 \quad \text{otherwise}. \end{aligned} \tag{64}$$

Then, to minimize

(65) $$S(\boldsymbol{\lambda}(n)) = E\left\{E_f \sum_{k=0}^{\infty} [\lambda_k(n)\hat{a}_k - a_k]^2 | a\right\}$$

we have

(66) $$\lambda_k(n) = \frac{n\sigma_k^2}{(n-1)\sigma_k^2 + E[E_f(\varphi_k^2(x)|a]}.$$

But if $f \sim f_0$, then it is reasonable to replace E_f by E_{f_0} to give

(67) $$E[E_{f_0}(\varphi_k^2(\mathbf{x}))\,|a] = 1.$$

Thus, we have

(68) $$\lambda_0(n) = 1$$

$$\lambda_k(n) = \frac{n}{n - 1 + (1/\sigma_k^2)} \quad \text{for} \quad k > 0.$$

Using the trigonometric basis

(69) $$\begin{gathered} \varphi_0(x) = 1 \qquad 0 \le x \le 1 \\ \varphi_{2k-1}(x) = \sqrt{2}\sin(2\pi kx) \\ \varphi_{2k}(x) = \sqrt{2}\cos(2\pi kx), \qquad k = 1, 2, \ldots \\ f_0(x) = 1, \qquad 0 \le x \le 1, \\ E[E_f(\varphi_k^2(\mathbf{x})|a)] = 1 \quad \text{for all} \quad k. \end{gathered}$$

Brunk gives several empirical alternatives for estimating $\lambda_k(n)$ when this basis is used. For example, let

(70) $$\xi_k = \left[\frac{1}{2}\left\{\hat{a}_{2k-1}^2 + \hat{a}_{2k}^2\right\} - \frac{1}{n}\right].$$

Then, if

(71) $$\gamma_k = \sigma_{2k-1}^2 = \sigma_{2k}^2, \qquad k = 1, 2, \ldots$$

let

(72) $$\hat{\gamma}_k = (1/4t_k^2)t_k + [1/n + (n-1)t_k'/n]^{-2}t_k',$$

where

$$\begin{aligned} t_k &= (1/5)^{k-1} \\ t_k' &= \max(\xi_k, 0), \qquad k = 1, 2, \ldots . \end{aligned}$$

Using the resulting estimates in (66), we obtain a sequence of damping factors $\{\lambda_k(n)\}$. Brunk notes that many procedures for choosing a sequence of γ_k, which decrease smoothly to 0 as k increases, appear to work better than any rule of the form

$$\begin{aligned} \lambda_k(n) &= 1 \quad \text{for} \quad k \le M \\ &= 0 \quad \text{for} \quad k > M. \end{aligned} \tag{73}$$

Let us now consider the following iterative algorithm for estimating an unknown density f with domain of positivity $[a, b]$. We assume a random sample $\{x_1, x_2, \ldots, x_n\}$ is given. Also we shall assume we have a prior guess as to the density—say $\hat{f}_0$ (the parameters characterizing $\hat{f}_0$ may be estimated by classical means, e.g., maximum likelihood). Let $\{\varphi_j^{(1)}\}$ be an appropriate orthonormal (with respect to $\hat{f}_0$) basis. Then we take as our estimate for f

$$\hat{f}_1(x) = \hat{f}_0(x) \sum_{j=0}^{N(1)} \hat{a}_j^{(1)} \varphi_j^{(1)}(x), \tag{74}$$

where

$$\hat{a}_0^{(1)} = 1$$

$$\hat{a}_j^{(1)} = \frac{1}{n} \sum_{i=1}^{n} \varphi_j^{(1)}(x_i), \qquad j > 1.$$

Using $\hat{f}_1$ as a new weight function, we obtain a new orthonormal basis $\{\varphi_j^{(2)}\}$. Then our second estimate for f is

$$\hat{f}_2(x) = \hat{f}_1(x) \sum_{j=0}^{N(2)} a_j^{(2)} \varphi_j^{(2)}(x), \tag{75}$$

where

$$\hat{a}_0^{(2)} = 1$$

$$\hat{a}_j^{(2)} = \frac{1}{n} \sum_{i=1}^{n} \varphi_j^{(2)}(x_i), \qquad j > 1.$$

If the algorithm converges to a result close to the true f, it is reasonable to hope that after a few iterations $N(m)$ can be taken to be small. Naturally, a crucial consideration here is the selection of an appropriate basis $\{\varphi_j^{(m)}\}$ at each stage of the process. We as yet have no good answers as to how this task may be accomplished and merely state it as an open problem.

As a procedure related to series estimation, we consider the following approach based on cubic B-splines. We shall assume that the data has been transformed to lie in $[-M(n), M(n)]$ where n is the size of a random sample $\{x_1, x_2, \ldots, x_n\}$ and $M(n)$ is an integer. Let

$$b_k(x) = \frac{1}{6}\sum_{j=-2}^{2}(-1)^{j+2}\binom{4}{j+2}(k+j-x)_+^3, \tag{76}$$

where

$$\begin{aligned}(y)_+ &= y \quad \text{if} \quad y > 0\\ &= 0 \quad \text{otherwise.}\end{aligned}$$

$b_k(x)$ is readily seen to be a probability density symmetric about k. Our estimator for the unknown density f is

$$f_b(x) = \sum_{k=-M(n)+2}^{M(n)-2} c_k b_k(x), \tag{77}$$

where $(c_{-M(n)+2}, c_{-M(n)+1}, \ldots, c_{+M(n)-2})$ is chosen to maximize $\prod_{i=1}^{n} f_b(x_i)$ subject to the constants

$$\sum c_k = 1, c_k \geq 0 \ \forall k.$$

Clearly, $M(n)$ should go to infinity as $n \to \infty$, but at a rate slower than n. Perhaps $M(n) = 0(\sqrt{n})$ is a practical value. Again, we have not investigated this matter and state it as an open problem.

2.5. *Kernel Estimators*

Of the methods used to estimate probability densities of unknown functional form, the most used is the histogram. Let us suppose we have a random sample $\{x_1, x_2, \ldots, x_n\}$ from an unknown absolutely continuous probability density with domain of positivity $[a, b]$. In the event that the unknown density, say $g(x)$, has infinite range, we shall content ourselves with estimating the truncated density

$$\begin{aligned} f(x) &= \frac{g(x)}{\int_a^b g(t)\,dt} \quad \text{for} \quad a \leq x \leq b \\ &= 0 \quad \text{otherwise.}\end{aligned} \tag{78}$$

In the following, we shall assume that those points outside $[a, b]$ have been discarded and that each of the $\{x_i\}_{i=1}^n$ is in the interval $[a, b]$. As a practical matter, if the density g has domain $[-\infty, \infty]$, then as the sample size increases we will take a to be smaller and smaller, and b to be larger and larger.

Let us partition $[a, b]$ by $a = t_0 < t_1 < \cdots < t_m = b$. Then we shall obtain an estimator f_H of the form

$$\begin{aligned} f_H(t) &= c_i \quad \text{for} \quad t_i \le t < t_{i+1}, \qquad i = 0, \ldots, m-1 \\ f_H(b) &= c_{m-1} \\ f_H(t) &= 0 \quad \text{for} \quad t \notin [a, b], \end{aligned} \tag{79}$$

where $f_H(t) \ge 0$ and $\int_a^b f_H(t)\, dt = 1$.

If q_i is the number of observations falling in the ith interval, then for $\hat{f}_H$ we shall use

$$\hat{c}_i = \frac{q_i}{n(t_{i+1} - t_i)} \quad \text{for} \quad i = 0, \ldots, m-1. \tag{80}$$

The intuitive appeal of $\hat{f}_H$ is clear. The number of observations falling into each of the intervals is a multinomial variate. Thus, q_i/n estimates $\int_{t_i}^{t_{i+1}} f(t)\, dt$. Since we have assumed that f is absolutely continuous, if $t_{i+1} - t_i$ is small, then $f(t) \sim f(t_i)$ for $t_i \le t < t_{i+1}$. Hence, q_i/n estimates $(t_{i+1} - t_i)f(t)$, and $\dfrac{q_i}{n(t_{i+1} - t)}$ estimates $f(t)$.

Theorem 2. Among estimates of the form (79) $\hat{f}_H$ uniquely maximizes the likelihood

$$\begin{aligned} L(c_0, \ldots, c_{m-1}) &= \prod_{j=1}^{n} f_H(x_j) \\ &= \prod_{i=0}^{m-1} c_i^{q_i} \end{aligned} \tag{81}$$

Proof. The feasible set $C \subset R^m$ must satisfy

$$\sum_{i=0}^{m-1} c_i(t_{i+1} - t_i) = 1 \tag{82}$$

and

$$c_i \ge 0.$$

But L is a continuous function of $\boldsymbol{c}$ and C is compact. Thus there exists a global maximizer $\boldsymbol{c}^*$. Let us assume, without loss of generality, that each interval contains at least one observation, since if this were not the case, say for the ith interval, a moments reflection should convince the reader that c_i^* must be equal to zero. Moreover, if $c_i^* = 0$ for any i, then $L(\boldsymbol{c}^*) = 0$. But $\boldsymbol{c}' = \frac{1}{m}\left(\frac{1}{t_1 - t_0}, \frac{1}{t_2 - t_1}, \ldots, \frac{1}{t_m - t_{m-1}}\right) \in C$ and $L(\boldsymbol{c}') > 0$. Hence, the inequality constraints are not active at the solution and from Appendix I.5 there exists $\lambda \in R$, such that

$$\nabla L(\boldsymbol{c}^*) = \lambda(t_1 - t_0, t_2 - t_1, \ldots, t_m - t_{m-1}). \tag{83}$$

From (81), (82), and (83), we have

$$q_i L(\boldsymbol{c}^*) = \lambda c_i^*(t_{i+1} - t_i) \tag{84}$$

$$\sum_{i=0}^{m-1} q_i L(\boldsymbol{c}^*) = \lambda$$

$$nL(\boldsymbol{c}^*) = \lambda.$$

Substituting for λ in (84), we have

$$q_i L(\boldsymbol{c}^*) = nL(\boldsymbol{c}^*) c_i^*(t_{i+1} - t_i).$$

So

$$c_i^* = \frac{q_i}{n(t_{i+1} - t_i)}. \tag{85}$$

But any maximizer must satisfy (85), and we have earlier shown that a maximizer exists. ∎

Theorem 3. Suppose that f has continuous derivatives up to order three except at the endpoints of $[a, b]$, and f is bounded on $[a, b]$. Let the mesh be equal spacing throughout $[a, b]$, so that $t_{i+1,n} - t_{i,n} = 2h_n$. Then if $n \to \infty$ and $h_n \to 0$, so that $nh_n \to \infty$, then for $x \in (a, b)$

$$\text{MSE}(\hat{f}_H(x)) = E[(\hat{f}_H(x) - f(x))^2] \to 0,$$

i.e., $\hat{f}_H(x)$ is a consistent estimator for $f(x)$.

Proof. For any h_n, we adopt the rule that $[a, b]$ is to be divided into $k = [(b - a)/(2h_n)]$ interior intervals, with two intervals of length $(b - a - 2kh_n)/2$ at the ends. Then for any h_n, x is unambiguously contained in an interval with a well-defined midpoint $x' = x'(x, h_n)$.

Expanding f about x', we have

$$(86)\qquad f(x) = f(x') + f'(x')(x - x') + \frac{1}{2}f''(x')(x - x')^2 + 0(x - x')^3,$$

where

$$x \in [t_k, t_{k+1}), \quad \text{and} \quad t_{k+1} - t_k = 2h_n \quad \text{for} \quad k = 0, 1, \ldots, m - 1.$$

Letting $p_k = \int_{t_k}^{t_{k+1}} f(x)\,dx$,
we have

$$p_k = 2h_n f(x') + \frac{h_n^3}{3} f''(x') + 0(h_n^4).$$

Now,

$$(87)\qquad E[(\hat{f}_H(x') - f(x'))^2] = \text{Var}(\hat{f}_H(x')) + \text{Bias}^2(\hat{f}_H(x')),$$

where

$$\begin{aligned}\text{Var}(\hat{f}_H(x')) &= E\left(\frac{q_k}{2nh_n} - \frac{p_k}{2h_n}\right)^2 = \frac{p_k(1 - p_k)}{4nh_n^2}\\ &= \frac{1}{4nh_n}\left[2f(x') - 4f^2(x')h_n + \frac{h_n^2}{3} f''(x') + 0(h_n^3)\right]\\ &= \frac{1}{2nh_n}\left[f(x') - 2h_n f^2(x') + \frac{h_n^2}{6} f''(x') + 0(h_n^3)\right]\end{aligned}$$

and

$$\begin{aligned}\text{Bias}^2(\hat{f}_H(x')) &= \left[\frac{p_k}{2h_n} - f(x')\right]^2\\ &= \frac{h_n^4}{36}(f''(x'))^2 + 0(h_n^5).\end{aligned}$$

Thus,

$$(88)\qquad \text{MSE}(\hat{f}_H(x')) = \frac{f(x')}{2nh_n} + \frac{h_n^4}{36}(f''(x'))^2 + 0\left(\frac{1}{n}\right) + 0(h_n^5).$$

Thus, if $nh_n \to \infty$, $n \to \infty$, $h_n \to 0$,

$$\text{MSE}(\hat{f}_H(x')) \to 0.$$

Now, let x be any point in $[t_k, t_{k+1})$. We recall that $\hat{f}_H(x) = \hat{f}_H(x')$. Then

$$\begin{aligned}\text{MSE}(\hat{f}_H(x)) = E[\hat{f}_H(x) - f(x))^2] &= E[(\hat{f}_H(x') - f(x') + f(x') - f(x))^2]\\ &\le 2E[(\hat{f}_H(x') - f(x'))^2] + 2[(f(x') - f(x))^2].\end{aligned}$$

Assuming the spacing is sufficiently fine that $|f(x') - f(x)|$ is no greater than $|f(x') - f(t_k)|$ or $|f(x') - f(t_{k+1})|$, we have

$$\text{MSE}(\hat{f}_H(x)) \leq \frac{f(x')}{nh_n} + 2|f'(x')|^2 h_n^2 + 0\left(\frac{1}{n}\right) + 0(h_n^3). \tag{89}$$

Thus if $nh_n \to \infty$, $n \to \infty$, $h_n \to 0$, $\text{MSE}(\hat{f}_H(x)) \to 0$. ■

Remark. It is interesting to note from (88) and (89) that if we select

$$h_n = \left[\frac{9f(x')}{2(f''(x'))^2}\right]^{1/5} n^{-1/5},$$

then the mean square error at x' is of order $n^{-4/5}$. However, this choice of $\{h_n\}$ gives a MSE away from x' as high as order $n^{-2/5}$. On the other hand, we may, viz. (89) select $h_n = \left[\frac{f(x')}{4(f'(x'))^2}\right]^{1/3} n^{-1/3}$ to obtain convergence throughout the kth interval of order $n^{-2/3}$.

Integrating (89) over $[a, b]$, we may minimize our bound on the integrated mean square error by selecting

$$h_n = \left[\frac{1}{4\int (f'(x))^2\,dx}\right]^{1/3} n^{-1/3} \tag{90}$$

to obtain

$$\int \text{MSE}(\hat{f}_H(x)) = \text{IMSE} \leq 3\left[\frac{1}{2}\int (f')^2\,dx\right]^{1/3} n^{-2/3} + 0\left(\frac{1}{n} + h_n^3\right). \tag{91}$$

Now the likelihood in (81) is, of course, the joint density of $(x_1, x_2, \ldots, x_n)$ given $\boldsymbol{c} = (c_0, c_1, \ldots, c_{m-1})$; i.e.,

$$L(\boldsymbol{c}) = L(x_1, x_2, \ldots, x_n|\boldsymbol{c}). \tag{92}$$

Let us consider as a prior distribution for $\boldsymbol{c}' = (c_0(t_1 - t_0), \ldots, c_{m-1}(t_m - t_{m-1}))$

$$P(c_0', c_1', \ldots, c_{m-1}') = \frac{\Gamma(\theta_0 + \cdots + \theta_{m-1})}{\Gamma(\theta_0)\cdots\Gamma(\theta_{m-1})} c_0'^{\theta_0 - 1} \cdots c_{m-1}'^{\theta_{m-1}-1} \tag{93}$$

$$\text{if } \sum_{i=0}^{m-1} c_i' = 1, \quad c_i' > 0, \quad i = 0, \ldots, m-1$$

$$= 0 \quad \text{otherwise.}$$

Then the joint density of $(\boldsymbol{x}, \boldsymbol{c})$ is given by

$$s(x_1, \ldots, x_n, c_0', \ldots, c_{m-1}') = \prod_{i=0}^{m-1} c_i'^{q_i + \theta_i - 1} \frac{\Gamma(\theta_0 + \cdots + \theta_{m-1})}{\Gamma(\theta_0)\cdots\Gamma(\theta_{m-1})} h_n^{-n}. \tag{94}$$

The marginal density of $\mathbf{x}$, then, is given by

$$u(x_1, \ldots, x_n) = \left[\prod_{i=0}^{m-1} \frac{\Gamma(\theta_i + q_i)}{\Gamma(\theta_i)}\right] \frac{\Gamma(\theta_0 + \cdots + \theta_{m-1})}{\Gamma(\theta_0 + \cdots + \theta_{m-1} + n)} h_n^{-n}. \tag{95}$$

The posterior density of $\mathbf{c}'$ is then given by

$$\begin{aligned} v(c'_0, \ldots, c'_{m-1} | \mathbf{x}) &= \frac{s(x_1, \ldots, x_n, c'_0, \ldots, c'_{m-1})}{u(x_1, \ldots, x_n)} \\ &= \frac{\Gamma(\theta_0 + \cdots + \theta_{m-1} + n)}{\Gamma(\theta_0 + q_0) \cdots \Gamma(\theta_{m-1} + q_{m-1})} \prod_{i=0}^{m-1} c_i'^{q_i + \theta_i - 1}. \end{aligned} \tag{96}$$

Now, the Dirichlet distribution with parameter $(\theta_0, \ldots, \theta_{m-1})$ has

$$E(c'_i) = \frac{\theta_i}{\sum_j \theta_j} \tag{97}$$

$$\mathrm{Var}(c'_i) = \frac{\theta_i(\theta_0 + \cdots + \theta_{m-1} - \theta_i)}{(\sum \theta_j)^2 (\sum \theta_j + 1)} \tag{98}$$

$$\mathrm{Cov}(c'_i, c'_j) = \frac{-\theta_i \theta_j}{(\sum \theta_k)^2 (\sum \theta_k + 1)}. \tag{99}$$

So the posterior mean of c' is given by

$$c'_i = \frac{\theta_i + q_i}{\sum_{j=0}^{m-1} \theta_j + n}, \tag{100}$$

giving

$$c_i = \frac{\theta_i + q_i}{\left(\sum_{j=0}^{m-1} \theta_j + n\right)(t_{i+1} - t_i)}, \qquad i = 0, \ldots, m-1. \tag{101}$$

Now a ready guess is available for the prior parameter $(\theta_0, \ldots, \theta_{m-1})$ via (97). We may replace $E(c'_i) = \dfrac{\theta_i}{\sum \theta_j}$ by our prior feelings as to the number of observations which will fall into the ith interval, i.e., let

$$\frac{\theta_i}{\sum \theta_j} = \int_{t_i}^{t_{i+1}} f_p(x)\, dx, \tag{102}$$

where f_p is our best guess as to the true f. This will give us $m - 1$ equations for $\{\theta_0, \theta_1, \ldots, \theta_{m-1}\}$. The additional equation may be obtained from our feelings as to the variance of c_i for some i. Clearly, for fixed spacing and $(\theta_0, \ldots, \theta_{m-1})$, as $n \to \infty$, the posterior mean of $\mathbf{c}$ is the maximum likelihood estimator given in (85). Note, moreover, that if we wish to use the posterior mean estimator in such a way that we obtain a consistent estimator f, we may easily do so by simply requiring that $\operatorname{Min}_i (t_{i+1} - t_i)n \to \infty$ and $\operatorname{Max}_i (t_{i+1} - t_i) \to 0$ as $n \to \infty$. The method for choosing prior values for $(\theta_0, \ldots, \theta_{m-1})$ for m of arbitrary size is that suggested in the discussion following (101).

We note, then, that the consistent histogram is a nonparametric estimator, in the sense that the number of characterizing parameters increases without limit as the sample size goes to infinity. It has the advantage over the classical series estimation of being more local. However, it is discontinuous, completely ad hoc, harder to update, and typically requires more parameters for a given sample size than the orthonormal series estimators. Clearly, the histogram should offer interesting possibilities for generalization.

In 1956 Murray Rosenblatt, in a very insightful paper [38], extended the histogram estimator of a probability density. Based on a random sample $\{x_i\}_{i=1}^n$ from a continuous but unknown density f, the Rosenblatt estimator is given by

$$\hat{f}_n(x) = \frac{\#\text{ sample points in } (x - h_n, x + h_n)}{2nh_n}, \tag{103}$$

where h_n is a real valued number constant for each n, i.e.,

$$\hat{f}_n(x) = \frac{F_n(x + h_n) - F_n(x - h_n)}{2h_n}, \tag{104}$$

where

$$F_n(x) = \frac{\#\text{ sample points} \le x}{n}.$$

To obtain the formula for the mean square error of $\hat{f}_n$, Rosenblatt notes that if we partition the real line into three intervals $\{x|x \le x_1\}$, $\{x|x_1 < x \le x_2\}$ and $\{x|x > x_2\}$, and let $Y_1 = F_n(x_1)$, $Y_2 = F_n(x_2) - F_n(x_1)$ and $Y_3 = 1 - F_n(x_2)$, then (nY_1, nY_2, nY_3) is a trinomial random variable with probabilities $(F(x_1), F(x_2) - F(x_1), 1 - F(x_2)) = (p_1, p_2, p_3)$, where F is the cumulative distribution function of x, i.e., $F(x) = \int_{-\infty}^{x} f(t)\,dt$. Thus, we have

$$E[F_n(x)] = F(x) \tag{105}$$

(106) $\mathrm{Cov}[F_n(x_1), F_n(x_2)]$

$$= \frac{1}{n^2} E[(nY_1 - nF(x_1))(n(Y_1 + Y_2) - n(F(x_2) + F(x_1))]$$

$$= \frac{1}{n^2} \mathrm{Cov}(nY_1, nY_2) + \frac{1}{n^2} \mathrm{Var}(nY_1)$$

$$= \frac{-1}{n^2} nF(x_1)(F(x_2) - F(x_1)) + \frac{1}{n^2} nF(x_1)(1 - F(x_1))$$

$$= -\frac{F(x_1)F(x_2)}{n} + \frac{F(x_1)}{n},$$

assuming without loss of generality that $x_1 < x_2$.

Thus, making no restrictions on x_1 and x_2,

(107)

$\mathrm{Cov}[\hat{f}_n(x_1), \hat{f}_n(x_2)]$

$$= E\left[\left(\frac{F_n(x_1 + h_n) - F_n(x_1 - h_n)}{2h_n} - \frac{F(x_1 + h_n) - F(x_1 - h_n)}{2h_n}\right)\right.$$
$$\left.\times \left(\frac{F_n(x_2 + h_n) - F_n(x_2 - h_n)}{2h_n} - \frac{F(x_2 + h_n) - F(x_2 - h_n)}{2h_n}\right)\right]$$

$$= \frac{1}{4h_n^2} \{\mathrm{Cov}[F_n(x_1 + h_n), F_n(x_2 + h_n)] - \mathrm{Cov}[F_n(x_1 + h_n), F_n(x_2 - h_n)]$$
$$- \mathrm{Cov}[F_n(x_1 - h_n), F_n(x_2 + h_n)] + \mathrm{Cov}[F_n(x_1 - h_n), F_n(x_2 - h_n)]\}$$

$$= \frac{1}{4h_n^2 n} [-F(x_1 + h_n)F(x_2 + h_n) + F(\min(x_1 + h_n, x_2 + h_n))$$
$$+ F(x_1 + h_n)F(x_2 - h_n) - F(\min(x_1 + h_n, x_2 - h_n))$$
$$+ F(x_1 - h_n)F(x_2 + h_n) - F(\min(x_1 - h_n, x_2 + h_n))$$
$$- F(x_1 - h_n)F(x_2 - h_n) + F(\min(x_1 - h_n, x_2 - h_n))].$$

If $x_1 = x_2 = x$, then

(108) $$\mathrm{Var}[\hat{f}_n(x)] = \frac{1}{4h_n^2 n} [-F^2(x + h_n) + F(x + h_n) + F(x - h_n)F(x + h_n)$$
$$- F(x - h_n) + F(x - h_n)F(x + h_n) - F(x - h_n)$$
$$- F^2(x - h_n) + F(x - h_n)]$$

$$= \frac{1}{4h_n^2 n}[F(x + h_n) - F(x - h_n) - (F(x + h_n) - F(x - h_n))^2].$$

Then, if we pick $x_1 < x_2$ and h_n sufficiently small that $x_1 + h_n < x_2 - h_n$

(109)

$$
\begin{aligned}
&\operatorname{Cov}[\hat{f}_n(x_1), \hat{f}_n(x_2)] \\
&\quad = E\left[\left(\frac{F_n(x_1+h_n)-F_n(x_1-h_n)}{2h_n} - \frac{F(x_1+h_n)-F(x_1-h_n)}{2h_n}\right)\right. \\
&\qquad \times \left.\left(\frac{F_n(x_2+h_n)-F_n(x_2-h_n)}{2h_n} - \frac{F(x_2+h_n)-F(x_2-h_n)}{2h_n}\right)\right] \\
&\quad = \frac{1}{4h_n^2}[\operatorname{Cov}(F_n(x_1+h_n), F_n(x_2+h_n)) - \operatorname{Cov}(F_n(x_1+h_n), F_n(x_2-h_n)) \\
&\qquad - \operatorname{Cov}(F_n(x_1-h_n), F_n(x_2+h_n)) + \operatorname{Cov}(F_n(x_1-h_n), F_n(x_2-h_n))] \\
&\quad = \frac{1}{4h_n^2}\left[-\frac{F(x_1+h_n)F(x_2+h_n)}{n} + \frac{F(x_1+h_n)}{n} + \frac{F(x_1+h_n)F(x_2-h_n)}{n}\right. \\
&\qquad - \frac{F(x_1+h_n)}{n} + \frac{F(x_1-h_n)F(x_2+h_n)}{n} - \frac{F(x_1-h_n)}{n} \\
&\qquad \left. - \frac{F(x_1-h_n)F(x_2-h_n)}{n} + \frac{F(x_1-h_n)}{n}\right] \\
&\quad = \frac{1}{4h_n^2 n}[-F(x_1+h_n)[F(x_2+h_n)-F(x_2-h_n)] \\
&\qquad + F(x_1-h_n)[F(x_2+h_n)-F(x_2-h_n)]] \\
&\quad = \frac{-1}{4h^2 n}[(F(x_2+h_n)-F(x_2-h_n))(F(x_1+h_n)-F(x_1-h_n))] \\
&\quad = \frac{-1}{n}\,\frac{F(x_2+h_n)-F(x_2-h_n)}{2h_n}\,\frac{F(x_1+h_n)-F(x_2-h_n)}{2h_n} \\
&\quad = \frac{-1}{n} f(x_1)f(x_2) - \frac{h_n^2}{6n}[f(x_1)f''(x_2) + f(x_2)f''(x_1)] + 0\left(\frac{h_n^3}{n}\right)
\end{aligned}
$$

(assuming f is thrice differentiable at x_1 and x_2).

Now,

(110)
$$
\begin{aligned}
\operatorname{MSE}(\hat{f}_n(x)) &= E[(\hat{f}_n(x) - f(x))^2] = \operatorname{Var}[\hat{f}_n(x)] + \operatorname{Bias}^2[\hat{f}_n(x)] \\
&= \frac{1}{4h_n^2 n}[F(x+h_n) - F(x-h_n) - (F(x+h_n) - F(x-h_n))^2] \\
&\quad + \left[\frac{1}{2h_n}(F(x+h_n) - F(x-h_n)) - f(x)\right]^2.
\end{aligned}
$$

But $F(x+h_n) - F(x-h_n) = 2h_n f(x) + \frac{h_n^3}{3} f''(x) + 0(h_n^4)$, (assuming f is thrice differentiable at x).

$$
\begin{aligned}
(111) \quad \text{MSE}(\hat{f}_n(x)) &= \frac{1}{4h_n^2 n}\left[2h_n f(x) + \frac{h_n^3}{3} f''(x) + 0(h_n^4)\right. \\
&\quad \left. - \left(2h_n f(x) + \frac{h_n^3}{3} f''(x) + 0(h_n^4)\right)^2\right] \\
&\quad + \frac{1}{4h_n^2}\left[2h_n f(x) + \frac{h_n^3}{3} f''(x) + 0(h_n^4) - 2h_n f(x)\right]^2 \\
&= \frac{f(x)}{2h_n n} + \frac{h_n^4}{36}(f''(x))^2 + o\left(\frac{1}{h_n n} + h_n^4\right).
\end{aligned}
$$

So, if $h_n \to 0$ as $n \to \infty$ in such a way that $nh_n \to \infty$, $\text{MSE}(\hat{f}_n(x)) \to 0$. Holding f, x, and n constant; we may minimize the first two terms in (111) using a straightforward argument, yielding

$$(112) \qquad \hat{h}_n = \left[\frac{9}{2}\frac{f(x)}{(f''(x))^2}\right]^{1/5} n^{-1/5},$$

with corresponding asymptotic (in n) mean square error of

$$(113) \qquad \text{MSE}(\hat{f}_n(x)) = \frac{5}{4} 9^{-1/5} 2^{-4/5} |f(x)|^{4/5} (|f''(x)|)^{2/5} n^{-4/5}.$$

Studies of a variety of measures of consistency of density estimates are given in [4, 28, 33, 42, 51, 61].

The comparison between the Rosenblatt kernel estimator and the histogram estimator is of some interest. In essence, Rosenblatt's approach is simply a histogram which, for estimating the density of x, say, has been shifted so that x lies at the center of a mesh interval. For evaluating the density at another point, say y, the mesh is shifted again, so that y is at the center of a mesh interval. The fact that the MSE of the Rosenblatt procedure decreases like $n^{-4/5}$ instead of as slowly as $n^{-2/3}$, as with the fixed grid histogram, demonstrates the advantage of having the point at which f is to be estimated at the center of an interval and is reminiscent of the advantages of central differences in numerical analysis. We note that the advantage of the "shiftable" histogram approach lies in a reduction of the bias of the resulting estimator.

The shifted histogram estimator has another representation, namely,

$$(114) \qquad \hat{f}_n(x) = \frac{1}{n}\sum_{j=1}^{n} \frac{1}{h_n} w\left(\frac{x - x_j}{h_n}\right),$$

where $w(u) = \frac{1}{2}$ if $|u| < 1$
$= 0$ otherwise,
and the $\{x_j\}_{j=1}^n$ are the data points.
Thus, it is clear that for all $\{x_j\}_{j=1}^n$

$$\int \hat{f}_n(x)\,dx = 1$$

and

$$\hat{f}_n(x) \geq 0.$$

A shifted histogram is, like the fixed grid histogram, always a bona-fide probability density function.

To pick a global value for h_n, we may examine the integrated mean square error

$$\int \mathrm{MSE}(\hat{f}_n(x))\,dx = \mathrm{IMSE} \sim \frac{1}{2h_n n} + \frac{h_n^4}{36}\int_{-\infty}^{\infty} (f''(x))^2\,dx + o\left(\frac{1}{h_n n} + h_n^4\right)$$

to give

$$h_n = \left[\frac{9}{2\int (f''(x))^2\,dx}\right]^{1/5} n^{-1/5}, \tag{115}$$

and thus

$$\mathrm{IMSE} \sim 2^{-4/5} 9^{-1/5} \frac{5}{4}\left[\int (f''(x))^2\,dx\right]^{1/5} n^{-4/5}. \tag{116}$$

Although Rosenblatt suggested generalizing (114) to estimators using different bases than step functions, the detailed explication of kernel estimators is due to Parzen [35]. We shall consider as an estimator for $f(x)$

$$\hat{f}_n(x) = \int_{-\infty}^{\infty} \frac{1}{h_n} K\left(\frac{x-y}{h_n}\right) dF_n(y) = \frac{1}{nh_n}\sum_{j=1}^{n} K\left(\frac{x-x_j}{h_n}\right), \tag{117}$$

where

$$\begin{cases} \int_{-\infty}^{\infty} |K(y)|\,dy < \infty \\ \sup\limits_{-\infty<y<\infty} |K(y)| < \infty \\ \lim\limits_{y\to\infty} |yK(y)| = 0 \end{cases} \tag{118}$$

and

$$K(y) \geq 0 \tag{119}$$

$$\int_{-\infty}^{\infty} K(y)\,dy = 1.$$

It is interesting to note that, if $k(\cdot)$ is an even function,

$$\hat{\mu} = \int x\hat{f}_n(x)\,dx = \frac{1}{n}\sum_{j=1}^{n}\int \frac{x}{h_n}K\left(\frac{x-x_j}{h_n}\right)dx$$
$$= \frac{1}{n}\sum_{j=1}^{n} x_j = \bar{x}$$

and

$$\hat{\sigma}^2 = \int (x-\hat{\mu})^2\hat{f}_n(x)\,dx$$
$$= \frac{1}{n}\sum_{j=1}^{n}(x_j-\bar{x})^2 + h_n^2\int x^2K(x)\,dx$$
$$= s^2 + h_n^2\int x^2K(x)\,dx.$$

Parzen's consistency argument is based on the following lemma of Bochner [2].

Lemma. Let K be a Borel function satisfying (118). Let $g \in L^1$ (i.e., $\int|g(y)|\,dy < \infty$). Let

$$g_n(x) = \frac{1}{h_n}\int K\left(\frac{y}{h_n}\right)g(x-y)\,dy, \tag{120}$$

where $\{h_n\}$ is a sequence of positive constants having $\lim_{n\to\infty} h_n = 0$. Then if x is a point of continuity of g,

$$\lim_{n\to\infty} g_n(x) = g(x)\int_{-\infty}^{\infty} K(y)\,dy. \tag{121}$$

Proof.

$$\left|g_n(x) - g(x)\int_{-\infty}^{\infty}K(y)\,dy\right|$$
$$= \left|\int_{-\infty}^{\infty}\{g(x-y)-g(x)\}\frac{1}{h_n}K\left(\frac{y}{h_n}\right)dy\right|$$
$$\le \sup_{|y|\le\delta}|g(x-y)-g(x)|\int|K(z)|\,dz$$
$$+ \int_{|y|\ge\delta}\frac{|g(x-y)|}{y}\frac{y}{h_n}K\left(\frac{y}{h_n}\right)dy + |g(x)|\int_{|y|\ge\delta}\frac{1}{h_n}K\left(\frac{y}{h_n}\right)dy$$
$$\le \sup_{|y|\le\delta}|g(x-y)-g(x)|\int_{-\infty}^{\infty}|K(z)|\,dz$$
$$+ \frac{1}{\delta}\sup_{|z|\ge\delta/h_n}|zK(z)|\int_{-\infty}^{\infty}|g(y)|\,dy + |g(x)|\int_{|z|\ge\delta/h_n}|K(z)|\,dz.$$

Now, as $n \to \infty$, since $h_n \to 0$ the second and third terms go to zero, since $g \in L^1$ and $\lim_{y\to\infty} |yK(y)| = 0$. Then, as $\delta \to 0$, the first term goes to zero since $K \in L^1$ and x is a point of continuity of g. ■

Thus, we immediately have

Theorem 4. The kernel estimator $\hat{f}_n$ in (117) subject to (118) and (119) is asymptotically unbiased if $h_n \to 0$ as $n \to \infty$, i.e.,

$$\lim_n E(\hat{f}_n(x)) = f(x).$$

Proof. Merely observe that

$$(122) \qquad \begin{aligned} E[\hat{f}_n(x)] &= \frac{1}{h_n}\int K\left(\frac{x - x_j}{h_n}\right) f(x_j)\,dx_j \\ &= \int K(y) f(x - h_n y)\,dy \\ &= f(x) + 0(h_n). \quad ■ \end{aligned}$$

More importantly, we have

Theorem 5. The estimator $\hat{f}_n$ in (117) subject to (118) and (119) is consistent if we add the additional constraint $\lim_{n\to\infty} nh_n \to \infty$.

Proof. We have

$$\operatorname{Var}[\hat{f}_n(x)] = \frac{1}{n}\operatorname{Var}\left[\frac{1}{h_n} K\left(\frac{x-y}{h_n}\right)\right]$$

(since $\operatorname{Var}(\bar{z}) = \frac{1}{n}\operatorname{Var}(z)$).

Now

$$(123) \qquad \begin{aligned} \frac{1}{n}\operatorname{Var}\left[\frac{1}{h_n} K\left(\frac{x-y}{h_n}\right)\right] &\le \frac{1}{n} E\left[\left(\frac{1}{h_n} K\left(\frac{x-y}{h_n}\right)\right)^2\right] \\ &= \frac{1}{h_n n}\left[\frac{1}{h_n}\int K^2\left(\frac{x-y}{h_n}\right) f(y)\,dy\right] \\ &\to 0 \quad \text{if} \quad \lim_{n\to\infty} h_n n = \infty. \end{aligned}$$

But

$$(124) \qquad \begin{aligned} \operatorname{MSE}[\hat{f}_n(x)] &= E[(\hat{f}_n(x) - f(x))^2] \\ &= \operatorname{Var}[\hat{f}_n(x)] + \operatorname{Bias}^2(\hat{f}_n(x)). \end{aligned}$$

We have just proved that the variance goes to zero if $\lim_{n\to\infty} h_n n = \infty$. That the bias goes to zero was proved in the preceding theorem. Thus, $\text{MSE}[\hat{f}_n(x)] \to 0$, i.e., $\hat{f}_n(x)$ is a consistent estimator for $f(x)$. ∎

We have already seen that for the Rosenblatt shifted histogram the optimal rate of decrease of the MSE is of the order of $n^{-4/5}$. We now address the question of the degree of improvement possible with the Parzen family of kernel estimators. Let us consider the Fourier transform of the kernel K, where K satisfies the conditions in (118) and (119). That is, let

$$k(u) = \int_{-\infty}^{\infty} e^{iuy} K(y)\, dy, \tag{125}$$

assuming k is absolutely integrable.
Then, letting

$$\varphi_n(u) = \int_{-\infty}^{\infty} e^{iux}\, dF_n(x) = \frac{1}{n}\sum_{k=1}^{n} e^{iux_k}, \tag{126}$$

we have

$$\begin{aligned} \hat{f}_n(x) &= \frac{1}{nh_n}\sum_{k=1}^{n} K\{(x - x_k)/h_n\} \\ &= \frac{1}{2\pi}\int e^{-iux} k(h_n u)\varphi_n(u)\, du \end{aligned} \tag{127}$$

by the Levy Inversion Theorem.
Now,

$$\begin{aligned} E[\varphi_n(x)] &= \frac{1}{n}\sum_{k=1}^{n} E(e^{iux_k}) \\ &= \varphi(u), \end{aligned} \tag{128}$$

where

$$\varphi(u) = E(e^{iux}).$$

Thus,

$$E[\hat{f}_n(x)] = \frac{1}{2\pi}\int_{-\infty}^{\infty} e^{-iux} k(h_n u)\varphi(u)\, du. \tag{129}$$

So

$$\text{Bias}[\hat{f}_n(x)] = \frac{1}{2\pi}\int_{-\infty}^{\infty} e^{-iux}[k(h_n u) - 1]\varphi(u)\, du. \tag{130}$$

Now, if there exists a positive r, such that

$$k_r = \lim_{u \to 0} \left[\frac{1 - k(u)}{|u|^r} \right] \tag{131}$$

is nonzero and finite, it is called the *characteristic exponent* of the transform k, and k_r is called the *characteristic coefficient*. But

$$\begin{aligned} k_r &= \lim_{u \to 0} \left[\frac{1 - k(u)}{|u|^r} \right] = \lim_{u \to 0} \left[\frac{1 - \int_{-\infty}^{\infty} e^{iux} K(x)\, dx}{|u|^r} \right] \\ &= \lim_{u \to 0} \frac{1}{|u|^r} \left[1 - \int_{-\infty}^{\infty} K(x)\, dx - iu \int_{-\infty}^{\infty} xK(x)\, dx - \cdots \right. \\ &\quad \left. - \frac{(iu)^{r-1}}{(r-1)!} \int_{-\infty}^{\infty} x^{r-1} K(x)\, dx - \frac{(iu)^r}{r!} \int_{-\infty}^{\infty} x^r K(x)\, dx + 0(u^{r+1}) \right]. \end{aligned} \tag{132}$$

Clearly, if k_r is to be finite and positive, we must have

$$\begin{cases} \int K(x)\, dx = 1 \\ \int_{-\infty}^{\infty} x^j K(x)\, dx = 0 \quad \text{for} \quad j = 1, 2, \ldots, r-1 \\ \int x^r K(x)\, dx \neq 0 \quad \text{or} \quad \infty. \end{cases} \tag{133}$$

If K is itself to be a probability density, and therefore nonnegative, $r > 2$ is impossible if we continue to require the conditions of (119). Clearly, $r = 2$ is the most important case. Examples of kernels with characteristic exponents of 2 include the Gaussian, double exponential, and indeed any symmetric density K having $x^2 K(x) \in L^1$.

For the general case of a characteristic exponent r,

$$\frac{\text{Bias}[\hat{f}_n(x)]}{h_n^r} = \frac{1}{2\pi} \int_{-\infty}^{\infty} e^{-iux} \frac{k(h_n u) - 1}{|h_n u|^r} |u|^r \varphi(u)\, du \to k_r f^{(r)}(x), \tag{134}$$

where

$$f^{(r)}(x) = \left. \frac{d^r f}{dx^r} \right|_x = -\frac{1}{2\pi} \int_{-\infty}^{\infty} e^{-iux} |u|^r \varphi(u)\, du. \tag{135}$$

But from (123) we have

$$\text{Var}[\hat{f}_n(x)] \sim \frac{f(x)}{nh_n} \int K^2(y)\, dy.$$

Thus,

$$\text{(136)} \qquad \operatorname{MSE}[\hat{f}_n(x)] \sim \frac{f(x)}{nh_n} \int_{-\infty}^{\infty} K^2(y)\,dy + h_n^{2r} k_r^2 |f^{(r)}(x)|^2.$$

We may now, given f and K (and hence r), easily find that h_n which minimizes the asymptotic mean square error by solving

$$\text{(137)} \qquad 2rh_n^{2r-1} k_r^2 |f^{(r)}(x)|^2 = \frac{f(x)}{nh_n^2} \int_{-\infty}^{\infty} K^2(y)\,dy,$$

giving

$$\text{(138)} \qquad h_n(x) = \left[f(x) \int_{-\infty}^{\infty} K^2(y)\,dy \Big/ \{n2rk_r^2 |f^{(r)}(x)|^2\} \right]^{1/(2r+1)}.$$

For this h_n, we have

$$\text{(139)} \quad \operatorname{MSE}_{\text{opt}} \sim (2r+1) \left\{ \frac{f(x)}{n2r} \int K^2(y)\,dy \right\}^{2r/(2r+1)} |k_r f^{(r)}(x)|^{2/(2r+1)}.$$

Thus we note that in the class of Parzen kernel estimators satisfying (118) the rate of decrease of the mean square error is of order $n^{-2r/(2r+1)}$. In practice, then, we should not expect a more rapid decrease of MSE for estimators of this class than $n^{-4/5}$—that obtainable by use of the shifted histogram.

A customary procedure is to obtain a global optimal h_n by minimizing the integrated mean square error

$$\text{(140)} \qquad \operatorname{IMSE}[\hat{f}_n] \sim \frac{1}{nh_n} \int_{-\infty}^{\infty} K^2(y)\,dy + h_n^{2r} k_r^2 \int_{-\infty}^{\infty} |f^{(r)}(x)|^2\,dx$$

to obtain

$$\text{(141)} \quad h_n = n^{-1/(2r+1)} \left[\frac{\int K^2(y)\,dy}{2rk_r^2} \right]^{1/(2r+1)} \left[\int |f^{(r)}(y)|^2\,dy \right]^{-1/(2r+1)}$$

$$= n^{-1/(2r+1)} \left[\frac{\int K^2(y)\,dy}{2r\left(\int y^r K(y)\,dy/r!\right)^2} \right]^{1/(2r+1)} \left[\int |f^{(r)}(y)|^2\,dy \right]^{-1/(2r+1)}$$

$$= n^{-1/(2r+1)} \alpha(K) \beta(f).$$

Since the Parzen approach assumes the functional form of K (and hence r) to be given, the evaluation of $n^{-1/(2r+1)}\alpha(K)$ is unlikely to cause as much difficulty as the determination of $\beta(f)$.

In the following, we shall generally restrict our considerations to kernels where $r = 2$. Below we show a table of useful kernels with their standard supports and their corresponding α values

Table 2.1

K_i		$\alpha(K_i)$
$K_1(y) = \frac{1}{2}$	$\lvert y\rvert \le 1$	1.3510
$K_2(y) = 1 - \lvert y\rvert$	$\lvert y\rvert \le 1$	1.8882
$K_3(y) = \frac{15}{16}(1 - y^2)^2$	$\lvert y\rvert \le 1$	2.0362
$K_4(y) = \frac{1}{\sqrt{2\pi}} e^{-y^2/2}$	$\lvert y\rvert < \infty$	0.7764

Although the Gaussian kernel K_4 is fairly popular, the quartic kernel estimator using K_3 is nearly indistinguishable in its smoothness properties and has distinct computational advantages due to its finite support.

In Figures 2.1 through 2.6, we display an empirical interactive technique for picking h_n using K_3. A random sample of size 300 has been generated from the bimodal mixture

$$(142)\quad f(x) = .75\frac{1}{\sqrt{2\pi}}\exp\left[-\frac{(x+1.5)^2}{2}\right] + .25\frac{3}{\sqrt{2\pi}}\exp\left[-\frac{(x-1.5)^2}{2/9}\right].$$

We note that for $h = 5$, the estimated density is overly smoothed. However, the user who does not know the functional form of f would not yet realize that the estimator is unsatisfactory. Now, as h is decreased to 2, the bimodality of f is hinted. Continuing the decrease of h, the shape of f becomes clearer. When we take h to .2, the noise level of the estimator becomes unacceptable. Such additional structure in f beyond that which we have already perceived for slightly larger h_n values is obscured by the fact that nh_n is too small. Accordingly, the user would return to the estimator obtained for $h_n = .4$.

We note here the fact that kernel estimators are not, in general, robust against poor choices of h_n. A guess of h_n a factor of 2 from the optimal may frequently double the integrated mean square error. Accordingly, we highly recommend the aforementioned interactive approach, where the user starts off with h_n values which are too large and then sequentially decreases h_n until overly noisy probability density estimates are obtained. The point where

```
BIMODAL  NORMAL WITH MEANS = -+  1.50 VARIANCE OF LEFT = 1 WITH WEIGHT AND VARIANCE OF RIGHT =   C.2500   0.1111
  SAMPLE  SIZE =   300 WITH    75 SAMPLES ON RIGHT       VS.
        KERNEL ESTIMATOR WITH KERNEL = QUAR WITH H =        5.00000000
        INTEGRATED MEAN SQUARE ERROR              0.124810934 E-01
        INTEGRATED SQUARE ERROR                   0.612855554 E-01
        MAXIMUM ABSOLUTE DIFFERENCE               0.196941674 E+00
```

ABSCISSA	ORDINATE *	ORDINATE O	
-5.00E+00	6.55E-04	4.15E-02	
-4.80E+00	1.29E-03	4.82E-02	
-4.60E+00	2.45E-03	5.51E-02	
-4.40E+00	4.46E-03	6.23E-02	
-4.20E+00	7.82E-03	6.96E-02	1.
-4.00E+00	1.31E-02	7.68E-02	2.
-3.80E+00	2.12E-02	8.40E-02	2.
-3.60E+00	3.30E-02	9.10E-02	2.
-3.40E+00	4.92E-02	9.79E-02	1.
-3.20E+00	7.05E-02	1.05E-01	6.
-3.00E+00	9.71E-02	1.11E-01	2.
-2.80E+00	1.29E-01	1.18E-01	9.
-2.60E+00	1.63E-01	1.24E-01	9.
-2.40E+00	2.00E-01	1.30E-01	9.
-2.20E+00	2.34E-01	1.36E-01	9.
-2.00E+00	2.64E-01	1.40E-01	9.
-1.80E+00	2.86E-01	1.45E-01	9.
-1.60E+00	2.98E-01	1.48E-01	9.
-1.40E+00	2.98E-01	1.51E-01	9.
-1.20E+00	2.86E-01	1.53E-01	9.
-1.00E+00	2.64E-01	1.54E-01	9.
-8.00E-01	2.34E-01	1.55E-01	9.
-6.00E-01	2.00E-01	1.54E-01	9.
-4.00E-01	1.63E-01	1.53E-01	7.
-2.00E-01	1.29E-01	1.51E-01	9.
0.0	9.72E-02	1.48E-01	5.
2.00E-01	7.07E-02	1.44E-01	4.
4.00E-01	5.05E-02	1.40E-01	2.
6.00E-01	4.08E-02	1.35E-01	1.
8.00E-01	5.42E-02	1.29E-01	7.
1.00E+00	1.10E-01	1.23E-01	8.
1.20E+00	2.07E-01	1.16E-01	9.
1.40E+00	2.91E-01	1.09E-01	9.
1.60E+00	2.88E-01	1.02E-01	9.
1.80E+00	2.01E-01	9.46E-02	6.
2.00E+00	9.78E-02	8.73E-02	5.
2.20E+00	3.33E-02	8.01E-02	1.
2.40E+00	7.96E-03	7.30E-02	1.
2.60E+00	1.36E-03	6.63E-02	
2.80E+00	1.78E-04	5.99E-02	
3.00E+00	2.40E-05	5.39E-02	
3.20E+00	5.45E-06	4.83E-02	
3.40E+00	1.86E-06	4.31E-02	
3.60E+00	6.74E-07	3.83E-02	
3.80E+00	2.38E-07	3.37E-02	
4.00E+00	8.08E-08	2.96E-02	
4.20E+00	2.64E-08	2.57E-02	
4.40E+00	8.26E-09	2.21E-02	
4.60E+00	2.49E-09	1.87E-02	
4.80E+00	7.20E-10	1.56E-02	
5.00E+00	2.00E-10	1.27E-02	

END OF PLOT.

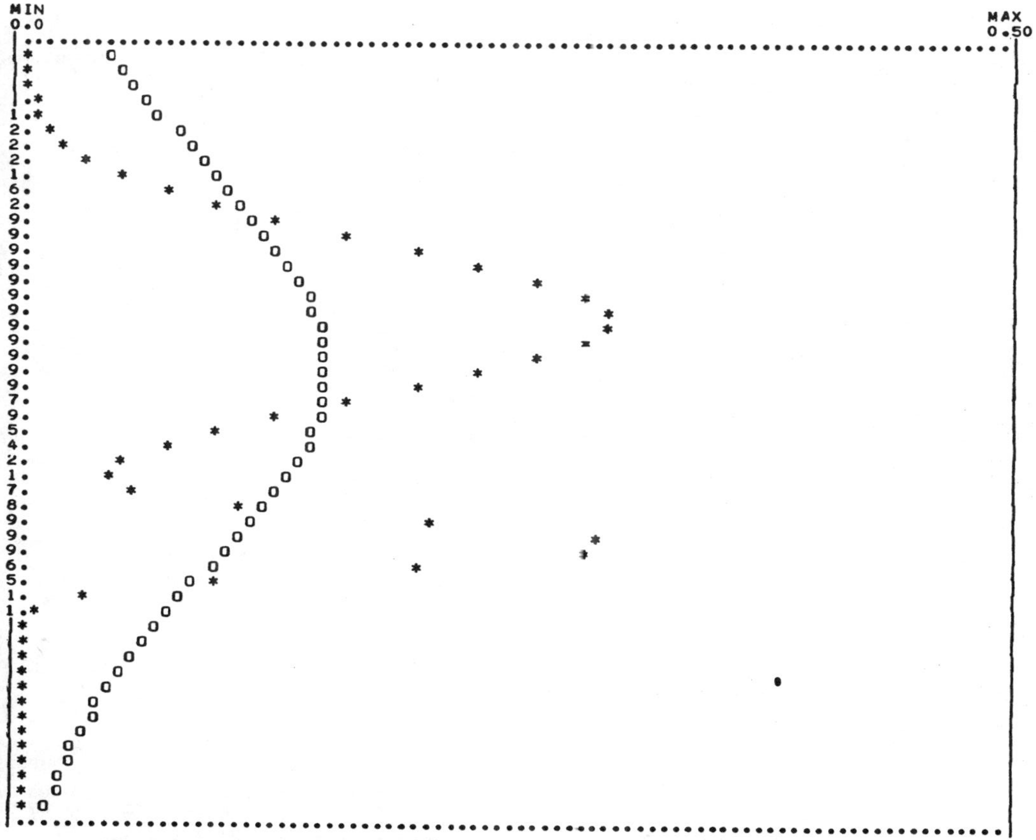

Figure 2.1. $n = 300$ bimodal quartic kernel $h_n = 5.0$.

```
BIMODAL NORMAL WITH MEANS = -+  1.50 VARIANCE OF LEFT = 1 WITH WEIGHT AND VARIANCE OF RIGHT =   0.2500   0.1111
 SAMPLE SIZE =   300 WITH   75 SAMPLES ON RIGHT      VS.
        KERNEL ESTIMATOR WITH KERNEL = QUAR WITH H =        2.00000000
        INTEGRATED MEAN SQUARE ERROR                0.522528961 E-02
        INTEGRATED SQUARE ERROR                     0.268322602 E-01
        MAXIMUM ABSOLUTE DIFFERENCE                 0.179653645 E+00

ABSCISSA        ORDINATE      MIN                                          MAX
                *        O    0.0                                          0.50
-5.00E+00 6.55E-04 4.44E-03
-4.80E+00 1.29E-03 6.88E-03
-4.60E+00 2.45E-03 1.00E-02
-4.40E+00 4.46E-03 1.46E-02
-4.20E+00 7.82E-03 2.11E-02 1.
-4.00E+00 1.31E-02 3.00E-02 2.
-3.80E+00 2.12E-02 4.18E-02 2.
-3.60E+00 3.30E-02 5.68E-02 2.
-3.40E+00 4.92E-02 7.49E-02 1.
-3.20E+00 7.05E-02 9.57E-02 6.
-3.00E+00 9.71E-02 1.18E-01 2.
-2.80E+00 1.29E-01 1.41E-01 9.
-2.60E+00 1.63E-01 1.65E-01 9.
-2.40E+00 2.00E-01 1.86E-01 9.
-2.20E+00 2.34E-01 2.06E-01 9.
-2.00E+00 2.64E-01 2.22E-01 9.
-1.80E+00 2.86E-01 2.33E-01 9.
-1.60E+00 2.98E-01 2.39E-01 9.
-1.40E+00 2.98E-01 2.39E-01 9.
-1.20E+00 2.86E-01 2.33E-01 9.
-1.00E+00 2.64E-01 2.21E-01 9.
-8.00E-01 2.34E-01 2.06E-01 9.
-6.00E-01 2.00E-01 1.88E-01 9.
-4.00E-01 1.63E-01 1.71E-01 7.
-2.00E-01 1.29E-01 1.56E-01 9.
 0.0      9.72E-02 1.46E-01 5.
 2.00E-01 7.07E-02 1.39E-01 4.
 4.00E-01 5.05E-02 1.35E-01 2.
 6.00E-01 4.08E-02 1.32E-01 1.
 8.00E-01 5.42E-02 1.31E-01 7.
 1.00E+00 1.10E-01 1.30E-01 8.
 1.20E+00 2.07E-01 1.29E-01 9.
 1.40E+00 2.91E-01 1.25E-01 9.
 1.60E+00 2.88E-01 1.20E-01 9.
 1.80E+00 2.01E-01 1.12E-01 6.
 2.00E+00 9.78E-02 1.01E-01 5.
 2.20E+00 3.33E-02 8.80E-02 1.
 2.40E+00 7.96E-03 7.30E-02 1.
 2.60E+00 1.36E-03 5.70E-02
 2.80E+00 1.78E-04 4.11E-02
 3.00E+00 2.40E-05 2.69E-02
 3.20E+00 5.45E-06 1.57E-02
 3.40E+00 1.86E-06 8.07E-03
 3.60E+00 6.74E-07 3.62E-03
 3.80E+00 2.38E-07 1.47E-03
 4.00E+00 8.08E-08 4.99E-04
 4.20E+00 2.64E-08 1.57E-04
 4.40E+00 8.26E-09 2.82E-05
 4.60E+00 2.49E-09 0.0
 4.80E+00 7.20E-10 0.0
 5.00E+00 2.00E-10 0.0
END OF PLOT.
```

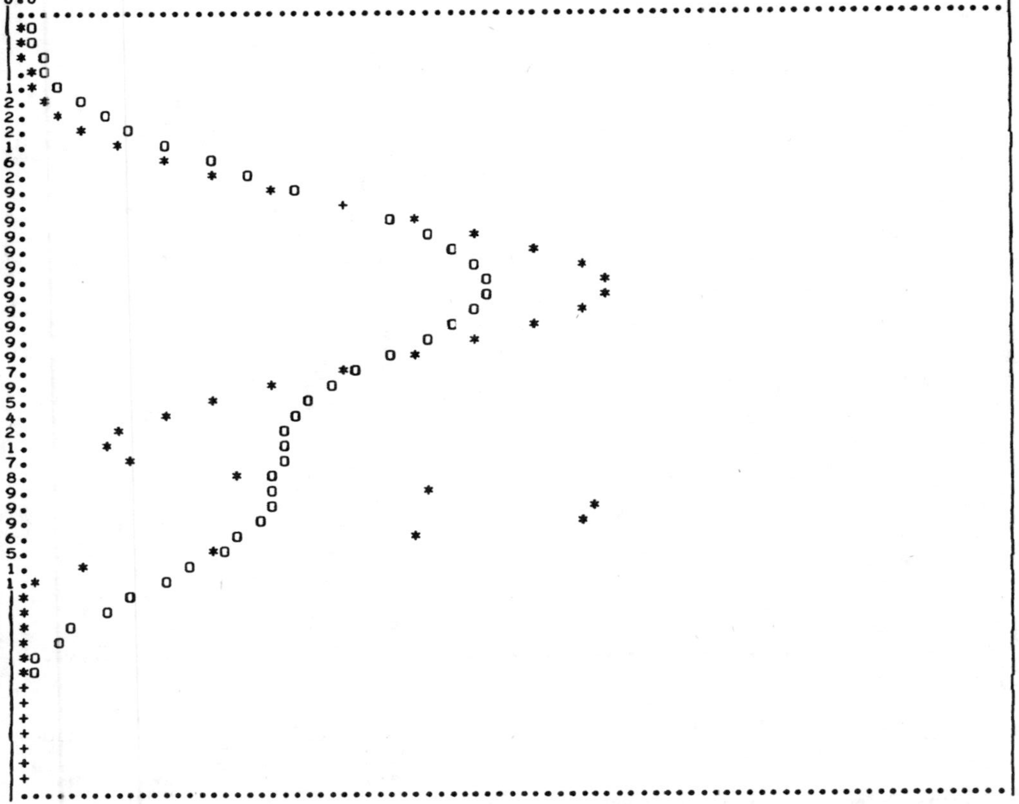

Figure 2.2. $n = 300$ bimodal quartic kernel $h_n = 2.0$.

```
BIMODAL NORMAL WITH MEANS = -+  1.50 VARIANCE OF LEFT = 1 WITH WEIGHT AND VARIANCE OF RIGHT =   0.2500   0.1111
 SAMPLE SIZE =   300 WITH    75 SAMPLES ON RIGHT      VS.
        KERNEL ESTIMATOR WITH KERNEL = QUAR WITH H =        0.80000001
        INTEGRATED MEAN SQUARE ERROR              0.865487382 E-03
        INTEGRATED SQUARE ERROR                   0.410439074 E-02
        MAXIMUM ABSOLUTE DIFFERENCE               0.840904713 E-01
```

ABSCISSA	ORDINATE *	ORDINATE O
-5.00E+00	6.55E-04	0.0
-4.80E+00	1.29E-03	2.35E-04
-4.60E+00	2.45E-03	1.65E-03
-4.40E+00	4.46E-03	6.00E-03
-4.20E+00	7.82E-03	1.31E-02
-4.00E+00	1.31E-02	2.05E-02
-3.80E+00	2.12E-02	2.76E-02
-3.60E+00	3.30E-02	3.48E-02
-3.40E+00	4.92E-02	4.18E-02
-3.20E+00	7.05E-02	5.80E-02
-3.00E+00	9.71E-02	8.66E-02
-2.80E+00	1.29E-01	1.24E-01
-2.60E+00	1.63E-01	1.67E-01
-2.40E+00	2.00E-01	2.11E-01
-2.20E+00	2.34E-01	2.50E-01
-2.00E+00	2.64E-01	2.77E-01
-1.80E+00	2.86E-01	2.88E-01
-1.60E+00	2.98E-01	2.87E-01
-1.40E+00	2.98E-01	2.80E-01
-1.20E+00	2.86E-01	2.67E-01
-1.00E+00	2.64E-01	2.50E-01
-8.00E-01	2.34E-01	2.28E-01
-6.00E-01	2.00E-01	2.01E-01
-4.00E-01	1.63E-01	1.67E-01
-2.00E-01	1.29E-01	1.34E-01
0.0	9.72E-02	1.07E-01
2.00E-01	7.07E-02	8.13E-02
4.00E-01	5.05E-02	6.42E-02
6.00E-01	4.08E-02	6.56E-02
8.00E-01	5.42E-02	9.36E-02
1.00E+00	1.10E-01	1.41E-01
1.20E+00	2.07E-01	1.90E-01
1.40E+00	2.91E-01	2.18E-01
1.60E+00	2.88E-01	2.10E-01
1.80E+00	2.01E-01	1.71E-01
2.00E+00	9.78E-02	1.16E-01
2.20E+00	3.33E-02	6.73E-02
2.40E+00	7.56E-03	3.27E-02
2.60E+00	1.36E-03	1.45E-02
2.80E+00	1.78E-04	5.21E-03
3.00E+00	2.40E-05	1.84E-03
3.20E+00	5.45E-06	3.94E-04
3.40E+00	1.86E-06	0.0
3.60E+00	6.74E-07	0.0
3.80E+00	2.38E-07	0.0
4.00E+00	8.08E-08	0.0
4.20E+00	2.64E-08	0.0
4.40E+00	8.26E-09	0.0
4.60E+00	2.49E-09	0.0
4.80E+00	7.20E-10	0.0
5.00E+00	2.00E-10	0.0

END OF PLOT.

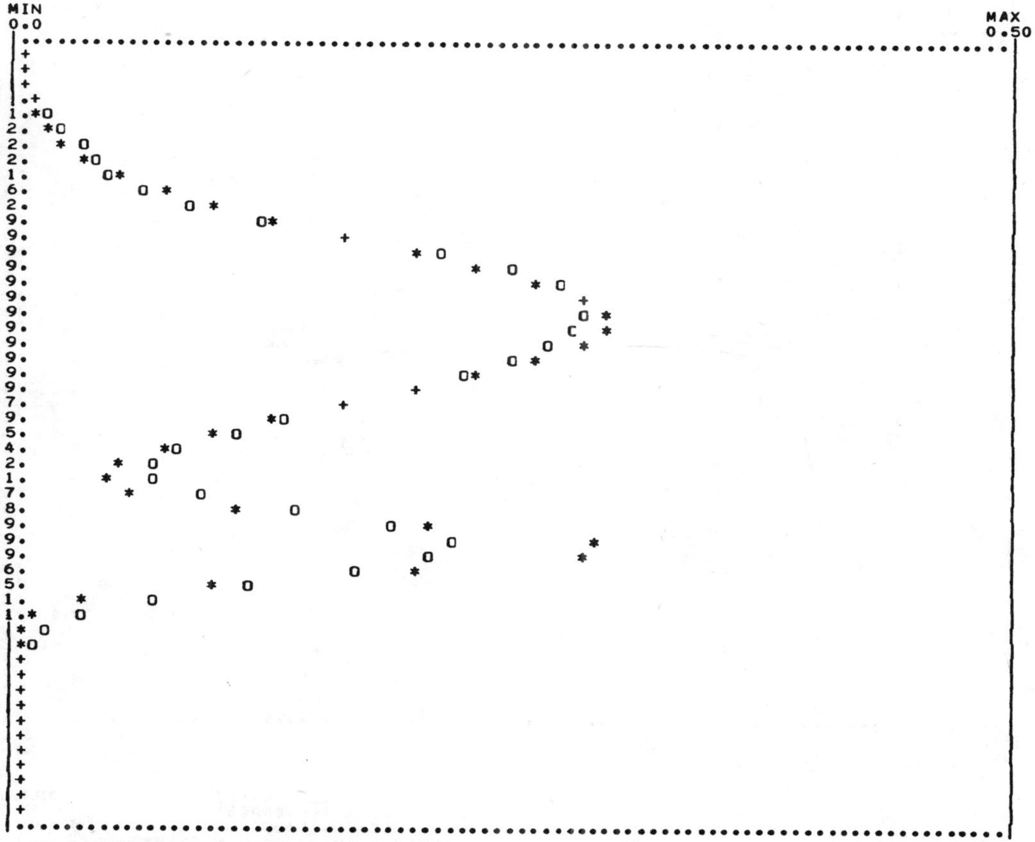

Figure 2.3. $n = 300$ bimodal quartic kernel $h_n = 0.8$.

```
BIMODAL NORMAL WITH MEANS = -+  1.50 VARIANCE OF LEFT = 1 WITH WEIGHT AND VARIANCE OF RIGHT =  0.2500  0.1111
 SAMPLE SIZE =  300 WITH   75 SAMPLES ON RIGHT     VS.
        KERNEL ESTIMATOR WITH KERNEL = QUAR WITH H =        0.60000002
        INTEGRATED MEAN SQUARE ERROR              0.503594056 E-03
        INTEGRATED SQUARE ERROR                   0.236694631 E-02
        MAXIMUM ABSOLUTE DIFFERENCE               0.574755073 E-01
```

ABSCISSA	ORDINATE *	O	
-5.00E+00	6.55E-04	0.0	
-4.80E+00	1.29E-03	0.0	
-4.60E+00	2.45E-03	5.32E-04	
-4.40E+00	4.46E-03	3.38E-03	
-4.20E+00	7.82E-03	1.15E-02	1.
-4.00E+00	1.31E-02	2.23E-02	2.
-3.80E+00	2.12E-02	2.96E-02	2.
-3.60E+00	3.30E-02	3.39E-02	2.
-3.40E+00	4.92E-02	3.97E-02	1.
-3.20E+00	7.05E-02	4.88E-02	6.
-3.00E+00	9.71E-02	7.83E-02	2.
-2.80E+00	1.29E-01	1.22E-01	9.
-2.60E+00	1.63E-01	1.69E-01	9.
-2.40E+00	2.00E-01	2.16E-01	9.
-2.20E+00	2.34E-01	2.51E-01	9.
-2.00E+00	2.64E-01	2.83E-01	9.
-1.80E+00	2.86E-01	3.03E-01	9.
-1.60E+00	2.98E-01	2.94E-01	9.
-1.40E+00	2.98E-01	2.73E-01	9.
-1.20E+00	2.86E-01	2.65E-01	9.
-1.00E+00	2.64E-01	2.57E-01	9.
-8.00E-01	2.34E-01	2.39E-01	9.
-6.00E-01	2.00E-01	2.01E-01	9.
-4.00E-01	1.63E-01	1.61E-01	7.
-2.00E-01	1.29E-01	1.34E-01	9.
0.0	9.72E-02	1.07E-01	5.
2.00E-01	7.07E-02	7.98E-02	4.
4.00E-01	5.05E-02	5.52E-02	2.
6.00E-01	4.08E-02	4.98E-02	1.
8.00E-01	5.42E-02	7.48E-02	7.
1.00E+00	1.10E-01	1.37E-01	8.
1.20E+00	2.07E-01	2.05E-01	9.
1.40E+00	2.91E-01	2.45E-01	9.
1.60E+00	2.88E-01	2.32E-01	9.
1.80E+00	2.01E-01	1.75E-01	6.
2.00E+00	9.78E-02	1.09E-01	5.
2.20E+00	3.33E-02	5.51E-02	1.
2.40E+00	7.96E-03	2.59E-02	1.
2.60E+00	1.36E-03	9.59E-03	
2.80E+00	1.78E-04	3.66E-03	
3.00E+00	2.40E-05	8.75E-04	
3.20E+00	5.45E-06	0.0	
3.40E+00	1.86E-06	0.0	
3.60E+00	6.74E-07	0.0	
3.80E+00	2.38E-07	0.0	
4.00E+00	8.08E-08	0.0	
4.20E+00	2.64E-08	0.0	
4.40E+00	8.26E-09	0.0	
4.60E+00	2.49E-09	0.0	
4.80E+00	7.20E-10	0.0	
5.00E+00	2.00E-10	0.0	

END OF PLOT.

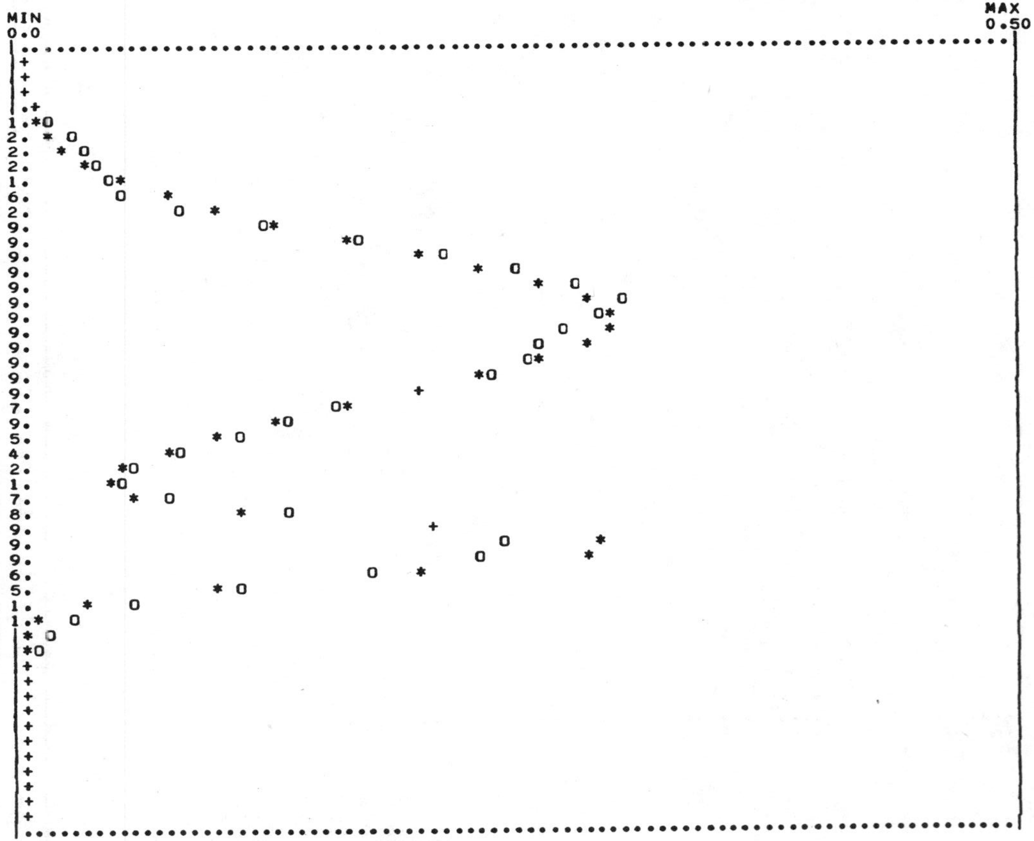

Figure 2.4. $n = 300$ bimodal quartic kernel $h_n = 0.6$.

```
BIMODAL NORMAL WITH MEANS = -+  1.50 VARIANCE OF LEFT = 1 WITH WEIGHT AND VARIANCE OF RIGHT =  0.2500  0.1111
 SAMPLE SIZE =  300 WITH   75 SAMPLES ON RIGHT    VS.
       KERNEL ESTIMATOR WITH KERNEL = QUAR WITH H =        0.39999998
       INTEGRATED MEAN SQUARE ERROR              0.497050816 E-03
       INTEGRATED SQUARE ERROR                   0.234523276 E-02
       MAXIMUM ABSOLUTE DIFFERENCE               0.500681400 E-01
```

ABSCISSA	ORDINATE *	ORDINATE O
-5.00E+00	6.55E-04	0.0
-4.80E+00	1.29E-03	0.0
-4.60E+00	2.45E-03	0.0
-4.40E+00	4.46E-03	1.63E-03
-4.20E+00	7.82E-03	8.17E-03
-4.00E+00	1.31E-02	2.38E-02
-3.80E+00	2.12E-02	3.48E-02
-3.60E+00	3.30E-02	3.22E-02
-3.40E+00	4.92E-02	3.43E-02
-3.20E+00	7.05E-02	5.09E-02
-3.00E+00	9.71E-02	6.22E-02
-2.80E+00	1.29E-01	1.17E-01
-2.60E+00	1.63E-01	1.86E-01
-2.40E+00	2.00E-01	2.10E-01
-2.20E+00	2.34E-01	2.49E-01
-2.00E+00	2.64E-01	2.89E-01
-1.80E+00	2.86E-01	3.14E-01
-1.60E+00	2.98E-01	3.07E-01
-1.40E+00	2.98E-01	2.63E-01
-1.20E+00	2.86E-01	2.47E-01
-1.00E+00	2.64E-01	2.80E-01
-8.00E-01	2.34E-01	2.54E-01
-6.00E-01	2.00E-01	1.87E-01
-4.00E-01	1.63E-01	1.59E-01
-2.00E-01	1.29E-01	1.36E-01
0.0	9.72E-02	1.07E-01
2.00E-01	7.07E-02	7.84E-02
4.00E-01	5.05E-02	5.39E-02
6.00E-01	4.08E-02	3.33E-02
8.00E-01	5.42E-02	5.83E-02
1.00E+00	1.10E-01	1.29E-01
1.20E+00	2.07E-01	2.20E-01
1.40E+00	2.51E-01	2.72E-01
1.60E+00	2.88E-01	2.50E-01
1.80E+00	2.01E-01	1.78E-01
2.00E+00	9.78E-02	9.71E-02
2.20E+00	3.33E-02	4.92E-02
2.40E+00	7.96E-03	1.82E-02
2.60E+00	1.36E-03	8.19E-03
2.80E+00	1.78E-04	2.58E-03
3.00E+00	2.40E-05	0.0
3.20E+00	5.45E-06	0.0
3.40E+00	1.86E-06	0.0
3.60E+00	6.74E-07	0.0
3.80E+00	2.38E-07	0.0
4.00E+00	8.08E-08	0.0
4.20E+00	2.64E-08	0.0
4.40E+00	8.26E-09	0.0
4.60E+00	2.49E-09	0.0
4.80E+00	7.20E-10	0.0
5.00E+00	2.00E-10	0.0

END OF PLOT.

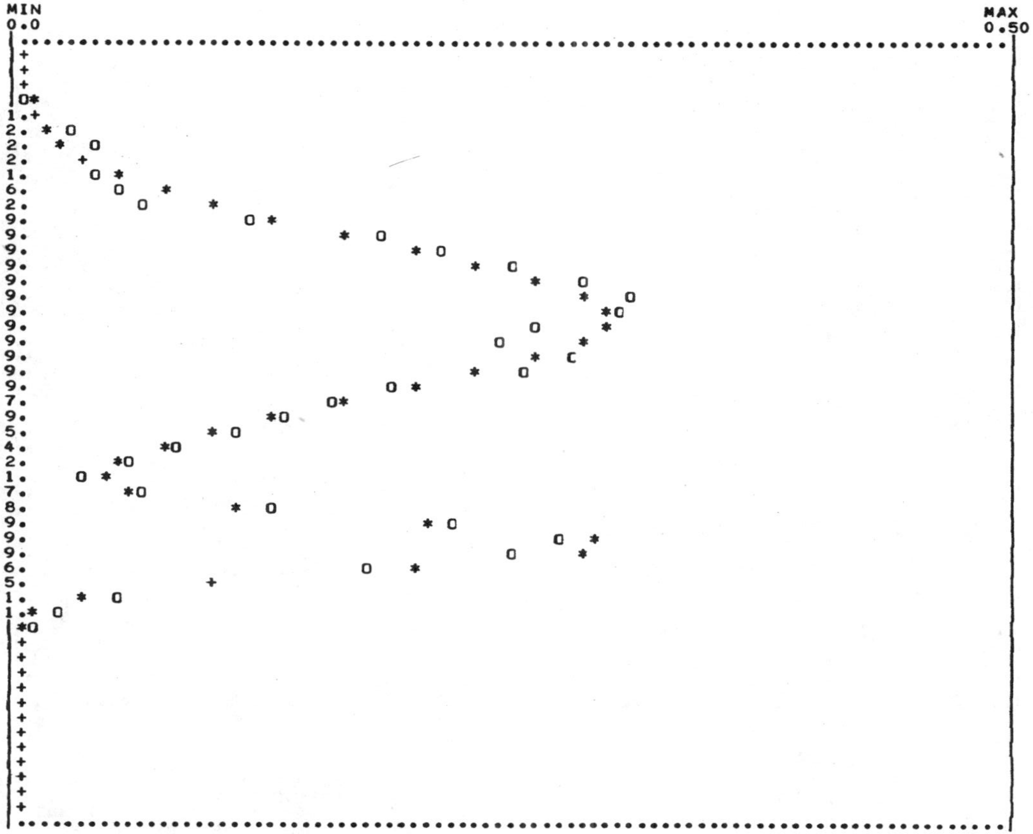

Figure 2.5. $n = 300$ bimodal quartic kernel $h_n = 0.4$.

```
BIMODAL NORMAL WITH MEANS = -+  1.50 VARIANCE OF LEFT = 1 WITH WEIGHT AND VARIANCE OF RIGHT =  0.2500  0.1111
 SAMPLE SIZE =   300 WITH    75 SAMPLES ON RIGHT      VS.
        KERNEL ESTIMATOR WITH KERNEL = QUAR WITH H =          0.19999999
        INTEGRATED MEAN SQUARE ERROR              0.148360804 E-02
        INTEGRATED SQUARE ERROR                   0.682390109 E-02
        MAXIMUM ABSOLUTE DIFFERENCE               0.936830044 E-01

ABSCISSA        ORDINATE
                *          O
-5.00E+00 6.55E-04 0.0
-4.80E+00 1.29E-03 0.0
-4.60E+00 2.45E-03 0.0
-4.40E+00 4.46E-03 0.0
-4.20E+00 7.82E-03 9.38E-03
-4.00E+00 1.31E-02 1.63E-02
-3.80E+00 2.12E-02 5.12E-02
-3.60E+00 3.30E-02 2.41E-02
-3.40E+00 4.92E-02 2.87E-02
-3.20E+00 7.05E-02 6.05E-02
-3.00E+00 9.71E-02 4.57E-02
-2.80E+00 1.29E-01 9.33E-02
-2.60E+00 1.63E-01 2.25E-01
-2.40E+00 2.00E-01 2.08E-01
-2.20E+00 2.34E-01 1.98E-01
-2.00E+00 2.64E-01 3.58E-01
-1.80E+00 2.86E-01 2.79E-01
-1.60E+00 2.98E-01 3.27E-01
-1.40E+00 2.98E-01 2.64E-01
-1.20E+00 2.86E-01 1.95E-01
-1.00E+00 2.64E-01 3.39E-01
-8.00E-01 2.34E-01 2.49E-01
-6.00E-01 2.00E-01 1.71E-01
-4.00E-01 1.63E-01 1.66E-01
-2.00E-01 1.29E-01 1.40E-01
 0.0      9.72E-02 9.60E-02
 2.00E-01 7.07E-02 9.00E-02
 4.00E-01 5.05E-02 4.38E-02
 6.00E-01 4.08E-02 3.04E-02
 8.00E-01 5.42E-02 4.49E-02
 1.00E+00 1.10E-01 1.11E-01
 1.20E+00 2.07E-01 2.50E-01
 1.40E+00 2.91E-01 2.71E-01
 1.60E+00 2.88E-01 2.88E-01
 1.80E+00 2.01E-01 1.54E-01
 2.00E+00 9.78E-02 1.03E-01
 2.20E+00 3.33E-02 4.53E-02
 2.40E+00 7.96E-03 9.00E-03
 2.60E+00 1.36E-03 1.29E-02
 2.80E+00 1.78E-04 0.0
 3.00E+00 2.40E-05 0.0
 3.20E+00 5.45E-06 0.0
 3.40E+00 1.86E-06 0.0
 3.60E+00 6.74E-07 0.0
 3.80E+00 2.38E-07 0.0
 4.00E+00 8.08E-08 0.0
 4.20E+00 2.64E-08 0.0
 4.40E+00 8.26E-09 0.0
 4.60E+00 2.49E-09 0.0
 4.80E+00 7.20E-10 0.0
 5.00E+00 2.00E-10 0.0
END OF PLOT.
```

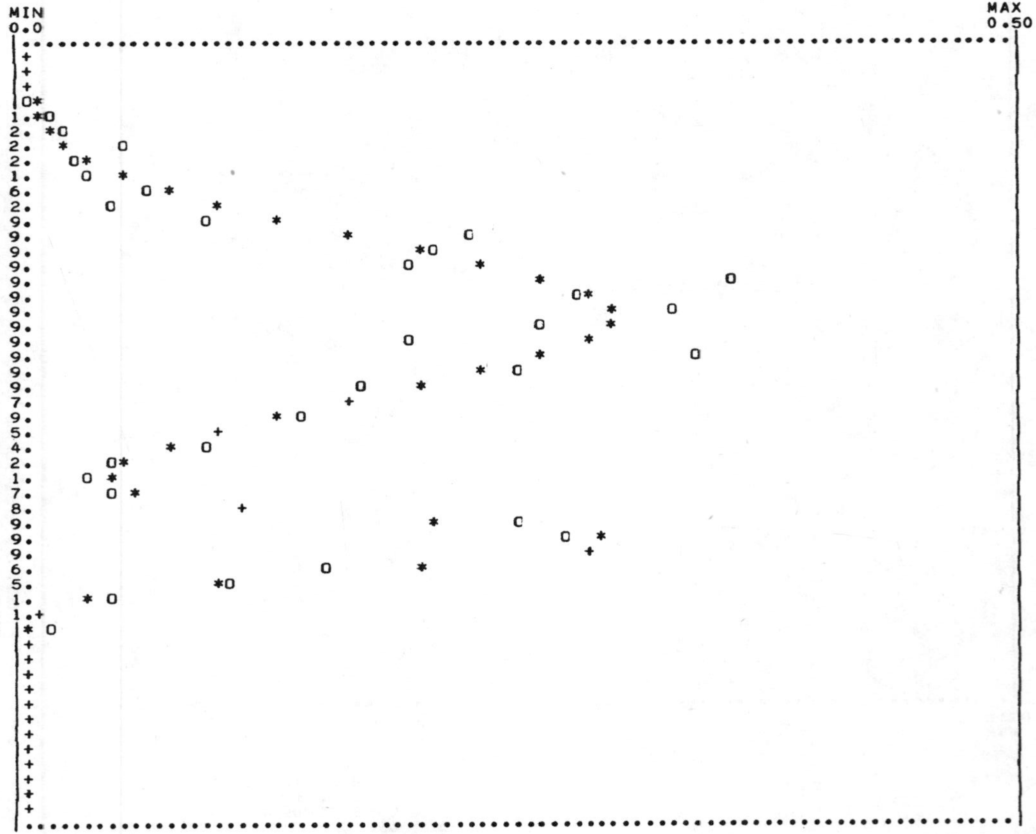

Figure 2.6. $n = 300$ bimodal quartic kernel $h_n = 0.2$.

further attempts at resolution by decreasing h_n drives one into a highly noisy estimate is generally fairly sharp and readily observable. However, for the user who is not really into time sharing, we give below an empirical algorithm which is frequently successful in selecting a satisfactory value of h_n, using a batch, i.e., one shot, approach.

Did we but know $\int_{-\infty}^{\infty} (f''(y))^2\, dy$, there would be no trouble in picking a good h_n for a particular kernel with characteristic exponent $r = 2$. It is tempting, since $\int_{-\infty}^{\infty} (f''(y))^2\, dy$ is sensitive, though not terribly so, to changes in f, to try functional iteration to obtain a satisfactory value for h_n. Thus, if we have a guess for h_n—say h_n^i—we may use (127) to obtain an estimate—say $\hat{f}_n^i$—for f. Then from (141), we have as our iterated guess

$$h_n^{i+1} = n^{-1/5}\alpha(K)\beta(\hat{f}_n^i). \tag{143}$$

More efficiently, since β is an explicit function of h_n, we may use Newton's method to find a zero of (143). Typically, we find that if Newton's method is employed, around five iterations are required to have $|h_n^{i+1} - h_n^i| < 10^{-5}$.

In Tables 2.2 and 2.3, we demonstrate the results of Monte Carlo simulations using (143). The distributions examined were the standard normal, the 50–50 mixture of two normal densities with a separation of means of three times their common standard deviation, a central t distribution with 5 degrees of freedom, and an $\mathscr{F}$ distribution with (10, 10) degrees of freedom. The average values of h_n^{∞}, together with the theoretically (asymptotically) optimal values of h_n, were computed from (143). In Table 2.3 we show integrated mean square error results obtained using the recursive technique in comparison with the kernel estimator using the theoretically optimal value of h_n. The average efficiency shown is simply the average of the ratio of the IMSE obtained using the theoretically optimal value of h_n divided

Table 2.2 Monte Carlo Results for Finding the Quasi-Optimal h_n (Each Row Represents 25 Samples)

Density	Sample size	Degenerate*	Mean	Std. dev.	Range	Theoretical
$N(0, 1)$	25	1	.54	.17	.20–.80	.56
$.5N(-1.5, 1) + .5N(1.5, 1)$	25	2	.77	.41	.09–1.35	.66
t_5	25	0	.59	.19	.25–.96	.41
$F_{10,10}$	25	2	.25	.09	.02–.42	.20
$N(0, 1)$	100	0	.35	.10	.09–.51	.42
$.5N(-1.5, 1) + .5N(1.5, 1)$	100	0	.43	.17	.12–.76	.50
t_5	100	0	.37	.09	.13–.54	.31
$F_{10,10}$	100	1	.15	.04	.05–.20	.15

* In less than 1% of the cases examined $h_n^{\infty} = 0$. These cases were excluded from the study.

Table 2.3 Integrated Mean Square Error, Using the Quasi-Optimal h_n vs. the Theoretically Optimal h

Density	Number of samples	Sample size	Quasi-optimal h_n Mean	Quasi-optimal h_n Std. dev.	Theoretical h_n Mean	Theoretical h_n Std. dev.	Average efficiency (%)
$N(0, 1)$	24	25	.0242	.0187	.0163	.119	67
$.5N(-1.5, 1)+.5N(1.5, 1)$	22*	25	.0233	.0306	.0095	.0070	41
t_5	25	25	.0230	.0109	.0210	.0091	91
$F_{10,10}$	22*	25	.0401	.0194	.0390	.0187	97
$N(0, 1)$	24*	100	.0071	.0042	.0050	.0028	70
$.5N(-1.5, 1)+.5N(1.5, 1)$	23*	100	.0053	.0024	.0036	.0021	68
t_5	25	100	.0085	.0099	.0069	.0039	81
$F_{10,10}$	24	100	.0217	.0221	.0168	.0094	77

* The largest one or two IMSE values were omitted because the corresponding quasi-optimal h_n was nearly zero.

by that obtained using the quasi-optimal value obtained by functional iteration.

Early on (1958) in the development of kernel density estimators, Whittle [60] proposed a Bayesian approach. Let us consider an estimator of the form

$$\hat{f}_n(x) = \frac{1}{n} \sum_{j=1}^{n} K_x(x_j). \tag{144}$$

To develop a somewhat more concise set of expressions, Whittle considers the sample size n to be a Poisson variate with expectation M. Letting

$$\varphi(x) = Mf(x), \tag{145}$$

we shall estimate $\varphi(x)$ with statistics, of the form

$$\hat{\varphi}(x) = \sum_{j=1}^{n} K_x(x_j). \tag{146}$$

Whittle assumes prior values on $\varphi(x)$ and $\varphi(x)\varphi(y)$ via

$$E_p[\varphi(x)] = \mu(x) \tag{147}$$

$$E_p[\varphi(x)\varphi(y)] = \mu(x, y). \tag{148}$$

The assumption of the existence of an appropriate measure on the space of densities is a nontrivial one [32].

Then a straightforward calculation shows that the Bayes risk is minimized when

$$\mu(y)K_x(y) + \int \mu(y, z)K_x(z)\, dz = \mu(y, x). \tag{149}$$

We then carry out the following normalization

$$\xi_x(y) = K_x(y)\sqrt{\frac{\mu(y)}{\mu(x)}} \tag{150}$$

$$\gamma(x, y) = \frac{\mu(x, y)}{\sqrt{\mu(x)\mu(y)}}. \tag{151}$$

Then (149) becomes

$$\xi_x(y) + \int \gamma(y, z)\xi_x(z)\, dz = \gamma(y, x). \tag{152}$$

Whittle suggests special emphasis on the case where $\gamma(y, x) = \gamma(y - x)$. This property, obviously motivated by second-order stationarity in time series analysis, might be appropriate, Whittle argues, in the case where $\mu(x)$ is rectangular. Then (152) becomes

$$\xi_x(y) + \int \gamma(y - z)\xi_x(z)\, dz = \gamma(y - x). \tag{153}$$

Now, if $\int\int|\gamma(y - x)|^2\, dx\, dy < \infty$, then (153) has a unique solution $\xi_x(y)$ in $L^2(-\infty, \infty)$, since $(I + \gamma*)$ is a Hilbert–Schmidt compact operator. Whittle solves for $\xi_x(y)$ by Fourier transforming (153) to obtain

$$\xi_x(y) = \frac{1}{2\pi}\int_{-\infty}^{\infty} e^{i\omega(y-x)}\frac{\Gamma(\omega)}{1 + \Gamma(\omega)}\, d\omega \tag{154}$$

where Γ is the Fourier transform of γ. He proposes the following form for γ

$$\gamma(x) = v(a + be^{-c|x|}). \tag{155}$$

(This causes some difficulty, since $\gamma \notin L^1(-\infty, \infty)$ if $a \neq 0$, and hence the Fourier transform does not exist.)
Then, solving for $\xi_x(y)$ we have

$$\xi_x(y) = \frac{vbc}{\theta}e^{-\theta|y-x|}, \tag{156}$$

where

$$\theta = (2vbc + c^2)^{1/2}. \tag{157}$$

But then $\xi_x(y)$ does not depend on a. Let us consider two possibilities for γ, namely,

$$\gamma_1(x) = v(a_1 + be^{-c|x|})$$

and

$$\gamma_2(x) = v(a_2 + be^{-c|x|}).$$

Both γ_1 and γ_2 yield the same $\xi_x(y)$. So from (153) we have

$$(va_1 - va_2) \int \xi_x(z)\, dz = va_1 - va_2. \tag{158}$$

Thus, any solution to (153) must have

$$\int \xi_x(z)\, dz = 1. \tag{159}$$

But this requires, from (156) and (157) that

$$\frac{2vbc}{\theta^2} = 1$$

or that

$$\frac{2vbc}{2vbc + c^2} = 1.$$

Thus, it must be true that $c^2 = 0$. Hence, it must be true that $\theta = 0$. Thus, we have shown

Theorem 6. (156) is not a solution to (153) for arbitrary a. In fact, it is immediately clear that in order for (156) to solve (153) we must have $a = 0$ (which, of course, makes γ an $L^1(-\infty, \infty)$ function).

We can also establish the "translation invariance" of $\xi_x(y)$, namely,

Theorem 7. $\xi_x(y) = \xi_0(y - x)$.

Proof.

$$\xi_x(y) = \frac{vbc}{\theta} e^{-\theta|y-x|} = \xi_0(y - x).$$

(This theorem is true for any second-order stationary γ.) ∎

Now, to return to the original problem of obtaining a (Bayesian) optimal form for $K_x(y)$, we use (150) and (156) to obtain

$$K_x(y) = \sqrt{\frac{\mu(x)}{\mu(y)}} \frac{vbc}{\theta} e^{-\theta|y-x|}. \tag{160}$$

But if $\hat{f}_n(x)$ is to be a probability density, we must have

$$1 = \int \hat{f}_n(x)\, dx = \frac{1}{n} \sum_{j=1}^{n} \int K_x(x_j)\, dx. \tag{161}$$

For the second-order stationarity assumption $\gamma(y, x) = \gamma(y - x)$ to be reasonable, Whittle suggests that one should assume $\mu(x) = \mu(y)$ for all (x, y). But this would imply

$$K_x(y) = \frac{vbc}{\theta} e^{-\theta|y-x|} \tag{162}$$

and that

$$\frac{1}{n}\sum_{j=1}^{n}\frac{vbc}{\theta}\int e^{-\theta|x_j-x|}\,dx=1. \tag{163}$$

If this is to hold for any possible set of sample points $\{x_j\}_{j=1}^{n}$, we must have

$$\frac{2vbc}{\theta^2}=1. \tag{164}$$

We have seen already that this is impossible. Thus, if (161) is to hold, the kernel in (162) gives an empty family of estimators.

Now if γ is any second-order stationary kernel, we are led to estimates of the form

$$\hat{f}_n(x)=\frac{1}{n}\sum_{j=1}^{n}\frac{\mu(x)}{\mu(x_j)}\,\xi_0(x_j-x).$$

If this is to be a density for any possible sample $\{x_j\}_{j=1}^{n}$, we must have ξ_0 to be a probability density and $\mu(x)=\mu(y)$ for all x and y. Thus, after much labor, we are led, simply, to a Parzen estimator. A more recent data-oriented approach is given by de Figueiredo [17].

If one finds it unreasonable or impractical to make rather strong prior assumptions about the unknown density f, it is tempting to try a minimax approach for optimal kernel design. Such an attack has been carried out by Wahba [53] motivated by her earlier paper [52] and a result of Farrell [15].

Let us assume that $f\in W_p^{(m)}(M)$, i.e., $[\int_{-\infty}^{\infty}|f^{(m)}(x)|^p\,dx]^{1/p}\le M$, where $p\ge 1$.

Since f is a density, there exists a $\Lambda<\infty$ such that $\sup_{-\infty<x<\infty} f(x)\le\Lambda$.

Next, we shall restrict out consideration to kernels which satisfy the following conditions:

$$\sup_{-\infty<y<\infty}|K(y)|<\infty \tag{165}$$

$$\int_{-\infty}^{\infty}|K(y)|\,dy<\infty \tag{166}$$

$$\lim_{|y|\to\infty}|yK(y)|=0 \tag{167}$$

$$\int_{-\infty}^{\infty}K(y)\,dy=1 \tag{168}$$

$$\int_{-\infty}^{\infty}y^jK(y)\,dy=0,\qquad j=1,2,\ldots,m-1 \tag{169}$$

$$\int_{-\infty}^{\infty}|y|^{m-1/p}|K(y)|\,dy<\infty. \tag{170}$$

Now, for the estimators of the form

$$\hat{f}_n(x) = \frac{1}{nh_n} \sum_{k=1}^{n} K\left(\frac{x - x_k}{h_n}\right),$$

we have

$$\text{(171)} \qquad \operatorname{Var}[\hat{f}_n(x)] = \frac{f(x)}{nh_n} \int_{-\infty}^{\infty} K^2(y)\, dy \left[1 + 0\left(\frac{1}{nh_n}\right)\right]$$

$$\leq B \frac{1}{nh_n}\left[1 + 0\left(\frac{1}{nh_n}\right)\right],$$

where

$$B = \Lambda \int_{-\infty}^{\infty} K^2(y)\, dy.$$

The bias term is given by

$$\text{(172)} \quad \operatorname{Bias}[\hat{f}_n(x)] = E[\hat{f}_n(x) - f(x)] = \int_{-\infty}^{\infty} K(-\xi) f(x + \xi h)\, d\xi - f(x).$$

Taking a Taylor's expansion for $f(x + \xi h)$ we have

$$\text{(173)} \quad f(x + \xi h) = f(x) + \sum_{i=1}^{m-1} \frac{(\xi h)^i}{i!} f^{(i)}(x) + \int_x^{x+\xi h} \frac{(x + \xi h - u)^{m-1}}{(m-1)!} f^{(m)}(u)\, du.$$

Using (168), (169), (170), and (173) in (172), we have

$$\text{(174)} \quad E[\hat{f}_n(x) - f(x)] = \int_{-\infty}^{\infty} K(-\xi) \int_x^{x+\xi h} \frac{(x + \xi h - u)^{m-1}}{(m-1)!} f^{(m)}(u)\, du\, d\xi.$$

Applying Hölder's inequality to the inner integral in (174), we have

$$\text{(175)} \qquad \operatorname{Bias}^2[\hat{f}_n(x)] \leq M^2 A h_n^{2m-2/p},$$

where

$$\text{(176)} \quad A = \frac{1}{[(m-1)!]^2[(m-1)q + 1]^{2/q}} \left[\int_{-\infty}^{\infty} |K(\xi)|\, |\xi|^{m-1/p}\, d\xi\right]^2,$$

with

$$\frac{1}{q} + \frac{1}{p} = 1.$$

Thus, ignoring the factor $\left(1 + 0\left(\frac{1}{nh_n}\right)\right)$ we have

$$\text{(177)} \qquad \operatorname{MSE}(\hat{f}_n(x)) \leq M^2 A h_n^{2m-2/p} + B \frac{1}{nh_n}.$$

The right-hand side of (177) is minimized for

$$h_n = \left[\frac{1}{2m - 2/p}\frac{B}{M^2 A}\right]^{1/(2m+1-2/p)} n^{-1/(2m+1-2/p)}. \tag{178}$$

Substituting this value back into (177), we have

$$\text{MSE}(\hat{f}_n(x)) \leq D n^{-(2m-2/p)/(2m+1-2/p)}(1 + o(1)), \tag{179}$$

where

$$D = \frac{(2m + 1 - 2/p)}{(2m - 2/p)^{(2m-2/p)}}(M^2 A B^{2m-2/p})^{1/(2m+1-2/p)}$$

and

$$AB^{2m-2/p} = \frac{1}{[(m-1)!\{((m-1)/(1-1/p)) + 1\}^{1-1/p}]^2} \times \left[\int_{-\infty}^{\infty} |K(\xi)|\,|\xi|^{m-1/p}\,d\xi\right]^2 \left[\Lambda \int_{-\infty}^{\infty} K^2(y)\,dy\right]^{2m-2/p}.$$

Thus, we know that for any density in $W_p^{(m)}(M)$ the mean square error is no greater than the right-hand side of (179). We do not know how tight this upper bound on the mean square error is. However, Wahba observes that, on the basis of the bound, we should pick K so as to minimize

$$\left[\int_{-\infty}^{\infty} |K(\xi)|\,|\xi|^{m-1/p}\,d\xi\right]\left[\int_{-\infty}^{\infty} K^2(\xi)\,d\xi\right]^{m-1/p} \tag{180}$$

This task has been undertaken by Kazakos [26], who notes that the upper bound is achieved for $W_p^2(M)$ by

$$\begin{aligned} K(y) &= \frac{(1 + a^{-1})}{2}[1 - |y|^a] \quad \text{for} \quad |y| < 1 \\ &= 0 \quad \text{otherwise,} \end{aligned} \tag{181}$$

where $a = 2 - 1/p$.
(We have here omitted a superfluous scale parameter used by Kazakos. The special case where $p = \infty$ gives the quadratic window of Epanechnikov [12] discussed also by Rosenblatt [39].)

We can now evaluate A, B, h_n and the Wahba upper bound on the mean square error:

$$\begin{aligned} A &= \frac{(1 + a^{-1})^2}{[(2-1)!]^2[(2-1)q + 1]^{2/q}}\left[\int_0^1 (1 - y^a)y^{2-1/p}\,dy\right]^2 \\ &= \frac{1}{(q+1)^{2/q}(2a+1)^2} \end{aligned} \tag{182}$$

$$(183) \qquad B = \Lambda \frac{(1 + a^{-1})^2}{2} \int_0^1 (1 - 2y^a + y^{2a})\, dy$$

$$= \Lambda \frac{a+1}{2a+1}.$$

h_n and the Wahba upper bound on the MSE may be obtained by substituting for A and B in (178) and (179) respectively.

Returning to (178), let us consider

$$(184) \quad h_n = \left[\frac{(q+1)^{2/q}}{4 - 2/p} \frac{\int K^2(y)\, dy}{\left[\int |K(y)|\, |y|^{2-1/p}\, dy\right]^2} \frac{\Lambda}{M^2} \right]^{1/(5-2/p)} n^{-1/(5-2/p)}.$$

In most practical cases, if f has infinite support it will be reasonable to assume that $f^{(2)}$ is bounded. To obtain the fastest rate of decrease in n, we shall take $p = \infty$, thus giving

$$(185) \qquad h_n^{(1)} = \left[\frac{\int K^2(y)\, dy}{\left[\int |K(y)| y^2\, dy\right]^2} \right]^{1/5} \left[\frac{\sup |f^{(2)}|^2}{\sup_x f(x)} \right]^{-1/5} n^{-1/5}$$

$$\left(= \left(\frac{15\Lambda}{M^2} \right)^{1/5} n^{-1/5} \text{ (using the kernel in (181))} \right).$$

Comparing (185) with the h_n obtained from the integrated mean square error criterion formula in (141), i.e.,

$$(186) \qquad h_n^{(2)} = \left[\frac{\int K^2(y)\, dy}{\left[\int |K(y)| y^2\, dy\right]^2} \right]^{1/5} \left[\int |f^{(2)}(y)|^2\, dy \right]^{-1/5} n^{-1/5}.$$

We note that the term in the first brackets of (185) and (186) are identical. The only difference is in the second bracket term. How reasonable is it to suppose that we will have reliable knowledge of $\sup|f^{(2)}|$ and $\sup|f|$? Probably it is not frequently the case that we shall have very good guesses as to these quantities. Of course, an iterative algorithm could be constructed to estimate these quantities from the data. However, data-based estimates for these two quantities violate the philosophy of minimax kernel construction. These suprema in (185) are taken with respect to all densities in $W_\infty^{(2)}(M)$. Even if we were content to replace these suprema using a kernel estimator based on the data, we would, of course, find them unreliable relative to the earlier mentioned estimate (143) for $\int (f^{(2)})^2\,dx$. Moreover, the kernels given in (181) suffer from rough edges. Scott's quartic kernel K_3 [43], a form of Tukey's biweight function [see § 2.3], given in Table 2.1, does not have this difficulty.

It is intuitively clear that if we actually know the functional form of the density f, a kernel estimator will be inferior to that obtained by the classical method of using standard techniques for estimating the parameters which characterize f, and substituting these estimates for the parameters into f. For example, let us consider the problem for estimating the normal density with unit variance but unknown mean μ on the basis of a random sample of size n. If $\bar{x}$ is the sample mean, then the classical estimate for f is given by:

$$\hat{f}_p = \frac{1}{\sqrt{2\pi}} \exp\left[-\frac{1}{2}(x-\bar{x})^2\right]. \tag{187}$$

Then the integrated mean square error is given by

$$\begin{aligned} \text{IMSE} &= \frac{1}{\Pi}\left[1-\sqrt{\frac{n}{n+\frac{1}{2}}}\right] \\ &= \frac{1}{\Pi}\left[\frac{1}{4(n+\frac{1}{2})} + 0\left(\frac{1}{n^2}\right)\right]. \end{aligned} \tag{188}$$

Now, from (140) and (141) we have that the integrated mean square error of a kernel estimator based on the (density) kernel K of characteristic exponent 2 is bounded below by

$$\text{IMSE} \ge \frac{5}{4}\left[\int K^2(y)\,dy\right]^{4/5}\left[\int y^2K(y)\,dy\right]^{2/5}\left[\int (f''(y))^2\,dy\right]^{1/5} n^{-4/5}. \tag{189}$$

We can attain this bound only if we actually know $\int |f''(y)|^2\,dy$. We have seen that using our suggested procedure in (143) for estimating $\int |f''(y)|^2\,dy$ actually increases IMSE values over this bound significantly. Nonetheless, let us examine the efficiency of the kernel estimator in the utopian case where we actually know $\int |f''(y)|^2\,dy$. We shall first use the Gaussian kernel K_4. In this case, we have

$$\text{IMSE} \ge .3329n^{-4/5}. \tag{190}$$

Thus, the Gaussian kernel estimator has an efficiency relative to the classical estimator, which gradually deteriorates to zero as sample size goes to infinity. However, for a sample of size 100, the efficiency bound is 17%; and it has only fallen to 13% when the sample size has increased to 400. Moreover, we must remember that a nonparametric density estimation technique should not be used if we really know the functional form of the unknown density. If the true form of the density is anything (having $f'' \in L^2$) other than $N(\mu, 1)$, then the efficiency of the Gaussian kernel estimator, relative to the classical estimator, assuming the true density is $N(\mu, 1)$, becomes infinite as the sample size goes to infinity.

Nevertheless, there has been great interest in picking kernels which give improved efficiencies relative to the classical techniques which assume the functional form of the density is known. In particular, there has been extensive work on attempts to bring the rate of decrease of the IMSE as close to n^{-1} as possible. The first kernel estimator, the shifted histogram of Rosenblatt [38], showed a rate of decrease of $n^{-4/5}$.

Looking at (189), one notices that if one concedes the practical expedient of using kernels with IMSE rate of decrease $n^{-4/5}$, it is still possible to select K so as to minimize $[\int K^2(y)\,dy]^{4/5}[\int y^2 K(y)\,dy]^{2/5}$. A straightforward variational argument [12] shows that this may be done (in the class of symmetric density kernels) by using

$$K_5(y) = 3/4(1 - y^2) \quad \text{if} \quad |y| \le 1 \qquad (191)$$
$$= 0 \quad \text{otherwise.}$$

This is, as stated earlier, the kernel obtained in (181) by considering the Wahba "lower bound" on the mean square error for $m = 2$, $p = \infty$. Then the lower bound on the integrated mean square is given by

$$\text{IMSE} \ge .3198 n^{-4/5}. \qquad (192)$$

But comparing (192) with (190) we note that the Epanechnikov kernel estimator gives only a 4% improvement in the lower bound over the Gaussian kernel estimator. As a matter of fact, all the practical kernel estimators with characteristic exponent 2 have lower bound IMSE efficiencies very close to that of the "optimal" Epanechnikov kernel. We note that the nondifferentiability of the Epanechnikov kernel at the endpoints may cause practical difficulties when one faces the real world situation where $\int |f''(x)|^2\,dx$ must be estimated from the data. The Gaussian kernel does not suffer from this difficulty, but it has the computational disadvantage of infinite support. A suitable compromise might be K_3 or the B-spline kernel [2].

Among the first attempts to develop kernels which have IMSE decrease close to the order of n^{-1} was the 1963 work of Watson and Leadbetter [55]. They sought to discover that estimator

$$\hat{f}(x) = \frac{1}{n} \sum_{j=1}^{n} K_n(x - x_i), \qquad (193)$$

which minimized

(194) $$E\left[\int (\hat{f}_n(x) - f(x))^2 \, dx\right].$$

As a first step, they assumed that one knows the density $f(x)$ *precisely* insofar as computation of K is concerned. A straightforward application of Parseval's theorem yields as the formula for the Fourier transform of K_n:

(195) $$\varphi_{K_n}(t) = \frac{|\varphi_f(t)|^2}{\frac{1}{n} + \left[\frac{n-1}{n}\right]|\varphi_f(t)|^2},$$

where φ_f is the characteristic function of the density f. We note that the constraint that K_n be a density has not been applied, and the K_n corresponding to $\varphi_{K_n}(t)$ need not be a probability density. Consequently, estimated densities using the Watson–Leadbetter procedure need not be nonnegative nor integrate to unity. However, the procedure provides lower bounds for the integrated mean square error. Certainly, one should not expect to do better in kernel construction (purely by the IMSE criterion) than the case where one assumes the true density to be known and relaxes the constraint that a kernel be a density.

Watson and Leadbetter attempt to bring their procedure into the domain of practical application by considering situations where one may not know φ_f precisely. Let it be known that

(196) $\lim_{|t|\to\infty} |t|^p|\varphi_f(t)| = C^{1/2} > 0$ with $p > \frac{1}{2}$, i.e., that the characteristic function of f decreases algebraically of degree $p > \frac{1}{2}$.
Then,

(197) $$\hat{\varphi}_{K_n}(t) = h(A_n t),$$

where

$$h(t) = (1 + |t|^{2p})^{-1}$$
$$A_n = [nC]^{-1/2p}$$

gives us strictly positive asymptotic efficiency relative to the estimator given in (195). The order of decrease of the IMSE of the kernel estimator using $\hat{\varphi}_{K_n}$ is $n^{-1+2/p}$. Unfortunately, the resulting estimator is not a probability density.

Next, suppose $\varphi_f(t)$ has exponential decrease of degree 1 and coefficient ρ, i.e.,

(198) $$|\varphi_f(t)| \le A \exp[-\rho|t|^1]$$

for some constant A and all t, and

$$\lim_{v\to\infty} \int_0^1 [1 + e^{2\rho v}|\varphi_f(vt)|^2]^{-1}\, dt = 0. \tag{199}$$

Then, if

$$\hat{\varphi}_{K_n}(t) = h(A_n e^{\alpha|t|}),$$

where

$$h \in L^2$$
$$\sup_t |h(t)| < B < \infty$$
$$|1 - h(t)| \le B_1|t| \quad \text{for} \quad |t| \le 1$$
$$A_n = Dn^{-b}$$
$$\alpha = 2\rho b$$
$$b > \tfrac{1}{2},$$

then the resulting kernel estimator has a positive asymptotic efficiency relative to the estimator given in (195). The rate of decrease of the integrated mean square error is $(\log n)n^{-1}$.

A related approach of Davis [11] based on "saturation" theorems of Shapiro [45] seeks to improve the rate of convergence of the mean square error by eliminating the requirement that $K \in L^1$ from the conditions in (118). We recall from (134) that for $K \in L^1$

$$\frac{\text{Bias}[\hat{f}_n(x)]}{h_n^r} = \frac{1}{2\Pi}\int_{-\infty}^{\infty} e^{-iux}\frac{k(h_n u)^{-1}}{|h_n u|^r}|u|^r\varphi(u)\, du \to k_r f^{(r)}(x), \tag{200}$$

where k_r is the characteristic coefficient of K and r is its characteristic exponent.
Now, for all $j = 1, 2, \ldots, r-1$

$$\int_{-\infty}^{\infty} x^j K(x)\, dx = 0. \tag{201}$$

If we require that the conditions of (118) be operative, then we can do no better than $r = 2$. But then we can do no better than

$$\text{Bias}^2[\hat{f}_n(x)] = 0(h_n^4). \tag{202}$$

But

$$\text{Var}[\hat{f}_n(x)] = 0\left(\frac{1}{nh_n}\right). \tag{203}$$

Thus, we are forced to sacrifice the rate of decrease in the variance in order to accommodate the Bias^2 term. By eliminating the condition that $K \in L^1$, we are enabled to consider kernels which do not have their MSE's decreasing so slowly as $n^{-4/5}$.

The suggested kernel of Davis is simply the sinc function

$$K(x) = \frac{\sin x}{\Pi x}. \tag{204}$$

The sinc function is, of course, the Fourier transform of a step function. Generally, sinc functions occur in practice when one takes the Fourier transform of a signal limited in time extent. It is a great villain in time series analysis due to its heavy negative and positive minor lobes. However, in this context, Davis considers it to be a convenient non-L^1 (though L^2) function which integrates to unity. As we shall shortly see, the sinc function estimator can give substantial negative values. Of course, this is not surprising, since nonnegativity has neither explicitly nor implicitly been imposed as a condition.

In the case where the characteristic function of the unknown density shows algebraic decrease of order p, the sinc function estimator has Bias^2 decrease of order h_n^{2p-2}. The variance decrease is of order $1/(h_n n)$. Thus a choice of $h_n = cn^{-1(2p-1)}$ gives a mean square error decrease of order $n^{-1+1/(2p-1)}$.

In the case where the characteristic function of the density decreases exponentially with degree r and coefficient ρ, we have that Bias^2 of the sinc function estimator decreases at least as fast as $\exp[-2p/h_n^r]/h_n^2$. The variance decrease is still of order $1/nh_n$. A choice of $h_n = (\log n/2\rho)^{-1/r}$ gives mean square error decrease of order $(\log n)/n$, i.e., very close to the n^{-1} rate. In Figures 2.7 and 2.8 we show the results of applying the Davis algorithm to the estimation of $N(0, 1)$ for samples of size 20 and 100 respectively. The substantial negative side lobes are disturbing. This difficulty is even more strikingly seen in Figure 2.9, where we apply the sinc function procedure to estimation of an F density with (10, 10) degrees of freedom on the basis of a sample of size 400.

It is very likely that the negativity of non-L^1 estimates would be lessened by using kernels like those constructed by Tukey for time series analysis, e.g.,

$$K(x) = \frac{1}{\Pi}\left[.5\frac{\sin x}{x} + .25\frac{\sin(x-a_n)}{(x-a_n)} + .25\frac{\sin(x+a_n)}{(x+a_n)}\right]. \tag{205}$$

```
NORMAL (0,1) SAMPLE OF SIZE  100          VS.
       KERNEL ESTIMATOR WITH KERNEL  = FIE. WITH H =      0.47140461
       INTEGRATED MEAN SQUARE ERROR                 0.795389526 E-03
       INTEGRATED SQUARE ERROR                      0.398049131 E-02
       MAXIMUM ABSOLUTE DIFFERENCE                  0.527679920 E-01
```

ABSCISSA	ORDINATE *	O	MIN -0.10
-5.00E+00	1.49E-06	-4.17E-03	
-4.80E+00	3.96E-06	-4.11E-03	
-4.60E+00	1.01E-05	-3.30E-03	
-4.40E+00	2.49E-05	-1.89E-03	
-4.20E+00	5.89E-05	-2.08E-04	
-4.00E+00	1.34E-04	1.39E-03	
-3.80E+00	2.92E-04	2.53E-03	
-3.60E+00	6.12E-04	2.99E-03	
-3.40E+00	1.23E-03	2.75E-03	
-3.20E+00	2.38E-03	2.08E-03	
-3.00E+00	4.43E-03	1.60E-03	
-2.80E+00	7.92E-03	2.24E-03	
-2.60E+00	1.36E-02	5.19E-03	
-2.40E+00	2.24E-02	1.18E-02	
-2.20E+00	3.55E-02	2.33E-02	1
-2.00E+00	5.40E-02	4.08E-02	1
-1.80E+00	7.90E-02	6.49E-02	2
-1.60E+00	1.11E-01	9.57E-02	2
-1.40E+00	1.50E-01	1.33E-01	5
-1.20E+00	1.94E-01	1.74E-01	5
-1.00E+00	2.42E-01	2.19E-01	3
-8.00E-01	2.90E-01	2.63E-01	5
-6.00E-01	3.33E-01	3.05E-01	5
-4.00E-01	3.68E-01	3.41E-01	5
-2.00E-01	3.91E-01	3.69E-01	9
0.0	3.99E-01	3.86E-01	9
2.00E-01	3.91E-01	3.91E-01	9
4.00E-01	3.68E-01	3.83E-01	6
6.00E-01	3.33E-01	3.63E-01	5
8.00E-01	2.90E-01	3.32E-01	7
1.00E+00	2.42E-01	2.93E-01	6
1.20E+00	1.94E-01	2.47E-01	3
1.40E+00	1.50E-01	1.99E-01	3
1.60E+00	1.11E-01	1.51E-01	4
1.80E+00	7.90E-02	1.06E-01	
2.00E+00	5.40E-02	6.71E-02	1
2.20E+00	3.55E-02	3.55E-02	1
2.40E+00	2.24E-02	1.21E-02	
2.60E+00	1.36E-02	-3.19E-03	
2.80E+00	7.92E-03	-1.11E-02	
3.00E+00	4.43E-03	-1.30E-02	
3.20E+00	2.38E-03	-1.06E-02	
3.40E+00	1.23E-03	-5.70E-03	
3.60E+00	6.12E-04	5.26E-05	
3.80E+00	2.92E-04	5.20E-03	
4.00E+00	1.34E-04	8.77E-03	
4.20E+00	5.89E-05	1.02E-02	
4.40E+00	2.49E-05	9.61E-03	
4.60E+00	1.01E-05	7.20E-03	
4.80E+00	3.96E-06	3.67E-03	
5.00E+00	1.49E-06	-2.32E-04	

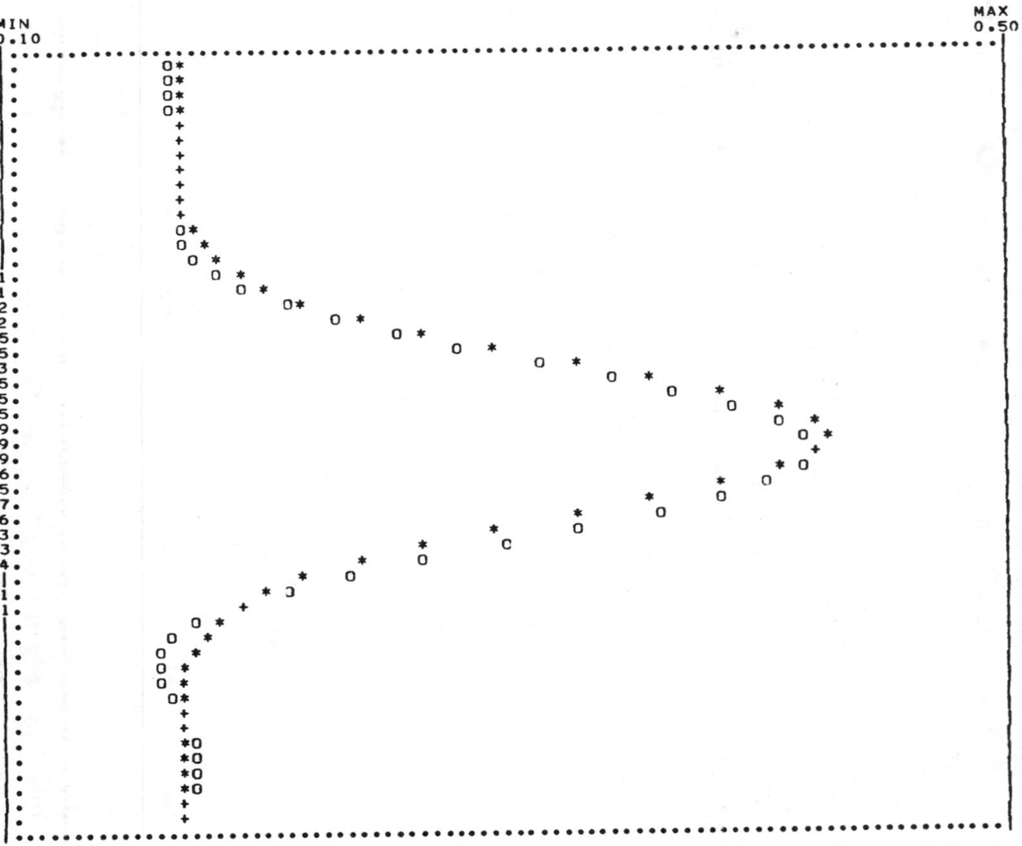

END OF PLOT.

Figure 2.7. $n = 20$ $N(0, 1)$ sinc kernel $h_n = 0.58$.

```
NORMAL (0,1) SAMPLE OF SIZE   20      VS.
       KERNEL ESTIMATOR WITH KERNEL = FIE. WITH H =        0.58447391
       INTEGRATED MEAN SQUARE ERROR                 0.153429294 E-02
       INTEGRATED SQUARE ERROR                      0.106731430 E-01
       MAXIMUM ABSOLUTE DIFFERENCE                  0.778741837 F-01
```

ABSCISSA	ORDINATE *	ORDINATE O
-5.00E+00	1.49E-06	1.26E-02
-4.80E+00	3.96E-06	1.75E-02
-4.60E+00	1.01E-05	2.05E-02
-4.40E+00	2.49E-05	2.11E-02
-4.20E+00	5.89E-05	1.91E-02
-4.00E+00	1.34E-04	1.44E-02
-3.80E+00	2.92E-04	7.39E-03
-3.60E+00	6.12E-04	-1.14E-03
-3.40E+00	1.23E-03	-1.02E-02
-3.20E+00	2.38E-03	-1.85E-02
-3.00E+00	4.43E-03	-2.47E-02
-2.80E+00	7.92E-03	-2.73E-02
-2.60E+00	1.36E-02	-2.49E-02
-2.40E+00	2.24E-02	-1.65E-02
-2.20E+00	3.55E-02	-1.19E-03
-2.00E+00	5.40E-02	2.13E-02
-1.80E+00	7.90E-02	5.06E-02
-1.60E+00	1.11E-01	8.62E-02
-1.40E+00	1.50E-01	1.27E-01
-1.20E+00	1.94E-01	1.70E-01
-1.00E+00	2.42E-01	2.15E-01
-8.00E-01	2.90E-01	2.59E-01
-6.00E-01	3.33E-01	2.99E-01
-4.00E-01	3.68E-01	3.33E-01
-2.00E-01	3.91E-01	3.59E-01
0.0	3.99E-01	3.76E-01
2.00E-01	3.91E-01	3.83E-01
4.00E-01	3.68E-01	3.79E-01
6.00E-01	3.33E-01	3.64E-01
8.00E-01	2.90E-01	3.40E-01
1.00E+00	2.42E-01	3.08E-01
1.20E+00	1.94E-01	2.70E-01
1.40E+00	1.50E-01	2.28E-01
1.60E+00	1.11E-01	1.84E-01
1.80E+00	7.90E-02	1.40E-01
2.00E+00	5.40E-02	9.93E-02
2.20E+00	3.55E-02	6.30E-02
2.40E+00	2.24E-02	3.25E-02
2.60E+00	1.36E-02	8.69E-03
2.80E+00	7.92E-03	-8.20E-03
3.00E+00	4.43E-03	-1.84E-02
3.20E+00	2.38E-03	-2.25E-02
3.40E+00	1.23E-03	-2.15E-02
3.60E+00	6.12E-04	-1.69E-02
3.80E+00	2.92E-04	-9.81E-03
4.00E+00	1.34E-04	-1.78E-03
4.20E+00	5.89E-05	5.99E-03
4.40E+00	2.49E-05	1.25E-02
4.60E+00	1.01E-05	1.69E-02
4.80E+00	3.96E-06	1.90E-02
5.00E+00	1.49E-06	1.86E-02

END OF PLOT.

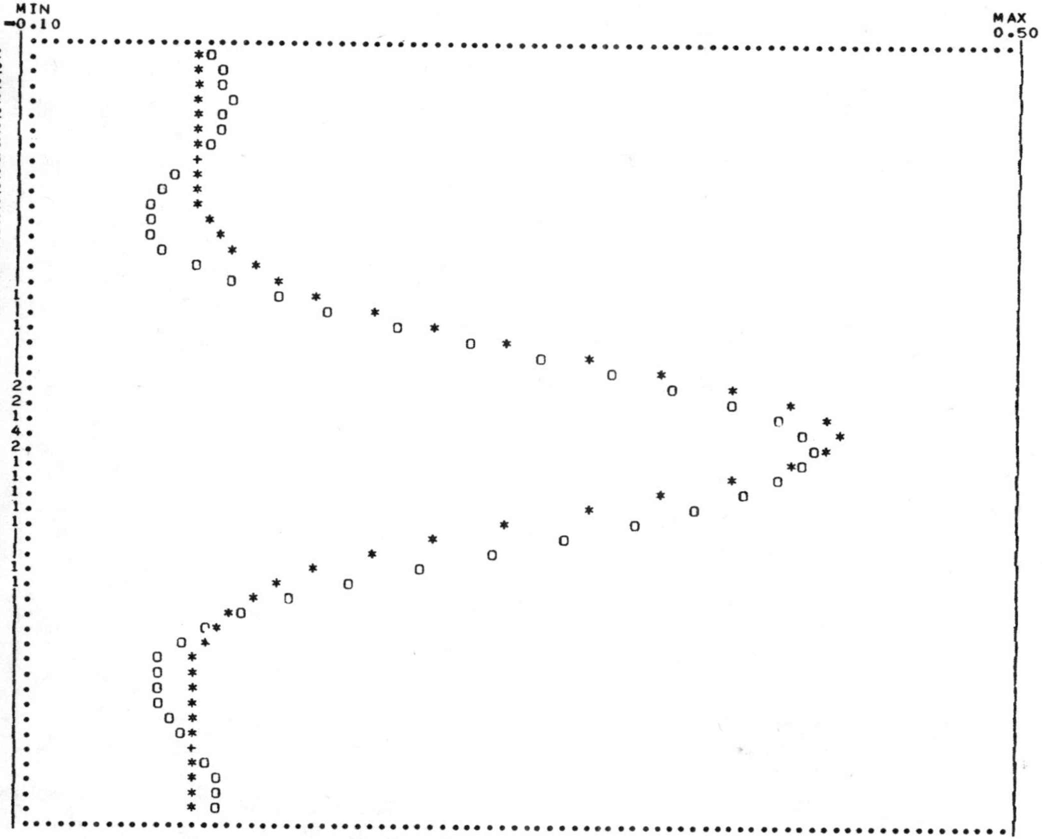

Figure 2.8. $n = 100$ $N(0, 1)$ sinc kernel $h_n = 0.47$.

```
F(10,10) SAMPLE OF SIZE  400        VS.
         KERNEL ESTIMATOR WITH KERNEL = FIE. WITH H =       0.41328561
         INTEGRATED MEAN SQUARE ERROR                   0.185997225 E-01
         INTEGRATED SQUARE ERROR                        0.605670996 E-01
         MAXIMUM ABSOLUTE DIFFERENCE                    0.275363922 E+00
```

ABSCISSA	ORDINATE *	ORDINATE O
-5.00E+00	0.0	2.24E-02
-4.80E+00	0.0	-3.99E-03
-4.60E+00	0.0	-3.42E-02
-4.40E+00	0.0	-6.14E-02
-4.20E+00	0.0	-7.82E-02
-4.00E+00	0.0	-7.77E-02
-3.80E+00	0.0	-5.47E-02
-3.60E+00	0.0	-7.09E-03
-3.40E+00	0.0	6.36E-02
-3.20E+00	0.0	1.52E-01
-3.00E+00	0.0	2.50E-01
-2.80E+00	1.63E-01	3.48E-01
-2.60E+00	5.58E-01	4.33E-01
-2.40E+00	7.43E-01	4.98E-01
-2.20E+00	7.23E-01	5.33E-01
-2.00E+00	6.15E-01	5.37E-01
-1.80E+00	4.92E-01	5.08E-01
-1.60E+00	3.82E-01	4.52E-01
-1.40E+00	2.92E-01	3.77E-01
-1.20E+00	2.23E-01	2.92E-01
-1.00E+00	1.71E-01	2.07E-01
-8.00E-01	1.31E-01	1.32E-01
-6.00E-01	1.01E-01	7.30E-02
-4.00E-01	7.87E-02	3.35E-02
-2.00E-01	6.17E-02	1.37E-02
0.0	4.87E-02	1.09E-02
2.00E-01	3.87E-02	1.99E-02
4.00E-01	3.10E-02	3.46E-02
6.00E-01	2.49E-02	4.89E-02
8.00E-01	2.02E-02	5.81E-02
1.00E+00	1.65E-02	5.93E-02
1.20E+00	1.36E-02	5.20E-02
1.40E+00	1.12E-02	3.76E-02
1.60E+00	9.30E-03	1.92E-02
1.80E+00	7.76E-03	4.31E-04
2.00E+00	6.51E-03	-1.50E-02
2.20E+00	5.49E-03	-2.44E-02
2.40E+00	4.65E-03	-2.66E-02
2.60E+00	3.95E-03	-2.18E-02
2.80E+00	3.37E-03	-1.19E-02
3.00E+00	2.89E-03	5.86E-04
3.20E+00	2.49E-03	1.25E-02
3.40E+00	2.15E-03	2.12E-02
3.60E+00	1.86E-03	2.49E-02
3.80E+00	1.62E-03	2.31E-02
4.00E+00	1.41E-03	1.64E-02
4.20E+00	1.23E-03	6.62E-03
4.40E+00	1.08E-03	-3.98E-03
4.60E+00	9.50E-04	-1.30E-02
4.80E+00	8.37E-04	-1.87E-02
5.00E+00	7.40E-04	-1.98E-02

END OF PLOT.

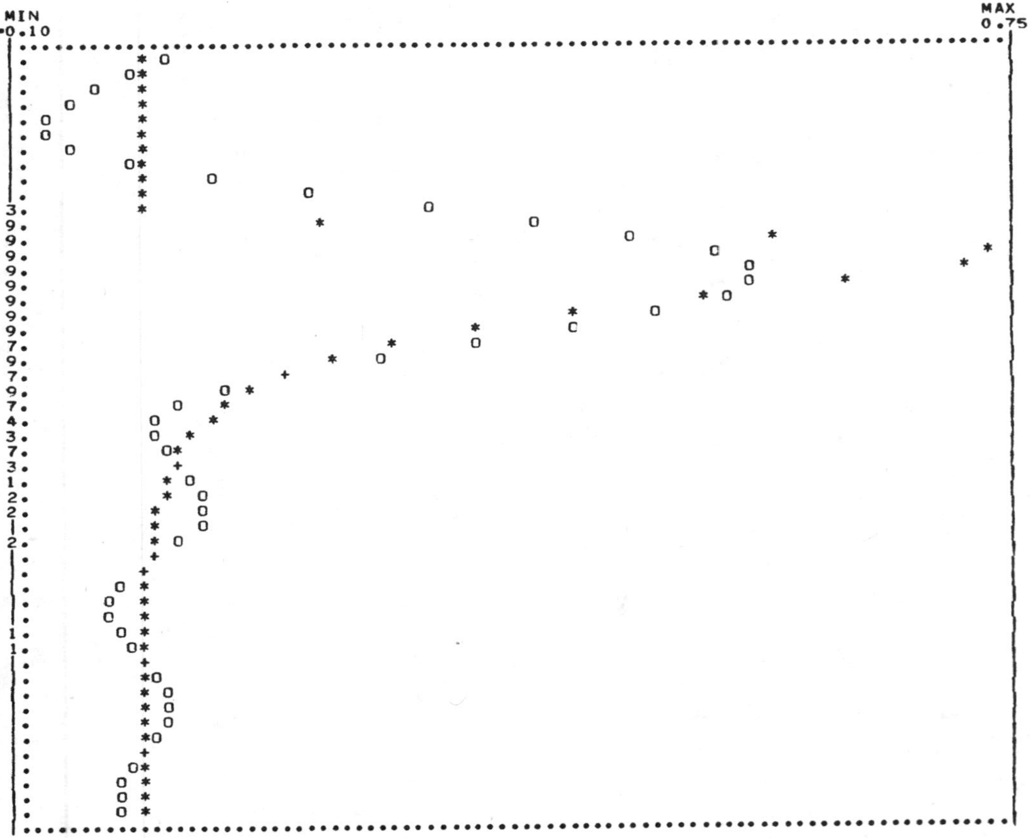

Figure 2.9. $n = 400$ $F_{10,10}$ sinc kernel $h_n = 0.41$.

In a sequence of papers [56, 57, 58, 59], Wegman has employed a maximum likelihood approach to obtain a modified histogram estimator for an unknown density f with domain $(a, b]$. His estimators are of the form

$$f(x|\mathbf{c}) = \sum_{j=0}^{k-1} c_j I_{(a_j,a_{j+1}]}^{(x)}, \tag{206}$$

where

$$\begin{aligned} I_{(c,d]}^{(x)} &= 1 \quad \text{if} \quad x \in (c, d] \\ &= 0 \quad \text{otherwise} \end{aligned}$$

$$c_j > 0 \qquad j = 0, 1, \ldots, k$$

$$\sum_{j=0}^{k-1} c_j(a_{j+1} - a_j) = 1$$

$$a_0 = a$$

$$a_k = b.$$

The criterion function to be maximized based on a sample of size n $\{x_1, x_2, \ldots, x_n\}$ is

$$L(\mathbf{c}) = \prod_{i=1}^{n} \hat{f}(x_i|\mathbf{c}) \tag{207}$$

subject to the constraint that at least $m(n)$ observations must fall into each of the k intervals where $k \le [n/m(n)]$ and $m(n)$ goes to infinity faster than $0(\sqrt{\log(\log n)})$.

In the case where $n = km(n)$, the solution is particularly simple, being analogous to that given in Theorem 2. Thus we have

$$\hat{f}(x) = \frac{1}{k} \sum_{j=0}^{k-1} \frac{1}{a_{j+1} - a_j} I_{(a_j,a_{j+1}]}^{(x)} \tag{208}$$

$$\begin{aligned} a_0 &= a \\ a_1 &= x_{(1m)} \\ a_2 &= x_{(2m)} \\ &\vdots \\ a_{k-1} &= x_{(\{k-1\}m)} \\ a_k &= b. \end{aligned}$$

Effectively, then, Wegman suggests that the interval widths of the histogram should vary across the data base in a manner inversely proportional to the density of data points in the interval.

This approach has implications for the selection of h_{in} in the case of a kernel density estimator:

$$\hat{f}_{in}(x) = \frac{1}{n} \sum_{i=1}^{n} K\{(x - x_i)/h_{in}\}/h_{in}. \tag{209}$$

We might, for example, use

$$h_{in}^k = \text{distance of } x_i \text{ to the } k^{th} \text{ nearest observation.}$$

Clearly, k should vary with n. Perhaps

$$k(n) = n^{1-\alpha} p, \tag{210}$$

where $0 < \alpha < 1$ and $0 < p < 1$
would be appropriate. For Parzen kernels, such conditions should insure consistency. For kernels with characteristic exponent of 2, $\alpha = .2$ may be (asymptotically) optimal. For densities with high frequency wiggles, small values of p should be used.

A recent approach of Carmichael and Parzen [8] attacks the probability density estimation problem using techniques motivated by the estimation of spectral densities of second order stationary time series. On the basis of a random sample $\{x_1, x_2, \ldots, x_n\}$ from an unknown density with domain $[-\pi, \pi]$, the estimate is

$$\hat{f}_p(x) = \frac{\hat{K}_p}{2\pi} \frac{1}{\left| \sum_{j=0}^{p} \hat{\alpha}_{jp} e^{ijx} \right|^2}, \tag{211}$$

where the parameters are obtained from the Yule–Walker equations

$$\begin{bmatrix} \hat{R}(0) & \ldots & \hat{R}(p-1) \\ \vdots & & \vdots \\ \hat{R}(1-p) & & \hat{R}(0) \end{bmatrix} \begin{bmatrix} \hat{\alpha}_{1p} \\ \vdots \\ \hat{\alpha}_{pp} \end{bmatrix} = - \begin{bmatrix} \hat{R}(-1) \\ \vdots \\ \hat{R}(-p) \end{bmatrix} \tag{212}$$

and

$$\sum_{j=0}^{p} \hat{\alpha}_{jp} \hat{R}(j) = \hat{K}_p, \qquad \alpha_{0p} = 1$$

with

$$\hat{R}(v) = \frac{1}{n} \sum_{j=1}^{n} \int_{-\pi}^{\pi} e^{ivx} \, \delta(x - x_j) \, dx.$$

Under very general conditions

$$|\hat{f}_p(x) - f(x)| \to 0 \text{ a.e. in } [-\pi, \pi] \tag{213}$$

provided that $\lim_{n\to\infty} p^3/n = 0$.

A lengthy paper of Boneva, Kendall, and Stefanov [6] uses the usual histogram as a starting point for a smoothed estimator of an unknown probability density function f with cdf F. Let the domain of the "histospline" estimator $[a, b]$ be divided into k equal width intervals with

$$[a, b] = \bigcup_{j=0}^{k-1} [a_j, a_{j+1}], \tag{214}$$

where

$$a_0 = a$$
$$a_k = b$$

and

$$a_{j+1} - a_j = \frac{b-a}{k}.$$

Then the histospline estimator $\hat{F}_H \in C^{(2)}[a, b]$ is the minimizer of

$$\int_a^b (\hat{F}'')^2 \, dx \tag{215}$$

subject to the constraint

$$F_H(a_j) = \frac{\text{number of samples} \le a_j}{n}.$$

The solution is a cubic spline [40]. The estimator of the unknown density is then

$$\hat{f}_H(x) = \hat{F}'_H(x). \tag{216}$$

Since the authors do not add the constraint $F'_H \ge 0$, their density estimator can go negative.

Schoenberg [41] has proposed a simple, less "wiggly" and nonnegative alternative to the histospline. To use this procedure, we first construct an ordinary histogram on the interval $[a, b]$. Thus, if we have k intervals of length h with endpoints $\{a = a_0, a_1, a_2, \ldots, a_k = b\}$, we have y_j for histogram height for the jth interval $I_j = [a_{j-1}, a_j)$, where $y_j =$ (number of observations in I_j)/nh. Let $A_j = \left(a_j - \frac{h}{2}, y_j\right)$ for $j = 1, 2, \ldots, k$. We also mark the extreme vertices $B_0 = (a, y_1)$ and $B_k = (b, y_k)$. These points are connected with the polygonal line $B_0A_1A_2 \cdots A_{k-1}A_kB_k$. The successive intersections $\{B_1, B_2, \ldots, B_{k-1}\}$ with the verticle segments of the histogram are marked. Clearly, $B_j = \left(a_j, \frac{y_j + y_{j+1}}{2}\right)$ for $j = 1, 2, \ldots, k-1$. The point A_j is equidistant from the verticle lines $x = a_{j-1}$ and $x = a_j$. We finally draw parabolic arcs $\{B_{j-1}B_j\}$ joining B_{j-1} to B_j, such that the parabola at B_{j-1} has tangent $B_{j-1}A_j$ and at B_j the tangent B_jA_j. Schoenberg's *splinegram* $s(x)$ is defined to be the quadratic spline, having as its graph the union of these arcs. A similar approach is given by Lii and Rosenblatt [30].

An algorithm of Good [21] examined by Good and Gaskins [22, 23] serves as the motivation for Chapters 4 and 5 of this study. We have already noted that in the case where the functional form of an unknown density f is known and characterized by a parameter θ we can assume a prior density h on this parameter and obtain, on the basis of a random sample $\{x_1, x_2, \ldots, x_n\}$ and the prior density the posterior density of θ via

$$(217) \qquad g(\theta|\mathbf{x}) = \frac{\prod_{j=1}^{n} f(x_j|\theta)h(\theta)}{\int_{\tau\in\Theta} \prod_{j=1}^{n} f(x_j|\tau)h(\tau)\,d\tau} = \prod_{j=1}^{n} f(x_j|\theta)h^*(\theta, \mathbf{x}).$$

It is then possible to use, for example, the posterior mode or posterior mean as an estimator for θ.

If we move to the case where we do not know the functional form of the unknown density, a Bayesian approach becomes more difficult [32]. To avoid these complications, Good suggests [21] a "Bayesian in mufti" algorithm. Let $\mathscr{F}$ be an appropriate space of probability densities. We seek to maximize in $\mathscr{F}$ the criterion functional

$$(218) \qquad w(f) = \sum_{j=1}^{n} \log f(x_j) - \Phi(f),$$

where

$$\Phi(f) \geq 0$$

and

$$\Phi(f_0) < \infty \qquad \text{where } f_0 \text{ is the true density.}$$

It is clear Good's approach is formally quasi-Bayesian, since

$$(219) \qquad \exp[w(f)] = \prod_{j=1}^{n} f(x_j) \exp[-\Phi(f)]$$

looks very similar to the expression (217).

We have the following argument from Good and Gaskins [22] to establish that their estimate $\hat{f}$ converges to the unknown density f_0 in the sense that for $a < b$

$$\int_a^b \hat{f}(x)\,dx \to \int_a^b f_0(x)\,dx.$$

For any $\hat{f}$

$$w(f_0) - w(\hat{f}) = \sum_{j=1}^{n} \log\left[\frac{f_0(x_j)}{\hat{f}(x_j)}\right] - \Phi(f_0) + \Phi(f) \tag{220}$$
$$\geq \sum_{j=1}^{n} \log\left[\frac{f_0(x_j)}{\hat{f}(x_j)}\right] - \Phi(f_0).$$

Thus,

$$E[w(f_0) - w(\hat{f})] \geq n \int \log\left[\frac{f_0(x)}{\hat{f}(x)}\right] f_0(x)\,dx - \Phi(f_0).$$

Now, since the logarithm is strictly concave, we have

$$-\int \log\left[\frac{f_0(x)}{\hat{f}(x)}\right] f_0(x)\,dx = \int \log\left[\frac{\hat{f}(x)}{f_0(x)}\right] f_0(x)\,dx < \log \int \frac{\hat{f}(x)}{f_0(x)} f_0(x)\,dx \tag{221}$$
$$= 0.$$

Thus, since the first term on the right is strictly positive and the second does not involve n, sooner or later, it is argued,

$$n \int \log\left[\frac{f_0(x)}{\hat{f}(x)}\right] f_0(x)\,dx - \Phi(f_0)$$

will be strictly positive. Then using Chebyshev's inequality, we have that in probability $w(f_0) - w(\hat{f})$ will ultimately be positive.

However, $\hat{f}(x)$ is not a fixed function but (assuming it exists) is the maximizer of the criterion functional in (218). It could be more accurately denoted $\hat{f}(x|\mathbf{x}, \Phi)$. We cannot, then, conclude, using the above argument, that

$$E[w(f_0) - w(\hat{f})] \geq \sum_{j=1}^{n} \underset{\mathbf{x}}{E} \log\left[\frac{f_0(x_j)}{\hat{f}(x_j|\mathbf{x}, \Phi)}\right] - \Phi(f_0) \tag{222}$$

is positive for any n, however large. We shall give a consistency theorem for a maximum penalized likelihood estimator in Chapter 5.

In order to guarantee that $\hat{f}$ be nonnegative, Good and Gaskins substitute γ^2 for f in (218), solve for the maximizer $\hat{\gamma}$, then obtain the solution $\hat{f} = \hat{\gamma}^2$. Thus, Good and Gaskins propose as an "equivalent" problem the maximization of

$$w(\gamma^2) = \sum_{j=1}^{n} \log \gamma^2(x_j) - \Phi(\gamma^2). \tag{223}$$

Assuming $\gamma \in L^2(-\infty, \infty)$, we may represent γ as a Hermite series:

$$\gamma_\infty(x) = \sum_{m=0}^{\infty} \gamma_m \varphi_m(x), \tag{224}$$

where the γ_m are real
and

$$\varphi_m = e^{-x^2/2}H_m(x)2^{-m/2}\pi^{-1/4}(m!)^{-1/2}$$

$$H_m(x) = (-1)^m e^{x^2}\frac{d^m}{dx^m}e^{-x^2}.$$

If $\gamma \in L^1(-\infty, \infty)$ and is of bounded variation in every finite interval, then the series converges to $\gamma(x)$ at all points of continuity of f.

To guarantee that $\int_{-\infty}^{\infty} \gamma_\infty^2(x)\,dx = 1$, the following condition is imposed:

$$\sum_{m=0}^{\infty} \gamma_m^2 = 1. \tag{225}$$

As penalty function, Good and Gaskins use

$$\Phi(\gamma) = 4\alpha \int_{-\infty}^{\infty} \gamma'^2(x)\,dx + \beta \int_{-\infty}^{\infty} \gamma''^2(x)\,dx. \tag{226}$$

Hence, it is necessary to assume that γ' and γ'' are also contained in $L^2(-\infty, \infty)$. Thus, for any r the criterion function to be maximized is

$$Q(\gamma) = \sum_{j=1}^{n} \log \gamma_r(x_j) - 4\alpha \int_{-\infty}^{\infty} (\gamma_r'(x))^2\,dx - \beta \int (\gamma_r''(x))^2\,dx, \tag{227}$$

where

$$\gamma_r(x) = \sum_{m=0}^{r} \gamma_m \varphi_m(x)$$

subject to

$$\sum_{m=0}^{r} \gamma_m^2 = 1.$$

Good and Gaskins use standard Lagrange multiplier techniques to attempt to find the maximizer of (227). This gives rise to the system of equations

$$\sum_{m=0}^{r} \gamma_m^2 = 1$$

$$\begin{aligned} 2\sum_{j=1}^{n} \varphi_k(x_j)\left\{\sum_{m=0}^{r} \gamma_m\varphi_m(x_j)\right\}^{-1} &- 4\alpha[2k\gamma_k - \{(k+1)(k+n)\}^{1/2}\gamma_{k+2} \\ &- \{(k-1)k\}^{1/2}\gamma_{k-2}] - \beta\Big[3k^2\gamma_k + 3k\gamma_k - (2k+3)\{(k+1)(k+2)\}^{1/2}\gamma_{k+2} \\ &- (2k-1)\{(k-1)k\}^{1/2}\gamma_{k-2} + \frac{1}{2}\{(k+1)(k+2)(k+3)(k+4)\}^{1/2}\gamma_{k+4} \\ &+ \frac{1}{2}\{(k-3)(k-2)(k-1)k\}^{1/2}\gamma_{k-4}\Big] - 2\lambda\gamma_k = 0; \qquad k = 0, 1, 2, \ldots, r. \end{aligned} \tag{228}$$

Practically, Good and Gaskins suggest using the approximation $\hat{\gamma}_R(x) \sim \hat{\gamma}_\infty(x)$ where R is a value beyond which the $\hat{\gamma}_j (j > R)$ are less than .001. There are a number of unanswered questions:

1) Under what conditions does the square of the solution to (223) solve (218)?
2) Under what conditions is the solution to (218) consistent?
3) Under what conditions does $\hat{\gamma}_m(x)$ have a limit as $m \to \infty$?
4) If $\hat{\gamma}_\infty(x)$ exists, under what conditions is it a solution to (223)?

Question (1) is addressed in Chapter 4. A modified version of question (2) is given in Chapter 5. The fact that in some small sample cases where we have been able to obtain closed form solutions—when the solution to (223) is the solution to (218)—these are at variance with the graphical representations of $\hat{f}$ given in [22] leads us to believe that questions (3) and (4) are nontrivial.

References

[1] Andrews, D. F., Bickel, P. J., Hampel, F. R., Huber, P. J., Rogers, W. H., and Tukey, J. W. (1972). *Robust Estimates of Location*. Princeton: Princeton University Press.

[2] Bennett, J. O., de Figueiredo, R. J. P., and Thompson, J. R. (1974). "Classification by means of B-spline potential functions with applications to remote sensing." *The Proceedings of the Sixth Southwestern Symposium on System Theory*, FA3.

[3] Bergstrom, Harald (1952). "On some expansions of stable distributions." *Arkiv for Matematik* 2: 375–78.

[4] Bickel, P. J., and Rosenblatt, M. (1973). "On some global measures of the deviations of density function estimates." *Annals of Statistics* 1: 1071–95.

[5] Bochner, Salomon (1960). *Harmonic Analysis and the Theory of Probability*, Berkeley and Los Angeles: University of California Press.

[6] Boneva, L. I., Kendall, D. G., and Stefanov, I. (1971). "Spline transformations: Three new diagnostic aids for the statistical data-analyst." *Journal of the Royal Statistical Society, Series B* 33: 1–70.

[7] Brunk, H. D. (1976). "Univariate density estimation by orthogonal series." Technical Report No. 51, Statistics Department, Oregon State University.

[8] Carmichael, Jean-Pierre (1976). "The Autoregressive Method: A Method of Approximating and Estimating Positive Functions." Ph. D. dissertation, State University of New York, Buffalo.

[9] Cramér, Harald (1928). "On the composition of elementary errors." *Skandinavisk Aktuarietidskrift* 11: 13–74, 141–80.

[10] ——— (1946). *Mathematical Methods of Statistics*. Princeton: Princeton University Press.

[11] Davis, K. B. (1975). "Mean square error properties of density estimates." *Annals of Statistics* 3: 1025–30.

[12] Epanechnikov, V. A. (1969). "Nonparametric estimates of a multivariate probability density." *Theory of Probability and Its Applications* 14: 153–58.

[13] Fama, Eugene F., and Roll, Richard (1968). "Some properties of symmetric stable distributions." *Journal of the American Statistical Association* 63: 817–36.

[14] ——— (1971). "Parameter estimates for symmetric stable distributions." *Journal of the American Statistical Association* 66: 331–38.

[15] Farrell, R. H. (1972). "On best obtainable asymptotic rates of convergence in estimates of a density function at a point." *Annals of Mathematical Statistics* 43: 170–80.

[16] Feller, William (1966). *An Introduction to Probability Theory and Its Applications*, II, New York: John Wiley and Sons.

[17] de Figueiredo, R. J. P. (1974). "Determination of optimal potential functions for density estimation and pattern classification." *Proceedings of the 1974 International Conference on Systems, Man, and Cybernetics* (C. C. White, ed.) IEEE Publication 74 CHO 908–4, 335–37.

[18] Fisher, R. A. (1937). "Professor Karl Pearson and the method of moments." *Annals of Eugenics* 7: 303–18.

[19] Galton, Francis (1879). "The geometric mean in vital and social statistics." *Proceedings of the Royal Society* 29: 365–66.

[20] ——— (1889). *Natural Inheritance*. London and New York: Macmillan and Company.

[21] Good, I. J. (1971). "A non-parametric roughness penalty for probability densities." *Nature* 229: 29–30.

[22] Good, I. J., and Gaskins, R. A. (1971). "Nonparametric roughness penalties for probability densities." *Biometrika* 58: 255–77.

[23] ——— (1972). "Global nonparametric estimation of probability densities. *Virginia Journal of Science* 23: 171–93.

[24] Graunt, John (1662). *Natural and Political Observations on the Bills of Mortality*.

[25] Johnson, N. L. (1949). "Systems of frequency curves generated by methods of translation." *Biometrika* 36: 149–76.

[26] Kazakos, D. (1975). "Optimal choice of the kernel function for the Parzen kernel-type density estimators."

[27] Kendall, Maurice G., and Stuart, Alan (1958). *The Advanced Theory of Statistics*, 1. New York: Hafner Publishing Company.

[28] Kim, Bock Ki, and Van Ryzin, J. (1974). "Uniform consistency of a histogram density estimator and model estimation." MRC Report 1494.

[29] Kronmal, R. A., and Tarter, M. E. (1968). "The estimation of probability densities and cumulatives by Fourier series methods." *Journal of the American Statistical Association* 63: 925–52.

[30] Lii, Keh-Shin, and Rosenblatt, M. (1975). "Asymptotic behavior of a spline estimate of a density function." *Computation and Mathematics with Applications* 1: 223–35.

[31] Lindeberg, W. (1922). "Eine neue Herleitung des Expenentialgesetzes in der Wahrscheinlichkeitsrechnung." *Mathematische Zeitschrift* 15: 211–26.

[32] de Montricher, G. M. (1973). *Nonparametric Bayesian Estimation of Probability Densities by Function Space Techniques*. Doctoral dissertation, Rice University, Houston, Texas.

[33] Nadaraya, E. A. (1965). "On nonparametric estimates of density functions and regression curves." *Theory of Probability and Its Applications* 10: 186–90.

[34] Ord, J. K. (1972). *Families of Frequency Distributions*. New York: Hafner Publishing Company.

[35] Parzen, E. (1962). "On estimation of a probability density function and mode." *Annals of Mathematical Statistics* 33: 1065–76.

[36] Pearson, Karl (1936). "Method of moments and method of maximum likelihood." *Biometrika* 28: 34–59.

[37] Rényi, Alfred (1970). *Foundations of Probability*. San Francisco: Holden Day.

[38] Rosenblatt, M. (1956). "Remarks on some nonparametric estimates of a density function." *Annals of Mathematical Statistics* 27: 832–35.

[39] ——— (1971). "Curve estimates." *Annals of Mathematical Statistics* 42: 1815–42.

[40] Schoenberg, I. J. (1946). "Contributions to the problem of approximation of equidistant data by analytical functions." *Quarterly of Applied Mathematics* 4: 45–99, 112–14.

[41] ——— (1972). "Notes on spline functions II on the smoothing of histograms." MRC Technical Report 1222.

[42] Schuster, Eugene F. (1970). "Note on the uniform convergence of density estimates." *Annals of Mathematical Statistics* 41: 1347–48.

[43] Scott, David W. (1976). *Nonparametric Probability Density Estimation by Optimization Theoretic Techniques.* Doctoral dissertation, Rice University, Houston, Texas.

[44] Scott, D. W., Tapia, R. A., Thompson, J. R. (1977) "Kernel density estimation revisited." *Nonlinear Analysis* 1: 339–72.

[45] Shapiro, J. S. (1969). *Smoothing and Approximation of Functions.* New York: Van Nostrand–Reinhold.

[46] Shenton, L. R. (1951). "Efficiency of the method of moments and the Gram-Charlier Type A distribution." *Biometrika* 38: 58–73.

[47] Tarter, M. E., and Kronmal, R. A. (1970). "On multivariate density estimates based on orthogonal expansions." *Annals of Mathematical Statistics* 41: 718–22.

[48] ——— (1976). "An introduction to the implementation and theory of nonparametric density estimation." *American Statistician* 30: 105–12.

[49] Tukey J. W. (1976). "Some recent developments in data analysis." Presented at 150th Meeting of the Institute of Mathematical Statistics.

[50] ——— (1977). *Exploratory Data Analysis.* Reading, Mass.: Addison–Wesley.

[51] Van Ryzin, J. (1969). "On strong consistency of density estimates." *Annals of Mathematical Statistics* 40: 1765–72.

[52] Wahba, Grace (1971). "A polynomial algorithm for density estimation." *Annals of Mathematical Statistics* 42: 1870–86.

[53] ——— (1975). "Optimal convergence properties of variable knot, kernel and orthogonal series methods for density estimation." *Annals of Statistics* 3: 15–29.

[54] Watson, Geoffrey S. (1969). "Density estimation by orthogonal series." *Annals of Mathematical Statistics* 40: 1496–98.

[55] Watson, G. S., and Leadbetter, M. R. (1963). "On the estimation of the probability density, I." *Annals of Mathematical Statistics* 34: 480–91.

[56] Wegman, Edward J. (1969). "A note on estimating a unimodal density." *Annals of Mathematical Statistics* 40: 1661–67.

[57] ——— (1970). "Maximum likelihood estimation of a unimodal density function." *Annals of Mathematical Statistics* 41: 457–71.

[58] ——— (1970). "Maximum likelihood estimation of a unimodal density, II." *Annals of Mathematical Statistics* 41: 2169–74.

[59] ———. "Maximum likelihood estimation of a probability density function." To appear in *Sankya, Series A*.

[60] Whittle, P. (1958). "On the smoothing of probability density functions." *Journal of the Royal Statistical Society* (B) 20: 334–43.

[61] Woodroofe, Michael (1970). "On choosing a delta-sequence." *Annals of Mathematical Statistics* 41: 1665–71.

CHAPTER

3 Maximum Likelihood Density Estimation

3.1. Maximum Likelihood Estimators

In this chapter we return to the classical maximum likelihood estimation procedure discussed in Chapter 2. We establish general existence and uniqueness results for the finite dimensional case, show that several popular estimators are maximum likelihood estimators, and, finally, show that the infinite dimensional case is essentially meaningless. This latter fact serves to motivate the maximum penalized likelihood density estimation procedures that will be presented in Chapter 4.

A good part of the material presented in this chapter originated in a preliminary version of [1]. At that time it was felt that the results were probably known and consequently were not included in the published version. However, we have not been able to find the material in this generality in the literature.

Consider the interval (a, b). As in the previous chapters we are again interested in the problem of estimating the (unknown) probability density function $f \in L^1(a, b)$ (Lebesgue integrable on (a, b)) which gave rise to the random sample $x_1, \ldots, x_n \in (a, b)$.

By the *likelihood* that $v \in L^1(a, b)$ gave rise to the random sample $x_1, \ldots, x_n$ we mean

$$L(v) = \prod_{i=1}^{n} v(x_i). \tag{1}$$

Let H be a manifold in $L^1(a, b)$, and consider the following constrained optimization problem

(2) maximize $L(v)$; subject to

$$v \in H, \int_a^b v(t)\,dt = 1 \qquad \text{and} \qquad v(t) \geq 0 \qquad \forall t \in (a, b).$$

By a *maximum likelihood estimate* based on the random sample $x_1, \ldots, x_n$ and corresponding to the manifold H, we mean any solution of problem (2).

We now restrict our attention to the special case when H is a finite dimensional subspace (linear manifold) of $L^1(a, b)$. In the following section we show that this special case includes several interesting examples. For $y \in R^n$ the notation $y \geq 0$ ($y > 0$) means $y_i \geq 0$ ($y_i > 0$), $i = 1, \ldots, n$. In this application $\langle\ ,\ \rangle$ denotes the Euclidean inner product in R^n (see Example I.1). The following three propositions will be useful in our analysis.

Proposition 1. Given the positive integers $q_1, \ldots, q_n$, define $f: R^n \to R$ by

$$f(y) = \prod_{i=1}^{n} y_i^{q_i}.$$

Also given $\alpha \in R^n$, such that $\alpha > 0$, define T by

$$T = \{y \in R^n : \langle \alpha, y \rangle = 1 \qquad \text{and} \qquad y \geq 0\}.$$

Then f has a unique maximizer in T which is given by y^*, where

$$y_i^* = \frac{q_i}{N\alpha_i} \qquad \text{and} \qquad N = \sum_{i=1}^{n} q_i.$$

Proof. Clearly, T is compact and f is continuous; hence by Theorem I.3 there exists a maximizer which we shall denote by y^*. Now, if for some $1 \leq i \leq n$, we have $y_i^* = 0$, then $f(y^*) = 0$; however, $y = \left(\frac{1}{\alpha_1}, \ldots, \frac{1}{\alpha_n}\right) \in T$ and $f(y) > 0$, which would be a contradiction. It follows that y^* is such that $y^* > 0$. From Theorem I.8 there exists λ, such that

$$\nabla f(y^*) = \lambda \alpha. \tag{3}$$

Taking the gradient of f and using (3) leads to

$$q_i f(y^*) = \lambda \alpha_i y_i^*, \qquad i = 1, \ldots, n. \tag{4}$$

From (4) and the fact that $\langle \alpha, y^* \rangle = 1$ we have

$$\lambda = \lambda \langle \alpha, y^* \rangle = f(y^*) \sum_{i=1}^{n} q_i = Nf(y^*). \tag{5}$$

Substituting this value for λ in (4) establishes the proposition; since we have proved that $\nabla f(y) = \lambda\alpha$ has a unique solution. ■

Proposition 2. Consider $f: R^n \to R$ defined by

$$f(y) = \prod_{i=1}^{n} y_i. \tag{6}$$

Let T be any convex subset of $R^n_+ = \{y \in R^n : y \geq 0\}$ which contains at least one element $y > 0$. Then f has at most one maximizer in T.

Proof. Since there is at least one element $y > 0$ in T any maximizer of f in T will not lie on the boundary of R^n_+. Therefore, maximizing f over T is equivalent to maximizing the log of f over T, which is in turn equivalent to minimizing the negative log over T. However, by part (ii) of Proposition I.16 it is easy to see that $-\log f(y)$ with $f(y)$ given by (6) is a strictly convex function on the interior of R^n_+. The proposition now follows from Theorem I.2. ■

Proposition 3. Let $\Phi_1, \ldots, \Phi_n$ be n linearly independent members of $L^1(a, b)$ and let

$$T_+ = \left\{\alpha \in R^n : \int_a^b \sum_{i=1}^{n} \alpha_i \Phi_i(t)\, dt = 1 \quad \text{and} \quad \sum_{i=1}^{n} \alpha_i \Phi_i(t) \geq 0 \ \forall t \in [a, b]\right\}. \tag{7}$$

Then T_+ is a convex compact subset of R^n.

Proof. If T_+ is empty, we are through, so suppose that $\alpha^* = (\alpha_1^*, \ldots, \alpha_n^*) \in T_+$. Clearly, T_+ is closed and convex. Suppose T_+ is not bounded. Then there exists γ_m such that $(1 - \lambda)\alpha^* + \lambda\gamma_m \in T_+$ for $0 \leq \lambda \leq m$. Let $\beta_m = (1 - \lambda_m)\alpha^* + \lambda_m\gamma_m$, where $\lambda_m = \|\gamma_m - \alpha^*\|^{-1}$. Observe that $\|\beta_m - \alpha^*\| = 1$. Let β be any limit point of β_m. Then,

$$(1 - \lambda)\alpha^* + \lambda\beta \in T_+, \qquad \forall\lambda \in [0, \infty). \tag{8}$$

Let $\delta = \beta - \alpha^*$. It follows from (8) that

$$\lambda \sum_{i=1}^{n} \delta_i \Phi_i(t) \geq -\sum_{i=1}^{n} \alpha_i^* \Phi_i(t), \qquad \forall\lambda \in [0, \infty);$$

hence,

$$\sum_{i=1}^{n} \delta_i \Phi_i(t) \geq 0 \qquad \forall t \in [a, b].$$

However,

$$\int_a^b \sum_{i=1}^{n} \delta_i \Phi_i(t)\, dt = 0, \qquad \text{so} \quad \sum_{i=1}^{n} \delta_i \Phi_i(t) = 0.$$

This contradicts the linear independence of $\Phi_1, \ldots, \Phi_n$ and proves the proposition. ■

We are now ready to prove our existence theorem.

Theorem 1 (Existence.) If H is a finite dimensional subspace of $L^1(a, b)$, then a maximum likelihood estimate based on $x_1, \ldots, x_n$ and corresponding to H exists.

Proof. Proposition 3 shows that in this case the constraint set of problem (2) is a convex compact subset of the finite dimensional space H. The existence of a solution of problem (2) now follows from Theorem I.3. ■

In addition to Theorem 1, we have the following weak form of uniqueness.

Theorem 2 (Uniqueness.) Suppose H is a finite dimensional subspace of $L^1(a, b)$, with the property that there exists at least one $\varphi \in H$ satisfying $\varphi(t) \geq 0\ \forall t \in [a, b]$ and $\varphi(x_i) > 0, i = 1, \ldots, n$ for the random sample $x_1, \ldots, x_n$. If φ_1 and φ_2 are maximum likelihood estimates based on $x_1, \ldots, x_n$ and corresponding to H, then

$$\varphi_1(x_i) = \varphi_2(x_i), \qquad i = 1, \ldots, n, \tag{9}$$

i.e., any two estimates must agree at the sample points.

Proof. The proof is a straightforward application of Proposition 2. ■

3.2. The Histogram as a Maximum Likelihood Estimator

In Chapter 2 we presented some results concerning the classical histogram and approaches for constructing generalized histograms. Let us return to the standard histogram. Consider a partition of the interval (a, b), say, $a = t_1 < t_2 < \cdots < t_{m+1} = b$. Let T_i denote the half-open half-closed interval $[t_i, t_{i+1})$ for $i = 1, \ldots, m$. Let $I(T_i)$ denote the indicator function of the interval T_i, i.e., $I(T_i)(x) = 0$, if $x \notin T_i$ and $I(T_i)(x) = 1$, if $x \in T_i$. For the random sample $x_1, \ldots, x_n \in (a, b)$, we let $M(T_i)$ denote the number of these samples which fall in the interval T_i. Clearly, $\sum_{i=1}^{m} M(t_i) = n$. Now, the classical theory (see, for example, Chapter 2 of Freund [2]) tells us that if we wish to construct a histogram with class intervals T_i, we make the heights of the rectangle with base T_i proportional to $M(T_i)$. This means that the histogram will have

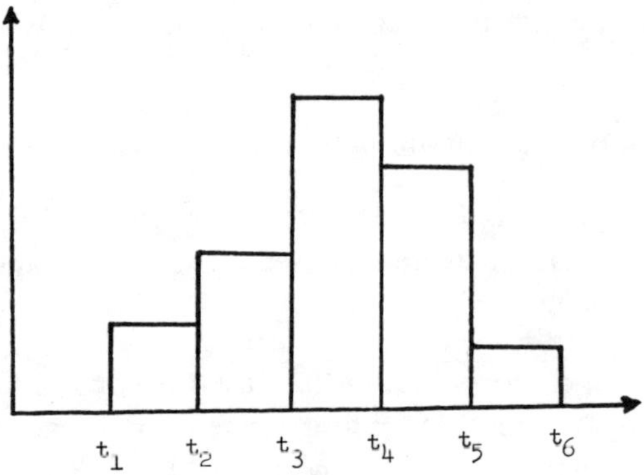

Figure 3.1

the form

$$f^* = \sum_{i=1}^{m} \frac{\alpha M(T_i)}{(t_{i+1} - t_i)} I(T_i) \tag{10}$$

for some constant of proportionality α. Since the area under the step function f^* must be equal to one, we must have

$$\sum_{i=1}^{m} \alpha M(T_i) = 1. \tag{11}$$

From (11) it follows that $\alpha = \frac{1}{n}$ and consequently the histogram estimator is given by

$$f^* = \sum_{i=1}^{m} \frac{M(T_i)}{n(t_{i+1} - t_i)} I(T_i). \tag{12}$$

The graph of f^* would resemble the graph in Figure 3.1.

In Chapter 2 we proved that the histogram estimator is actually a maximum likelihood estimator. For the sake of unity, and also to demonstrate the usefulness of Theorem 1, we will show that this result is actually a straightforward consequence of Theorem 1.

Proposition 4. The histogram (12) is the unique maximum likelihood estimate based on the random sample $x_1, \ldots, x_n$ and corresponding to the subspace of $L^1(a, b)$ defined by $S_0(t_1, \ldots, t_m) = \left\{ \sum_{i=1}^{m} y_i I(T_i) : y_i \in R \right\}$.

Proof. By Theorem 1 a maximum likelihood estimate

$$v^* = \sum_{i=1}^{m} y_i^* I(T_i) \tag{13}$$

exists. Observe that if $M(T_i) = 0$, then y_i^* in (13) is also equal to zero. For if $y_i^* > 0$, then we could modify v^* in (13) by setting $y_i^* = 0$ and increasing some $y_j^*(j \neq i)$, so that the constraints of problem (2) would still be satisfied, but the likelihood functional would be increased. However, this would lead to a contradiction, since v^* maximizes the likelihood over this constraint set. We therefore lose no generality by assuming $M(T_i) > 0 \ \forall i$. Let $q_i = M(T_i)$ and $\alpha_i = t_{i+1} - t_i$ for $i = 1, \ldots, m$. The constraints of problem (2) take the form $y \geq 0$ and $\langle \alpha, y \rangle = 1$ for $y_1 I(T_1) + \cdots + y_m I(T_m) \in S_0(t_1, \ldots, t_m)$. Proposition 1 now applies and says that (12) and (13) are the same. The uniqueness follows from Theorem 2, since the conditions are satisfied by this choice of H and (9) implies two maximum likelihood estimates would coincide on all intervals which contain sample points. And, as before, they would both be zero on intervals which contain no sample points.

Let $\{X_1, \ldots, X_m\}$ where $X_i < X_{i+1}$ denote the distinct samples in $\{x_1, \ldots, x_n\}$. Also, for each X_i, $i = 1, \ldots, m$, let q_i denote the number of samples equal to X_i so that $\sum_{i=1}^{m} q_i = n$. Finally, let X_0, X_{m+1} be two real numbers such that $a \leq X_0 < X_1$ and $X_m < X_{m+1} \leq b$.

Proposition 5. The maximum likelihood estimate based on $x_1, \ldots, x_n \in (a, b)$ and corresponding to the linear manifold

$$S_0(X_1, \ldots, X_m) = \left\{ \sum_{i=1}^{m} y_i I([X_i, X_{i+1})) : y_i \in R \right\} \subset L^1(a, b)$$

exists and is uniquely given by

$$v^* = \sum_{i=1}^{m} y_i^* I([X_i, X_{i+1})), \tag{14}$$

where

$$y_i^* = \frac{q_i}{n(X_{i+1} - X_i)}. \tag{15}$$

Proof. The proof is similar to the proof of Proposition 4 and follows from Proposition 1 by setting $\alpha_i = X_{i+1} - X_i$. ∎

Let $S_1(X_0, \ldots, X_{m+1})$ denote the subspace of $L^1(a, b)$ consisting of continuous functions which are linear on $[X_i, X_{i+1}]$, $i = 0, \ldots, m$ and vanish outside the interval (X_0, X_{m+1}).

Proposition 6. The maximum likelihood estimate based on $x_1, \ldots, x_n \in (a, b)$ and corresponding to the subspace $S_1(X_0, \ldots, X_{m+1})$ exists and is uniquely given by that v^* which satisfies

$$v^*(X_i) = \frac{2q_i}{n(X_{i+1} - X_{i-1})}, \qquad i = 1, \ldots, m. \tag{16}$$

Proof. If we associate $v \in S_1(X_0, \ldots, X_{m+1})$ with $y \in R^m$, where $y_i = v(X_i)$, $i = 1, \ldots, m$, then we may again consider solving problem (2) by Proposition 1. Toward this end, let $\alpha_i = \frac{1}{2}(X_{i+1} - X_{i-1})$. The constraints of problem (2) become $y \geq 0$ and $\langle \alpha, y \rangle = 1$. The result now follows from Proposition 1 in the same way that Propositions 4 and 5 did. Again we have uniqueness because two members of $S_1(X_0, \ldots, X_{m+1})$ which agree at $X_0, \ldots, X_{m+1}$ must coincide. ■

The notation $S_i(t_1, \ldots, t_m)$, $i = 1, 2$, used above actually denotes the class of polynomial splines of degree i with knots at the points $t_1, \ldots, t_m$. For example, when $i = 0$ we are working with functions which are piecewise constant, and for $i = 1$ we are working with continuous functions which are piecewise linear. It follows that the maximum likelihood estimate given by Proposition 5 is a histogram and resembles the histogram in Figure 3.1. Moreover, the maximum likelihood estimate given by Proposition 6 can be thought of as a generalized histogram and its graph would resemble the graph given in Figure 3.2.

The notion of the histogram as a polynomial spline of degree 0 could be extended not only to splines of degree 1 as in Proposition 6 but to splines of arbitrary degree, using either the proportional area approach or the maximum likelihood approach. However, in general, such an approach would create severe problems, including the facts that we could not obtain

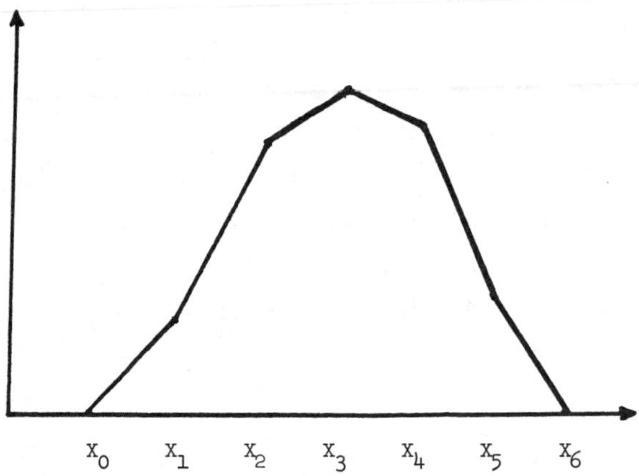

Figure 3.2

the estimate in closed form and it would be essentially impossible to enforce the nonnegativity constraint. For splines of degree 0 or 1 the nonnegativity constraint is a trivial matter, since they cannot "wiggle."

3.3. The Infinite Dimensional Case

In general, if the manifold H in problem (2) is infinite dimensional, then a maximum likelihood estimate will not exist. To see this observe that a solution is idealized by

$$f^*(x) = \frac{1}{n} \sum_i \delta(x - x_i), \tag{17}$$

where δ is the Dirac delta mass at the origin. The linear combination of Dirac spikes f^* satisfies the constraints of problem (2) and gives a value of $+\infty$ for the likelihood. These comments are formal, since δ does not exist as a member of $L^1(a, b)$. However, they allow us to conclude that in any infinite dimensional manifold $H \subset L^1(a, b)$, which has the property that it is possible to approximate Dirac delta spikes, e.g., given $t^* \in (a, b)$ there exists $f_m \in H$, such that f_m integrates to one $f_m(t) \geq 0 \ \forall t \in (a, b)$ and $\lim_{m \to \infty} f_m(t^*) = +\infty$, the likelihood will be unbounded and maximum likelihood estimates will not exist. Moreover, most infinite dimensional manifolds in $L^1(a, b)$ will have this property. This is certainly true of the continuous functions, the differentiable functions, the infinitely differentiable functions, and the polynomials. Of course, we could consider pathological examples of infinite dimensional manifolds in $L^1(a, b)$, for which a maximum likelihood estimate exists. As an extreme case, let H be all continuous functions on (a, b) which vanish at the samples $x_1, \ldots, x_n$. In this case, any member of H, which is nonnegative and integrates to one is a maximum likelihood estimate based on $x_1, \ldots, x_n$.

The fact that in general the maximum likelihood estimate does not exist for infinite dimensional H has a very important interpretation in terms of finite dimensional H. Specifically, for large (but finite) dimensional H the maximum likelihood estimate must necessarily lead to unsmooth (contains spikes) estimates and a numerically ill-posed problem. Granted, if it is known a priori that the unknown probability density function is a member of the finite dimensional manifold H, then the maximum likelihood approach would probably give satisfactory results. However, this prior knowledge is known only in very special cases.

The above observations lead us to the following criticism of maximum likelihood estimation in general: for small dimensional H there is no flexibility and the solution will be greatly influenced by the subjective choice of H; while for large dimensional H the solution must necessarily be unsmooth and the optimization problem must necessarily create numerical problems. For example, notice that for a fixed random sample $x_1, \ldots, x_n$ the histogram estimate f^* given by (12) has the unfortunate property that $f^*(x) \to +\infty$ for $x \in \{x_1, \ldots, x_n\}$ and $f^*(x) \to 0$ for $x \notin \{x_1, \ldots, x_n\}$ as the number of intervals T_i goes to infinity and the length of T_i goes to zero. Whether or not a reasonable estimate is obtained is completely dependent on the delicate art of choosing T_i, $i = 1, \ldots, m$ properly, i.e., the interplay between the intervals T_i and the random sample $x_1, \ldots, x_n$. Indeed, this is a fundamental part of the consistency proof given in Section 2.5.

For the generalized histogram estimators given in Proposition 5 and Proposition 6 we are not free to choose the dimension of the problem (i.e., the linear manifold H) independent of the data, $x_1, \ldots, x_n$. Moreover, whether the estimate develops spikes as the sample size n becomes large, depends on the quantity $n \min_i (X_{i+1} - X_i)$, respectively, $n \min_i (X_{i+1} - X_{i-1})$, which the user has no control over.

In numerical analysis it is standard to handle an infinite dimensional problem by restricting the problem to a finite dimensional subspace (e.g., discrete mesh) and then using the solution of the approximate problem as an approximation to the solution of the original problem. However, this approach is usually predicated on the security that the infinite dimensional problem is well-posed and that as the dimension of the finite dimensional subspace becomes infinite (e.g., mesh size approaches zero) the approximate problem approaches the original problem in some reasonable manner. It is then a straightforward matter to establish convergence of the approximate solution to the true solution. In general, this is the situation with numerical methods for the solution of differential and integral equations. However, in many statistical applications the cart is put before the horse and considerable effort is spent in showing that the finite dimensional problem is well-posed, however, the asymptotic infinite dimensional problem is either ignored, meaningless or ill-posed. This latter phenomenon we choose to call *dimensional instability*. We further feel that dimensional stability is distinct from the numerical analysts' use of stability and the statisticians' use of robustness and that in some sense it implies these latter two notions. This notion probably merits further investigation.

In this section we have argued that the maximum likelihood density estimation problem as posed in Section 1 of this chapter is dimensionally unstable. The purpose of Chapter 4 is to describe one general approach for introducing dimensional stability into the maximum likelihood density estimation problem. In Chapter 5 we show how the theory developed in Chapter 4 can be used to develop a numerically efficient algorithm.

References

[1] de Montricher, G. M., Tapia, R. A., and Thompson, J. R. (1975). "Nonparametric maximum likelihood estimation of probability densities by penalty function methods." *Annals of Statistics* 3:1329–48.

[2] Freund, J. E. (1962). *Mathematical Statistics*. Englewood Cliffs, New Jersey: Prentice–Hall.

CHAPTER

4 Maximum Penalized Likelihood Estimation

4.1 Maximum Penalized Likelihood Estimators

The material in this chapter is taken primarily from de Montricher, Tapia, and Thompson [1] and relies heavily on the theory of mathematical optimization in Hilbert space. A fairly complete (for the purposes of this chapter) treatment of mathematical optimization theory can be found in Appendix I. The reader not familiar with the theory will benefit by reading this appendix before embarking on the present material.

As in Chapter 3, let H be a manifold in $L^1(a, b)$ and consider a functional $\Phi: H \to R$. Given the random sample $x_1, \ldots, x_n \in (a, b)$ the *Φ-penalized likelihood* of $v \in H$ is defined by

$$\hat{L}(v) = \prod_{i=1}^{n} v(x_i) \exp(-\Phi(v)). \tag{1}$$

Consider the constrained optimization problem:

(2) maximize $\hat{L}(v)$; subject to

$$v \in H, \qquad \int_a^b v(t)\,dt = 1 \qquad \text{and} \qquad v(t) \geq 0, \qquad \forall t \in (a, b).$$

The general form of the penalized likelihood (1) is due to Good and Gaskins [2]. Their specific suggestions are analyzed in Sections 3 and 4.

Any solution to problem (2) is said to be a *maximum penalized likelihood estimate* based on the random sample $x_1, \ldots, x_n$ corresponding to the manifold H and the penalty function Φ. Motivated by the observations made in Chapter 3 we will be particularly interested in the case when H is an infinite dimensional Hilbert space (see Appendix I.1). In the case that H

is a Hilbert space, a very natural penalty function to use is $\Phi(v) = \|v\|^2$ where $\|\circ\|$ denotes the norm on H. Consequently, when H is a Hilbert space and we refer to the penalized likelihood functional on H or to the maximum penalized likelihood estimate corresponding to H with no reference to the penalty functional Φ, we are assuming that Φ is the square of the norm in H. The Hilbert space inner product will be denoted by $\langle \circ, \circ \rangle$ so that $\langle x, x \rangle = \|x\|^2$.

For problem (2) to make sense we would like H to have the property that for $x_1, \ldots, x_n \in (a, b)$ there exists at least one $v \in H$, such that

$$\int_a^b v(t)\,dt = 1, \qquad v(t) \geqq 0 \qquad \forall t \in (a, b) \tag{3}$$

and

$$v(x_i) > 0 \qquad i = 1, \ldots, n.$$

Proposition 1. Suppose that H is a reproducing kernel Hilbert space and D is a closed convex subset of $\{v \in H : v(x_i) \geqq 0, i = 1, \ldots, n\}$ with the property that D contains at least one function which is positive at the samples $x_1, \ldots, x_n$. Then the penalized likelihood functional (1) has a unique maximizer in D.

Proof. The $\langle \circ, \circ \rangle$-penalized likelihood functional $\hat{L}$ is clearly continuous when $\langle \circ, \circ \rangle$ represents the inner product in a reproducing kernel Hilbert space. By assumption there exists at least one $v \in D$, such that $\hat{L}(v) > 0$. Hence, maximizing $\hat{L}$ over D is equivalent to minimizing $J = -\log \hat{L}$ over D. A straightforward calculation gives the second derivative of J as

$$J''(v)(\eta, \eta) = \sum_{i=1}^{n} \frac{\eta(x_i)\eta(x_i)}{v(x_i)^2} + 2\langle \eta, \eta \rangle. \tag{4}$$

The proposition now follows from Theorem 7 in Appendix I.4. ∎

Theorem 1. Suppose H is a reproducing kernel Hilbert space, integration over (a, b) is a continuous functional and there exists at least one $v \in H$ satisfying (3). Then the maximum penalized likelihood estimate corresponding to H exists and is unique.

Proof. The proof follows from Proposition 1, since the constraints in (2) give a closed convex subset of $\{v \in H : v(x_i) \geqq 0, i = 1, \ldots, n\}$. ∎

The nonnegativity constraint in problem (2) is, in general, impossible to enforce when working with algorithms which deal with continuous densities which are not piecewise linear. For this reason, and others, we often find examples in the statistical literature where a problem with a nonnegativity constraint is solved by working with an equivalent problem stated in terms of the square root of the unknown density. In this latter problem, the nonnegativity constraint is redundant. Specifically, given $H \subset L^1(a, b)$ and $J: H \to R$ consider the following two problems:

(5) maximize $J(v)$; subject to

$$v \in H, \qquad \int_a^b v(t)\,dt = 1 \qquad \text{and} \qquad v(t) \geq 0 \qquad \forall t \in (a, b),$$

and

(6) maximize $J(u^2)$; subject to

$$u^2 \in H \qquad \text{and} \qquad \int_a^b u^2(t)\,dt = 1.$$

The following proposition is obvious.

Proposition 2.

(i) If v^* solves problem (5), then $u^* = \sqrt{v^*}$ solves problem (6).

(ii) If u^* solves problem (6), then $v^* = (u^*)^2$ solves problem (5).

Part of the price one pays for no longer having to work with the nonnegativity constraint is that the integral constraint is now nonlinear. In the following, when we consider the Good and Gaskins maximum penalized likelihood estimators, we will be dealing with constrained optimization problems of the form

(7) maximize $J(v)$; subject to

$$\sqrt{v} \in \hat{H}, \qquad \int_a^b v(t) = 1 \qquad \text{and} \qquad v(t) \geq 0 \qquad \forall t \in (a, b),$$

where J is defined on $H \subset L^1(a, b)$ and $\hat{H}$ has the property that $w \in \hat{H}$ implies $w^2 \in H$. In order to avoid the nonnegativity constraint in problem (7), Good and Gaskins [2] suggested that one work with the analogous version of problem (6); namely,

(8) maximize $J(u^2)$; subject to

$$u \in \hat{H} \qquad \text{and} \qquad \int_a^b u^2(t)\,dt = 1.$$

However, in this case there is a very subtle distinction between problems (7) and (8), and they are not always equivalent. Specifically, we have the following relationship between these two problems.

Proposition 3.

If u^* solves problem (8), then $v^* = (u^*)^2$ solves problem (7) if and only if we have the additional condition that $|u^*| \in \hat{H}$.

That the additional condition in Proposition 3 is needed is clear from the requirement in problem (7) that $\sqrt{v} \in \hat{H}$. In the sequel, $\hat{H}$ will be a Hilbert space. Clearly, the condition that $u^* \in \hat{H}$ implies $|u^*| \in \hat{H}$ is trivially satisfied when $\hat{H} = R^n$. Moreover, if u^* is a nonnegative function, then this condition is also trivially satisfied. However, if u^* is a member of the Sobolev space $H^s(-\infty, \infty)$ (see Example 4 of Appendix I.1) or the restricted Sobolev space $H_0^s(a, b)$ (see Example 5 of Appendix I.1) which has a simple zero, then $|u^*|$ will also be a member of the space if and only if $s \leq 1$. Consequently, if we are working with $H^s(-\infty, \infty)$ or $H_0^s(a, b)$ with $s > 1$, and we can show that in our particular application problem (8) has a solution which takes on both positive and negative values, then the Good and Gaskins transformation is not valid. We will show in Section 4.4 that this is exactly the case with one of the estimators proposed by Good and Gaskins.

In Section 4.2 we will analyze the maximum penalized likelihood estimators proposed by de Montricher, Tapia, and Thompson [1]. The two estimators proposed by Good and Gaskins in [2] will be analyzed in Sections 4.3 and 4.4. The existence, uniqueness, and characterization results presented for these two estimators were established in [1] and were not considered by Good and Gaskins in [2].

Before going on to Section 4.2 we introduce a mathematical tool which will be extremely useful in the following three sections. For $f, g \in H_0^1(a, b)$ (see Appendix I.1) expressions of the form $\int_a^b f^{(j)}(t)g^{(k)}(t)\,dt$ make sense if $0 \leq j, k \leq 1$. Moreover, the integration by parts $\int_a^b f^{(j)}(t)g'(t)\,dt = -\int_a^b f^{(j+1)}(t)g(t)\,dt$ makes sense if and only if $j = 0$. However, by considering the $L^2(a, b)$ function f' as a distribution on (a, b) (see Horvath [3]) and interpreting derivatives as distributional derivatives we can write

$$\int_a^b f'(t)g'(t)\,dt = -\int_a^b f''(t)g(t)\,dt,$$

as long as we remember that f'' is a distribution and not a member of $L^2(a, b)$. This formal manipulation (which can be rigorously justified) offers us a very useful tool. For example, since $H_0^1(a, b)$ is a reproducing kernel Hilbert space (see Proposition 10 of Appendix I.2), we have by the Riesz representation theorem (see Theorem 1 of Appendix I.1) that there exists a unique $v_0 \in H_0^1(a, b)$, such that $v(0) = \int_a^b v_0'(t)v'(t)\,dt$ for all $v \in H_0^1(a, b)$ (we have assumed that $0 \in (a, b)$). If we proceed formally to integrate by parts, we obtain $v(0) = \int_a^b v_0'(t)v'(t)\,dt = -\int_a^b v_0''(t)v(t)\,dt$, and it follows that as a distribution $-v_0''$ is equal to δ the Dirac delta distribution. Recalling that δ is the distributional derivative of the distribution associated with $H(t)$, the Heaviside unit function ($H(t) = 0$ if $t \leq 0$ and $H(t) = 1$ if $t > 0$), it follows that $v_0'(t) = -H(t) + C_0$ and $v_0(t) = -S(t) + C_0 t + C_1$, where $S(t)$ is the continuous linear spline $S(t) = 0$, if $t \le 0$ and $S(t) = t$, if $t \ge 0$. Now, choosing the constants C_0 and C_1, so that $v_0(a) = v_0(b) = 0$, we see that $v_0 \in H_0^1(a, b)$ and v_0 is the Riesz representer of the evaluation functional $v \to v(0)$.

The above remarks clearly generalize to $H_0^s(a, b)$ and to $H^s(-\infty, \infty)$ (see Example 5 or Appendix I.1).

4.2. The de Montricher–Tapia–Thompson Estimator

Consider the restricted Sobolev space given in Example 5 of Appendix I.1; namely,

$$H_0^s(a, b) = \{f : f^{(j)} \in L^2(a, b), j = 0, \ldots, s \quad \text{and} \quad f^{(j)}(a) = f^{(j)}(b) = 0, j = 0, \ldots, s - 1\},$$

with inner product

$$\langle f, g \rangle = \int_a^b f^{(s)}(t) g^{(s)}(t)\, dt. \tag{9}$$

Theorem 2. The maximum penalized likelihood estimate corresponding to the Hilbert space $H_0^s(a, b)$ exists and is unique. Moreover, if the estimate is positive in the interior of an interval, then in this interval it is a polynomial spline (monospline) of degree $2s$ and of continuity class $2s - 2$ with knots exactly at the sample points.

Proof. The existence and uniqueness follow from Theorem 1, since Proposition 10 of Appendix I.2 shows that $H_0^s(a, b)$ is a reproducing kernel Hilbert space, Proposition 11 of Appendix I.2 shows that integration over the interval $[a, b]$ gives a linear functional which is continuous on $H_0^s(a, b)$ and there obviously exist members of $H_0^s(a, b)$ which satisfy (3).

When no confusion can arise, we will delete the variable of integration in definite integrals. Consider an interval $I_+ = [\alpha, \beta] \subset [a, b]$. Let $I_- = \{t \in [a, b] : t \notin [\alpha, \beta]\}$. Define the two functionals J_+ and J_- on $H_0^s(a, b)$ by

$$J_+(v) = -\sum_i \log v(x_i) + \int_{I_+} [v^{(s)}]^2,$$

and

$$J_-(v) = -\sum_i \log v(x_i) + \int_{I_-} [v^{(s)}]^2,$$

where the summation in the first formula is taken over all i, such that $x_i \in I_+$, and the summation in the second formula is taken over all i, such that $x_i \in I_-$. It should be clear that

$$J(v) = J_+(v) + J_-(v),$$

where as before $J(v) = -\log \hat{L}(v)$ and $\hat{L}$ is the penalized likelihood in $H_0^s(a, b)$. Let v_* denote the maximum penalized likelihood estimate for the samples $x_1, \ldots, x_n$. Suppose v_* is positive on the interval I_+. We claim that v_* solves the following constrained optimization problem:

(10) minimize $J_+(v)$; subject to

$$v \in H_0^s(a, b), v^{(m)}(\alpha) = v_*^{(m)}(\alpha), v^{(m)}(\beta) = v_*^{(m)}(\beta), \qquad m = 0, \ldots, s-1,$$

$$\int_{I_+} v = \int_{I_+} v_* \quad \text{and} \quad v(t) \geqq 0, \qquad t \in I_+.$$

To see this, observe that if v_+ satisfies the constraints of problem (10) and $J_+(v_*) > J_+(v_+)$, then the function v^* defined by

$$v^*(t) \begin{cases} = v_+(t), & t \in I_+ \\ = v_*(t), & t \in I_- \end{cases}$$

satisfies the constraints of problem (2), with $H_0^s(a, b)$ playing the role of H and $J(v_*) = J_+(v_*) + J_-(v_*) > J_+(v_+) + J_-(v_*) = J(v^*)$, which in turn implies that $\hat{L}(v_*) < \hat{L}(v^*)$; however, this contradicts the optimality of v_* with respect to problem (2). Now, define the functional G on $H_0^s(\alpha, \beta)$ by

$$G(v) = J(v_* + v) \quad \text{for} \quad v \in H_0^s(\alpha, \beta).$$

Consider the constrained optimization problem

(11) minimize $G(v)$; subject to

$$v \in H_0^s(\alpha. \beta) \quad \text{and} \quad \int_{I_+} v = 0.$$

Suppose $v \neq 0$ is a solution of problem (11). Then, $(1-t)v_* + t(v_* + v) = v_* + tv$ satisfies the constraints of problem (10) for $t > 0$ and sufficiently small. By the strict convexity of J (see (4) and Appendix I.3) we have

$$J(v_* + tv) < (1-t)J(v_*) + tJ(v_* + v) \leq J(v_*), \tag{12}$$

which contradicts the optimality of v_* with respect to problem (10). It follows that the zero function is the unique solution of problem (11). From the theory of Lagrange multipliers (see Appendix I.5), we must have

$$\nabla G(0) + \lambda v_0 = 0, \tag{13}$$

where λ is a real number, $\nabla G(0)$ is the gradient of G at 0 (see Appendix I.3) and v_0 is the gradient of the functional $v \to \int_{I_+} v$ in the space $H_0^s(\alpha. \beta)$. Clearly, in this case v_0 is merely the Riesz representer of the functional $v \to \int_{I_+} v$ (see Appendix I.1). Specifically,

$$\int_{I_+} v_0^{(s)} v^{(s)} = \int_{I_+} v. \tag{14}$$

Integrating by parts (in the distribution sense) we see that $v_0^{(2s)} = (-1)^s$; hence v_0 is a polynomial of degree $2s$ in $[\alpha, \beta]$. A straightforward calculation shows that

$$\nabla G(0) = -\left(\sum_i \frac{v_i}{v_*(x_i)} - 2v_*\right), \tag{15}$$

where the summation is taken over i, such that $x_i \in I_+$ and v_i is the Riesz representer of the functional $v \to v(x_i)$ in $H_0^s(\alpha, \beta)$, i.e.,

$$\int_{I_+} v^{(s)} v_i^{(s)} = v(x_i). \tag{16}$$

As before integrating by parts (in the distribution sense) we see that $v_i^{(2s)} = (-1)^s \delta_i$ where δ_i is the Dirac distribution at the point x_i. It follows that v_i is a polynomial spline of degree $2s - 1$ and of continuity class $2s - 2$, with a knot exactly at the sample point x_i. From (15) and (16) we have that v_* restricted to the interval $[\alpha, \beta]$ is a polynomial spline of degree $2s$ and of continuity class $2s - 2$, with knots exactly at the sample points in $[\alpha, \beta]$. A simple continuity argument takes care of the case when v_* is only positive on the interior of $[\alpha, \beta]$. Schoenberg [4] defines a monospline to be the sum of a polynomial of degree $2s$ and a polynomial spline of degree $2s - 1$. This proves the theorem. ∎

In Chapter 5 we will consider numerical algorithms based on Theorem 2 with $H_0^s(a, b)$ replaced by a finite dimensional subspace of $H_0^s(a, b)$.

4.3. *The First Estimator of Good and Gaskins*

Motivated by information theoretic considerations Good and Gaskins [2] consider the maximum penalized likelihood estimate corresponding to the penalty function

$$\Phi_1(v) = \alpha \int_{-\infty}^{\infty} \frac{v'(t)^2}{v(t)}\, dt \qquad (\alpha > 0). \tag{17}$$

They did not define the manifold H, but it is obvious from the constraints that must be satisfied and the fact that

$$\frac{1}{4}\Phi_1(v) = \alpha \int_{-\infty}^{\infty} \left[\frac{dv^{1/2}}{dt}\right]^2 dt \tag{18}$$

what the underlying manifold H should be, namely, $v^{1/2} \in H^1(-\infty, \infty)$ where $H^1(-\infty, \infty)$ is the Sobolev space given in Example 4 of Appendix I.1, namely, $H^1(-\infty, \infty) = \{f : R \to R : f'$ exists almost everywhere and f,

$f' \in L^2(-\infty, \infty)\}$ with inner product

(19) $$\langle f, g\rangle = \int_{-\infty}^{\infty} f(t)g(t)\,dt + \int_{-\infty}^{\infty} f'(t)g'(t)\,dt.$$

The situation here is going to be very delicate, because it is possible to show that the integration functional is *not* continuous in $H^1(-\infty, \infty)$. In the present application, problem (2) takes the form

(20) $$\text{maximize } L_1(v) = \prod_{i=1}^{n} v(x_i)\exp(-\Phi_1(v)); \text{ subject to}$$

$$v^{1/2} \in H^1(-\infty, \infty), \int_{-\infty}^{\infty} v(t)\,dt = 1 \quad \text{and} \quad v(t) \geqq 0 \quad \forall t \in (-\infty, \infty).$$

In an effort to avoid the nonnegativity constraint in problem (20), Good and Gaskins considered working with $v^{1/2}$ instead of v. Specifically, if we let $u = v^{1/2}$, then restating problem (20) in terms of u we obtain

(21) $$\text{maximize } \prod_{i=1}^{n} u(x_i)^2 \exp\left(-4\alpha \int_{-\infty}^{\infty} u'(t)^2\,dt\right); \text{ subject to}$$

$$u \in H^1(-\infty, \infty) \quad \text{and} \quad \int_{-\infty}^{\infty} u(t)^2\,dt = 1.$$

Problem (21) is solved for u^*, and then $v^* = (u^*)^2$ is accepted as the solution to problem (20). This transformation was discussed in Section 4.1, and Proposition 3 shows that since we are working with $H^1(-\infty, \infty)$, it is valid.

Problem (21) cannot possibly have a unique solution. To see this, notice that if u^* is a solution, then so is $-u^*$. Adding the nonnegativity constraint to problem (21) and restating in the form obtained by taking the square root of the objective functional (since it is nonnegative) we arrive at the following constrained optimization problem:

(22) $$\text{maximum } \hat{L}(v) = \prod_{i=1}^{n} v(x_i)\exp(-\Phi(v)); \text{ subject to}$$

$$v \in H^1(-\infty, \infty), \int_{-\infty}^{\infty} v(t)^2\,dt = 1 \quad \text{and} \quad v(t) \geqq 0, \quad \forall t \in (-\infty, \infty)$$

where

(23) $$\Phi(v) = 2\alpha \int_{-\infty}^{\infty} v'(t)^2\,dt$$

and α is given in (17).

Proposition 4.

(i) If v solves problem (20), then $v^{1/2}$ solves problem (21) and problem (22).

(ii) If u solves problem (21), then $|u|$ solves problem (22) and u^2 solves problem (20).

(iii) If v solves problem (22), then v solves problem (21) and v^2 solves problem (20).

Proof. The proof follows from Proposition 3 and the fact that if $v \geqq 0$, then

$$\Phi(v^{1/2}) = \tfrac{1}{2}\Phi_1(v) \tag{24}$$

and

$$\hat{L}_1(v) = \hat{L}(v^{1/2})^2. \quad \blacksquare \tag{25}$$

Corollary 1. If problem (22) has a unique solution, then problem (20) has a unique solution; and although problem (22) cannot have a unique solution, it will have solutions, and the square of any of these solutions will give the unique solution of problem (20).

The remainder of this section is dedicated to demonstrating that problem (22) has a unique solution which is a positive exponential spline with knots only at the sample points. The same will then be true of Good's and Gaskins's first maximum penalized likelihood estimate.

Along with problem (22) we will consider the constrained optimization problem obtained by only requiring nonnegativity at the sample points:

(26) maximize $\hat{L}(v)$; subject to

$$v \in H^1(-\infty, \infty), \quad \int_{-\infty}^{\infty} v(t)^2\, dt = 1 \quad \text{and} \quad v(x_i) \geqq 0, \quad i = 1, \ldots, n.$$

Given $\lambda > 0$ and α in problem (22), we may also consider the constrained optimization problem:

(27) maximize $\hat{L}_\lambda(v) = \prod_{i=1}^{n} v(x_i) \exp(-\Phi_\lambda(v))$; subject to

$$v \in H^1(-\infty, \infty), \quad \int_{-\infty}^{\infty} v(t)^2\, dt = 1 \quad \text{and} \quad v(x_i) \geqq 0, \quad i = 1, \ldots, n,$$

where

$$\Phi_\lambda(v) = 2\alpha \int_{-\infty}^{\infty} v'(t)^2\, dt + \lambda \int_{-\infty}^{\infty} v(t)^2\, dt. \tag{28}$$

Our study of problem (27) will begin with the study of the following constrained optimization problem:

(29) maximize $\hat{L}_\lambda(v)$; subject to

$$v \in H^1(-\infty, \infty) \quad \text{and} \quad v(x_i) \geqq 0, \quad i = 1, \ldots, n,$$

where $\hat{L}_\lambda$ is given by problem (27). Let $L^2 = L^2(-\infty, \infty)$.

Proposition 5. Problem (29) has a unique solution. Moreover, if v_λ denotes this solution, then

(i) v_λ is an exponential spline with knots at the sample points $x_1, \ldots, x_n$;
(ii) $v_\lambda(t) > 0$, $\forall t \in (-\infty, \infty)$; and
(iii) $\|v_\lambda\|_{L^2} \geqq (n/4\lambda))^{1/2}$.

Proof. From Proposition 12 of Appendix I.2 $H^1(-\infty, \infty)$ is a reproducing kernel space. Also $\|v\|_\lambda^2 = \Phi_\lambda(v)$ gives a norm equivalent to the original norm on $H^1(-\infty, \infty)$. The existence of v_λ now follows from Proposition 1 with $D = \{v \in H^1(-\infty, \infty): v(x_i) \geqq 0, i = 1, \ldots, n\}$. We will denote the Φ_λ inner product by $\langle\ ,\ \rangle_\lambda$. Let v_i be the representer in the Φ_λ inner product of the continuous linear functional given by point evaluation at the point x_i, $i = 1, \ldots, n$, i.e.,

$$\langle v_i, \eta\rangle_\lambda = \eta(x_i), \qquad \forall \eta \in H^1(-\infty, \infty). \tag{30}$$

Equivalently,

$$2\alpha \int_{-\infty}^{\infty} v_i'(t)\eta'(t)\,dt + \lambda \int_{-\infty}^{\infty} v_i(t)\eta(t)\,dt = \eta(x_i), \qquad \forall \eta \in H^1(-\infty, \infty). \tag{31}$$

Integrating by parts (in the distribution sense) gives

$$\int_{-\infty}^{\infty} [-2\alpha v_i''(t) + \lambda v_i(t)]\eta(t)\,dt = \eta(x_i), \qquad \forall \eta \in H^1(-\infty, \infty); \tag{32}$$

hence,

$$-2\alpha v_i'' + \lambda v_i = \delta_i, \qquad i = 1, \ldots, n \tag{33}$$

where $\delta_i(t) = \delta_0(t - x_i)$ and δ_0 denotes the Dirac distribution, i.e., $\int_{-\infty}^{\infty} \delta_0(t)\eta(t)\,dt = \eta(0)$. If we let v_0 be the solution of (33) for $i = 0$, then

$$\begin{aligned} v_0(t) &= \frac{1}{2(2\alpha\lambda)^{1/2}} \exp((\lambda(2\alpha))^{1/2}t), \qquad t < 0 \\ &= \frac{1}{2(2\alpha\lambda)^{1/2}} \exp(-(\lambda/(2\alpha))^{1/2}t), \qquad t > 0 \end{aligned} \tag{34}$$

and $v_i(t) = v_0(t - x_i)$ for $i = 1, \ldots, n$. Since v_λ is the maximizer we have that $v_\lambda(x_i) > 0$, $i = 1, \ldots, n$. We necessarily have that the derivative of $\hat{L}_\lambda$ at v_λ must be the zero functional; equivalently the gradient of $\hat{L}_\lambda$, or for that matter the gradient of $\log \hat{L}_\lambda$, must vanish at v_λ, since $\hat{L}_\lambda$ and $\log \hat{L}_\lambda$ have the same maxima. A calculation similar to that used in the proof of Proposition 1 gives

$$\nabla_\lambda \log \hat{L}_\lambda(v) = 2v - \sum_{i=1}^{n} \frac{v_i}{v(x_i)}, \tag{35}$$

where ∇_λ denotes the gradient. It follows from (35) that

$$v_\lambda = \frac{1}{2}\sum_{i=1}^{n} \frac{v_i}{v_\lambda(x_i)}. \tag{36}$$

Properties (i) and (ii) are now immediate. Since $\langle v_i, v_\lambda \rangle_\lambda = v_\lambda(x_i)$, from (36) we have

$$\|v_\lambda\|_\lambda^2 = n/2. \tag{37}$$

A straightforward calculation shows that

$$v_i'(t)v_j'(t) \leqq \frac{\lambda}{2\alpha} v_i(t)v_j(t), \qquad \text{for} \quad i, j = 1, \dots, n. \tag{38}$$

So

$$\begin{aligned} v'(t)^2 &= \frac{1}{4}\left[\sum_i \left(\frac{v_i'(t)}{v_\lambda(x_i)}\right)^2 + \sum_{i,j} \frac{v_i'(t)v_j'(t)}{v_\lambda(x_i)v_\lambda(x_j)}\right] \\ &\leqq \frac{\lambda}{8\alpha}\left[\sum_i \left(\frac{v_i(t)}{v_\lambda(x_i)}\right)^2 + \sum_{i,j} \frac{v_i(t)v_j(t)}{v_\lambda(x_i)v_\lambda(x_j)}\right] = \frac{\lambda}{2\alpha} v_\lambda(t)^2. \end{aligned} \tag{39}$$

Integrating in t gives

$$2\alpha\|v_\lambda'\|_{L^2(-\infty,\infty)}^2 \leqq \lambda\|v_\lambda\|_{L^2(-\infty,\infty)}^2. \tag{40}$$

By definition of the Φ_λ-norm and (37) we have property (iii). This proves the proposition. ■

Proposition 6. Problem (26) has a unique solution.

Proof. Let $B = \{v \in H^1(-\infty, \infty): \int_{-\infty}^{\infty} v(t)^2\,dt \leqq 1 \text{ and } v(x_i) \geqq 0, i = 1, \dots, n\}$. Clearly, B is closed and convex. If $\hat{L}_\lambda$ is given by (27), then by Proposition 1 it will have a unique maximizer in B; say μ_λ. Now, by property (iii) of Proposition 5, if we choose $0 < \lambda < \frac{1}{4}$, then, v_λ, the unique solution of problem (29), will be such that $\|v_\lambda\|_{L^2(-\infty,\infty)} > 1$. We will show that for this range of λ, $\|\mu_\lambda\|_{L^2(-\infty,\infty)} = 1$. Consider $v_\theta = \theta v_\lambda + (1-\theta)\mu_\lambda$. We know that $-\log \hat{L}_\lambda$ is a strictly convex functional (see the proof of Proposition 1). Moreover, $\log \hat{L}_\lambda(v_\lambda) \geqq \log \hat{L}_\lambda(\mu_\lambda)$; hence $\log \hat{L}_\lambda(v_\theta) \geqq \log \hat{L}_\lambda(\mu_\lambda)$ for $0 < \theta < 1$. Now suppose $\|\mu_\lambda\|_{L^2(-\infty,\infty)} < 1$ and consider

$$g(\theta) = \|v_\theta\|_{L^2(-\infty,\infty)}. \tag{41}$$

We have $g(0) < 1$ and $g(1) > 1$. So for some $0 < \theta_0 < 1$, $g(\theta_0) = 1$ and $\log \hat{L}_\lambda(\mu_\lambda) \leqq \log \hat{L}_\lambda(v_{\theta_0})$. This is a contradiction, since μ_λ is the unique maximizer of $\hat{L}_\lambda$ in B; hence, $\|\mu_\lambda\|_{L^2(-\infty,\infty)} = 1$. This shows that μ_λ is the unique solution of problem (27) for $0 < \lambda < \frac{1}{4}$. However, the term $\lambda \int_{-\infty}^{\infty} v(t)^2 \, dt$ is constant over the constraint set in problems (26) and (27); hence, problems (26) and (27) have the same solutions for any $\lambda > 0$. This proves the proposition, since we have demonstrated that problem (24) has a unique solution for at least one λ. ■

Proposition 7. Problem (22) has a unique solution which is positive and an exponential spline with knots at the points $x_1, \ldots, x_n$.

Proof. If we can demonstrate that $\tilde{v}$, the unique solution of problem (26), has these properties we will be through. Let $G(v) = \log \hat{L}(v)$ where $\hat{L}$ is given in problem (22) and let

$$g(v) = \int_{-\infty}^{\infty} v(t)^2 \, dt \tag{42}$$

for $v \in H^1(-\infty, \infty)$. Clearly, $\tilde{v}(x_i) > 0$ for $i = 1, \ldots, n$; hence, from the theory of Lagrange multipliers there exist λ, such that $\tilde{v}$ satisfies the equations

$$G'(v) - \lambda g'(v) = 0 \qquad \text{and} \qquad g(v) = 1. \tag{43}$$

Using $L^2(-\infty, \infty)$ gradients (in the sense of distributions) (43) is equivalent to

$$-4\alpha v'' + 2\lambda v = \sum_{i=1}^{n} \frac{\delta_i}{v(x_i)} \qquad \text{and} \qquad g(v) = 1, \tag{44}$$

where δ_i is the Dirac distribution at the point x_i. Since we have already established that problem (26) has a unique solution, it follows that (44) must have a unique solution in $H^1(-\infty, \infty)$; namely, $\tilde{v}$. If $\lambda \leqq 0$, then any solution of the first equation in (44) would be a sum of trigonometric functions and could not possibly satisfy the constraint $g(v) = 1$, i.e., cannot be contained in $L^2(-\infty, \infty)$. It follows that $\lambda > 0$. Now observe that

$$G - \lambda g = \log \hat{L}_\lambda,$$

where $\hat{L}_\lambda$ is given by problem (27); hence, if $\tilde{v}$ satisfies (43) (from the first equation alone) it must also be a solution of problem (29) for this λ and therefore has the desired properties according to Proposition 5. This proves the proposition. ■

Theorem 3. The first maximum penalized likelihood estimate of Good and Gaskins exists and is unique; specifically, the maximum penalized likelihood estimate corresponding to the penalty function

$$\Phi(v) = \alpha \int_{-\infty}^{\infty} \frac{v'(t)^2}{v(t)} \, dt \qquad (\alpha > 0)$$

and the manifold

$$H = \{v : v \geqq 0 \quad \text{and} \quad v^{1/2} \in H^1(-\infty, \infty)\}$$

exists and is unique. Moreover, the estimate is positive and an exponential spline with knots only at the sample points.

Proof. The proof follows from Proposition 4 and Proposition 7. ∎

4.4. The Second Estimator of Good and Gaskins

Consider the functional $\Phi : H^2(-\infty, \infty) \to R$ defined by

$$\Phi(v) = \alpha \int_{-\infty}^{\infty} v'(t)^2\, dt + \beta \int_{-\infty}^{\infty} v''(t)^2\, dt \tag{45}$$

for some $\alpha \geqq 0$ and $\beta > 0$. By a *second maximum penalized likelihood estimate of Good and Gaskins* we mean any solution of the following constrained optimization problem:

$$\text{(46)} \quad \text{maximize } \hat{L}_1(v) = \prod_{i=1}^{n} v(x_i) \exp(-\Phi(v^{1/2}));\ \text{subject to}$$

$$v^{1/2} \in H^2(-\infty, \infty), \quad \int_{-\infty}^{\infty} v(t)\, dt = 1 \quad \text{and} \quad v(t) \geqq 0 \quad \forall t \in (-\infty, \infty).$$

As in the first case (described in the previous section), Good and Gaskins suggest avoiding the nonnegativity constraint by calculating the solution of problem (46) from the following constrained optimization problem:

$$\text{(47)} \quad \text{maximize } \prod_{i=1}^{n} v(x_i)^2 \exp(-\Phi(v));\ \text{subject to}$$

$$v \in H^2(-\infty, \infty) \quad \text{and} \quad \int_{-\infty}^{\infty} v(t)^2\, dt = 1,$$

where Φ is given by (45).

Along with problem (47) we consider the constrained optimization problem:

$$\text{(48)} \quad \text{maximize } \hat{L}(v) = \prod_{i=1}^{n} v(x_i) \exp(-\tfrac{1}{2}\Phi(v));\ \text{subject to}$$

$$v \in H^2(-\infty, \infty), \int_{-\infty}^{\infty} v(t)^2\, dt = 1 \quad \text{and} \quad v(x_i) \geqq 0, \quad i = 1, \ldots, n.$$

Problem (48) was obtained from problem (47) by taking the square root of the functional to be maximized (since it is nonnegative) and requiring nonnegativity at the sample points; hence, the two problems only differ by the nonnegativity constraints at the sample points. This simple difference will allow us to establish uniqueness of the solution of problem (48); whereas problem (47) cannot have a unique solution. We will presently demonstrate that the solutions of problem (47) and problem (48) are not necessarily nonnegative. It will then follow that we cannot obtain the solution of problem (46) by considering problem (47) and problem (48). If we naively use v_*^2, where v_* solves problem (48), as an estimate for the probability density function giving rise to the random sample $x_1, \ldots, x_n$, then, clearly, v_*^2 will be nonnegative and integrate to one and is therefore a probability density; however, the estimate obtained in this manner will not in the strict sense of our definition be the maximum penalized likelihood estimate corresponding to problem (46), i.e., the second maximum penalized likelihood estimate of Good and Gaskins. For this reason we will refer to this latter estimate as the *pseudo maximum penalized likelihood estimate of Good and Gaskins.*

The next six propositions are needed to show that the second maximum penalized likelihood estimate and the pseudo maximum penalized likelihood estimate of Good and Gaskins exist, are unique, and are distinct.

Given $\lambda > 0$ consider the constrained optimization problem:

(49) maximize $\hat{L}_\lambda(v) = \prod_{i=1}^{n} v(x_i) \exp(-\Phi_\lambda(v))$; subject to

$$v \in H^2(-\infty, \infty), \quad \int_{-\infty}^{\infty} v(t)^2\, dt = 1 \quad \text{and} \quad v(x_i) \geqq 0, \quad i = 1, \ldots, n,$$

where

$$\Phi_\lambda(v) = \Phi(v) + \lambda \int_{-\infty}^{\infty} v(t)^2\, dt$$

with $\Phi(v)$ given by (45).

As before we also consider the constrained optimization problem obtained by dropping the integral constraint:

(50) maximize $\hat{L}_\lambda(v)$; subject to

$$v \in H^2(-\infty, \infty), \qquad \text{and} \qquad v(x_i) \geqq 0, \qquad i = 1, \ldots, n.$$

Proposition 8. Problem (49) has a unique solution. Moreover, if v_λ denotes this solution, then

$$\|v_\lambda\|_{L^2(-\infty,\infty)} \to +\infty \qquad \text{as} \quad \lambda \to 0.$$

Proof. By Proposition 11 of Appendix I.2 the Sobolev space $H^2(-\infty, \infty)$ is a reproducing kernel Hilbert space. Moreover, if

$$\|v'\|_\lambda^2 = \Phi_\lambda(v),$$

then an integration by parts gives

$$\begin{aligned} \|v'\|_{L^2}^2 &= |\langle v, v''\rangle_{L^2}| \leqq \|v\|_{L^2}\|v''\|_{L^2} \\ &\leqq \tfrac{1}{2}[\|v\|_{L^2}^2 + \|v''\|_{L^2}^2], \end{aligned} \tag{51}$$

where L^2 denotes $L^2(-\infty, \infty)$; hence, $\|\circ\|_\lambda$ is equivalent to the original norm on $H^2(-\infty, \infty)$. The existence and uniqueness of v_λ now follows from Proposition 1.

We must now show that $\|v_\lambda\|_{L^2} \to +\infty$ as $\lambda \to 0$. From the fundamental theorem of calculus we have

$$\begin{aligned} v(x)^2 &= \int_{-\infty}^x \frac{dv(t)^2}{dt}\, dt = 2\int_{-\infty}^x v(t)v'(t)\, dt \\ &\leqq 2\|v\|_{L^2}\|v'\|_{L^2}. \end{aligned} \tag{52}$$

Also, $\|v''\|_{L^2} \leqq \|v\|_\lambda/\beta^{1/2}$, so that from (51) and (52)

$$v(x)^2 \leqq 2\|v\|_{L^2}^{3/2}(\|v\|_\lambda/\beta^{1/2})^{1/2}. \tag{53}$$

Evaluating (53) at x_i, taking logs (since $v(x_i) \geqq 0$) and summing over i gives

$$\sum_{i=1}^n \log v(x_i) \leqq \frac{n}{4}\log\left(\frac{4}{\beta^{1/2}}\|v\|_\lambda\right) + \frac{3n}{4}\log(\|v\|_{L^2}). \tag{54}$$

Hence, from (54) we see that

$$\log \hat{L}_\lambda(v) \leqq \frac{3n}{4}\log(\|v\|_{L^2}) + \frac{n}{4}\log\left(\frac{4}{\beta^{1/2}}\|v\|_\lambda\right) - \|v\|_\lambda^2. \tag{55}$$

In a manner exactly the same as that used to establish (37), we have that $\|v_\lambda\|_\lambda^2 = n/2$. Hence, from (55) and the fact that $\log \hat{L}_\lambda(v) \leqq \log \hat{L}_\lambda(v_\lambda)$ we obtain

$$\log \hat{L}_\lambda(v) \leqq \frac{3n}{4}\log(\|v_\lambda\|_{L^2}) + \frac{n}{8}\log(8n/\beta) - \frac{n}{2}, \tag{56}$$

for any $v \in \{u \in H^2(-\infty, \infty): u(x_i) \geqq 0, i = 1, \ldots, n\}$.

Let a and b be such that

$$a < \min_i(x_i) \qquad \text{and} \qquad \max_i(x_i) < b.$$

Given $\lambda > 0$ and ϵ and δ define the function θ_λ in the following piece-wise fashion:

$$\begin{aligned}\theta_\lambda(t) &= \lambda^\epsilon \exp(-(t-a)^2/2\sigma^2) && \text{for} \quad t \in (-\infty, a)\\ &= \lambda^\epsilon && \text{for} \quad t \in [a, b]\\ &= \lambda^\epsilon \exp(-(t-b)^2/2\sigma^2) && \text{for} \quad t \in (b, +\infty),\end{aligned}$$

where $\sigma = \lambda^\delta$. Straightforward calculations can be used to show

$$\log\left(\prod_{i=1}^{n} \theta_\lambda(x_i)\right) = \epsilon n \log(\lambda),$$

$$\begin{aligned}\|\theta_\lambda\|_{L_2}^2 &= (b-a)\lambda^{2\epsilon} + ((\pi\lambda)^{1/2})^{2\epsilon-\delta},\\ \|\theta_\lambda'\|_{L_2}^2 &= ((2\pi\lambda)^{1/2})^{2\epsilon-\delta},\\ \|\theta_\lambda''\|_{L_2}^2 &= 2((2\pi\lambda)^{1/2})^{2\epsilon-3\delta},\end{aligned}$$

and

(57) $$\|\theta_\lambda\|_\lambda^2 = (b-a)\lambda^{2\epsilon+1} + ((\pi\lambda)^{1/2})^{2\epsilon+\delta+1} + 4\alpha((2\pi\lambda)^{1/2})^{2\epsilon-\delta} + 2\beta((2\pi\lambda)^{1/2})^{2\epsilon-3\delta}.$$

If we want $\|\theta_\lambda\|_\lambda^2 \to 0$ as $\lambda \to 0$, it is sufficient to choose all exponents of λ in (57) positive. If we also want

$$\log\left(\prod_{i=1}^{n} \theta_\lambda(x_i)\right) \to +\infty \quad \text{as} \quad \lambda \to 0,$$

we should choose $\epsilon < 0$. This leads to the inequalities

(58) $$\begin{aligned}2\epsilon + 1 &> 0\\ 2\epsilon + \delta + 1 &> 0\\ 2\epsilon - \delta &> 0\\ 2\epsilon - 3\delta &> 0\\ \epsilon &< 0.\end{aligned}$$

The system of inequalities (58) has solutions; specifically, $\epsilon = -\frac{1}{32}$ and $\delta = -\frac{1}{8}$ is one such solution. With this choice of ϵ and δ we see that $\log \hat{L}_\lambda(\theta_\lambda) \to +\infty$ as $\lambda \to 0$. It follows from (56) by choosing $v = \theta_\lambda$ that $\|v_\lambda\|_{L_2} \to +\infty$ as $\lambda \to 0$. This proves the proposition. ∎

Theorem 4. Problem (48) has a unique solution, i.e., the pseudo maximum penalized likelihood estimate of Good and Gaskins exists and is unique.

Proof. By Proposition 8 there exists $\lambda > 0$, such that if v_λ is the unique solution of problem (49), then $\|v_\lambda\|_{L_2} > 1$. Now, if $B = \{v \in H^2(-\infty, \infty)$: $\int_{-\infty}^{\infty} v(t)^2\, dt \leqq 1$ and $v(x_i) \geqq 0$, $i = 1, \ldots, n\}$, then B is closed and convex.

The proof of the theorem is now exactly the same as the proof of Proposition 6. ■

By the change of unknown function $v \to v^{1/2}$ we see that problem (46) is equivalent to the following constrained optimization problem:

(59) maximize $\hat{L}(v) = \prod_{i=1}^{n} v(x_i) \exp(-\frac{1}{2}\Phi(v))$; subject to

$$v \in H^2(-\infty, \infty), \quad \int_{-\infty}^{\infty} v(t)^2\, dt = 1 \quad \text{and} \quad v(t) \geqq 0 \quad \forall t \in (-\infty, \infty),$$

where $\Phi(v)$ is given by (45).

In turn, for $\lambda > 0$ problem (59) is equivalent to

(60) maximize $\hat{L}_\lambda(v)$: subject to

$$v \in H^2(-\infty, \infty), \quad \int_{-\infty}^{\infty} v(t)^2\, dt = 1 \quad \text{and} \quad v(t) \geqq 0 \quad \forall t \in (-\infty, \infty),$$

where $\hat{L}_\lambda$ is defined in problem (49).

As in the previous two cases, we also consider the constrained optimization problem:

(61) maximize $\hat{L}_\lambda(v)$; subject to

$$v \in H^2(-\infty, \infty) \quad \text{and} \quad v(t) \geqq 0 \quad \forall t \in (-\infty, \infty),$$

where $\hat{L}_\lambda(v)$ is defined in problem (49).

Proposition 9. Problem (61) has a unique solution. Moreover, if v_λ^+ denotes this solution, then

$$\|v_\lambda^+\|_{L_2} \to +\infty \quad \text{as} \quad \lambda \to 0.$$

Proof. The existence of v_λ^+ follows from Proposition 1 as in the proof of Proposition 8. Let us first show that

$$\|v_\lambda^+\|_\lambda \leqq (n/2)^{1/2}. \tag{62}$$

Let $J_\lambda = -\log \hat{L}_\lambda$.

Clearly, for all nonnegative η in $H^2(-\infty, \infty)$ we have

$$J_\lambda'(v_\lambda^+)(\eta - v_\lambda^+) \geqq 0. \tag{63}$$

To see this, suppose that for some η (63) does not hold. Then by the definition of the Gâteaux derivative (see Appendix I.3) we will have

$$J_\lambda(v_\lambda^+ + t(\eta - v_\lambda^+)) < J(v_\lambda^+) \tag{64}$$

for $t > 0$ sufficiently small. However, $v_\lambda^+ + t(\eta - v_\lambda^+) = (1 - t)v_\lambda^+ + t\eta \geq 0$

for $t > 0$ sufficiently small. Hence (64) contradicts the optimality of v_λ^+ with respect to problem (61). A straightforward calculation shows that

$$J'_\lambda(v)(\eta) = \sum_{i=1}^{n} \frac{\eta(x_i)}{v(x_i)} - 2\langle v, \eta\rangle_\lambda;$$

hence,

$$J'_\lambda(v_\lambda^+)(v_\lambda^+) = n - 2\|v_\lambda\|_\lambda^2. \tag{65}$$

Now, choosing $\eta = 0$ in (63) and using (65) we arrive at (62). The functions θ_λ defined in the proof of Proposition 8 satisfy the constraints of this problem; hence,

$$\log \hat{L}_\lambda(\theta_\lambda) \leqq \log \hat{L}_\lambda(v_\lambda^+). \tag{66}$$

From (55) and (62) we have

$$\log \hat{L}_\lambda(\theta_\lambda) \leqq \frac{3n}{4} \log(\|v_\lambda^+\|_{L_2}) + \frac{n}{8} \log\left(\frac{8n}{\beta}\right) + \frac{n}{2}. \tag{67}$$

The proof now follows from (67), since $\log \hat{L}_\lambda(\theta_\lambda) \to +\infty$ as $\lambda \to 0$. ∎

Theorem 5. The second maximum penalized likelihood estimate of Good and Gaskins exists and is unique.

Proof. Using Proposition 9, the argument used to prove Theorem 4 shows that problem (60) has a unique solution, which is also the unique solution of problem (59). ∎

Theorem 6. The second maximum penalized likelihood estimate and the pseudo maximum likelihood estimate of Good and Gaskins are distinct.

Proof. We will show that it is possible for problem (48) to have solutions which are not nonnegative. Toward this end let $n = 1$, $x_1 = 0$, $\alpha = 0$, and $\beta = 2$. Let $G(v) = \log \hat{L}(v)$, i.e.,

$$G(v) = \log v(0) - \int_{-\infty}^{\infty} v''(t)^2 \, dt$$

and let

$$g(v) = \int_{-\infty}^{\infty} v(t)^2 \, dt.$$

As in the proof of Proposition 7, using the theory of Lagrange multipliers, we see that the solutions of problem (48) in this case are exactly the solutions of

$$v^{(iv)} + \lambda v = \frac{\delta_1}{2v(0)} \qquad \text{and} \qquad g(v) = 1, \tag{68}$$

where δ_1 is defined in the proof of Proposition 7. If we let $\tilde{v}$ denote the Fourier transform of v, then taking the Fourier transform of the first expression in (68) gives

$$\tilde{v}(w) = [2v(0)(\lambda + 16\pi^4 w^4)]^{-1}.$$

Since $\|\tilde{v}\|_{L^2(-\infty,\infty)} = \|v\|_{L^2(-\infty,\infty)} = 1$ we must have

$$\int_{-\infty}^{\infty} \frac{dw}{(\lambda + 16\pi^4 w^4)^2} = 4v(0)^2. \tag{69}$$

For the integral in (69) to exist we must have $\lambda > 0$. Now, the inverse Fourier transform of $(\lambda + 16\pi^4 w^4)^{-1}$ is given by v, where

$$v(t) = \frac{e^{bt}}{8b^3}[\cos bt - \sin bt], \qquad t \leqq 0 \tag{70}$$

$$= \frac{e^{-bt}}{8b^3}(\cos bt + \sin bt], \qquad t > 0$$

with $b = \lambda^{1/4}/2^{1/2}$. From (70) $v(0) = (8b^3)^{-1}$ and from (69) $v(0)^2 = \frac{1}{4}\lambda^{-7/4K}$, where $K = \|(1 + 16\pi^4 w^4)^{-1}\|_{L^2(-\infty,\infty)}$. Hence, $\lambda^{1/4} = 2K$ and $b = 2^{1/2}K$. It follows that v is not nonnegative. Moreover, $|v|$ does not have a continuous derivative, so $|v| \notin H^2(-\infty, \infty)$. ∎

Corollary 1. The second maximum penalized likelihood estimate of Good and Gaskins cannot be obtained by solving problem (47).

Proof. Observe that the solution constructed in the proof of Theorem 6 is also a solution of problem (47) for this example. ∎

References

[1] de Montricher, G. M., Tapia, R. A., and Thompson, J. R. (1975). "Nonparametric maximum likelihood estimation of probability densities by penalty function methods." *Annals of Statistics* 3: 1329–48.

[2] Good, I. J., and Gaskins, R. A. (1971). "Nonparametric roughness penalties for probability densities." *Biometrika* 58: 255–77.

[3] Horvath, J. (1966). *Topological Vector Spaces and Distributions.* Reading, Massachusetts: Addison-Wesley.

[4] Schoenberg, I. J. (1968). "Monosplines and quadrature formulae," in *Theory and Application of Spline Functions* (T.N.E. Greville, ed.). New York: Academic Press.

CHAPTER

5
Discrete Maximum Penalized Likelihood Estimation

5.1 Discrete Maximum Penalized Likelihood Estimators

Our examination of penalized likelihood density estimations will now be brought to the state of practical computation. The arguments below are due mainly to Scott [5] and to Scott, Tapia, and Thompson [7]. As before, given the random sample $x_1, \ldots, x_n$ we would like to estimate, on an interval (a, b), the unknown density function f which gave rise to this random sample. From Chapter 4 our estimates are required to integrate to one on (a, b) and have support in (a, b); hence it follows that if the support of the unknown density f is not contained in (a, b), then we will actually be estimating the density function $\bar{f}$, which is close to f in the following sense

$$\bar{f}(x) = \begin{cases} \dfrac{f(x)}{\int_a^b f(x)\,dx} & \text{if} \quad a < x < b \\ 0 & \text{otherwise.} \end{cases} \tag{1}$$

Notice that f and $\bar{f}$ will coincide if and only if (a, b) contains the support of f. For the random sample $x_1, \ldots, x_n$ it is assumed that all observations outside (a, b) have been censored.

We will construct approximations to the maximum penalized likelihood estimate corresponding to the Hilbert space $H_0^1(a, b)$ (see Theorem 2 of Chapter 4) by solving finite dimensional versions of problem (2) in Chapter 4, which are reasonable approximations to the infinite dimensional problem. The terminology "discrete" is used because our finite dimensional versions of problem (4.2) will be obtained by working with the finite dimensional linear manifolds which arise by introducing a discrete mesh on

(a, b) and then considering the polynomial spline spaces of functions which are piece-wise constant or piece-wise linear on this mesh. Specifically, for a given positive integer m consider the uniform mesh $a = t_0 < t_1 < \cdots < t_m = b$, where $t_i = a + ih$, $i = 0, \ldots, m$, with $h = (b - a)/m$. Let T_i denote the half-open half-closed interval $[t_{i-1}, t_i)$, for $i = 1, \ldots, m$, and let $I(T_i)$ denote the indicator function of the interval T_i, i.e., $I(T_i)(x) = 0$, if $x \notin T_i$ and $I(T_i)(x) = 1$, if $x \in T_i$. Let $S_0(t_0, \ldots, t_m)$ denote the functions with support in $[t_0, t_m)$, equivalently $[a, b)$, and which are constant on the intervals T_i. Finally, let $S_1(t_0, \ldots, t_m)$ denote the continuous functions with support in (t_0, t_m) and which are linear on the intervals T_i. A typical member of $S_0(t_0, \ldots, t_m)$ would have a graph similar to that given by Figure 3.1 in Section 3.2 of Chapter 3, while a typical member of $S_1(t_0, \ldots, t_m)$ would have a graph similar to that given by Figure 3.2 in Section 3.2 of Chapter 3.

Since each $s_0 \in S_0(t_0, \ldots, t_m)$ has the representation

$$s_0(x) = \sum_{i=1}^{m} y_{i-1} I(T_i)(x), \tag{2}$$

where

$$y_i = s_0(t_i), \qquad i = 0, \ldots, m-1, \tag{3}$$

it follows that $S_0(t_0, \ldots, t_m)$ is isomorphic to R^m. Moreover, assuming the correspondence (3), we see that

$$s_0(x) \geq 0 \quad \forall x \in (a, b) \Leftrightarrow y_i \geq 0, \qquad i = 0, \ldots, m-1 \tag{4}$$

and

$$\int_a^b s_0(x)\, dx = 1 \Leftrightarrow h \sum_{i=0}^{m-1} y_i = 1. \tag{5}$$

Furthermore, since each $s_1 \in S_1(t_0, \ldots, t_m)$ has the representation

$$s_1(x) = \sum_{i=1}^{m} [y_{i-1} + h^{-1}(x - t_{i-1})(y_i - y_{i-1})] I(T_i)(x), \tag{6}$$

where

$$y_i = s_1(t_i), \qquad i = 1, \ldots, m-1, \tag{7}$$

and by assumption $y_0 = s_1(t_0) = 0$ and $y_m = s_1(t_m) = 0$. It follows that $S_1(t_0, \ldots, t_m)$ is isomorphic to R^{m-1}. Moreover, assuming the correspondence (6), we see that

$$s_1(x) \geq 0 \quad \forall x \in (a, b) \Leftrightarrow y_i \geq 0, \qquad i = 1, \ldots, m-1 \tag{8}$$

and

$$\int_a^b s_1(x)\,dx = 1 \Leftrightarrow h \sum_{i=1}^{m-1} y_i = 1. \tag{9}$$

Let us now consider the $H_0^1(a, b)$ inner product defined on $S_i(t_0, \ldots, t_m)$, $i = 0, 1$, i.e., for s_0 given by (2) and s_1 given by (6) we are interested in the quantities

$$\|s_i\|_H^2 = \int_a^b s_i'(x)^2\,dx, \qquad i = 0, 1. \tag{10}$$

For $i = 1$, this creates no problem, since $S_1(t_0, \ldots, t_m) \subset H_0^1(a, b)$ and it follows that

$$\|s_1\|_s^2 = h^{-1} \sum_{i=1}^{m} (y_i - y_{i-1})^2. \tag{11}$$

However, since $S_0(t_0, \ldots, t_m) \not\subset H_0^1(a, b)$ the expression (10) is meaningless, but (11), even in this case, makes perfectly good sense, provided we interpret y_m to be zero. It therefore seems reasonable that if we wish to consider $S_i(t_0, \ldots, t_m)$, $i = 0, 1$ as approximations of the Hilbert space $H_0^1(a, b)$ we should work with the inner product

$$\langle s, \bar{s} \rangle_S = \alpha h^{-1} \sum_{i=1}^{m} (y_i - y_{i-1})(\bar{y}_i - \bar{y}_{i-1}) \qquad (\alpha > 0) \tag{12}$$

where $s, \bar{s} \in S_i(t_0, \ldots, t_m)$ and $y_i, \bar{y}_i$ are given by (2) and (3) or (6) and (7). Observe that (12) is actually a discrete Sobolev inner product approximating the continuous Sobolev inner product. Moreover, the constant α in (12) amounts to a scaling of the discrete Sobolev norm and will be important in our applications. We are now in a position to state our finite dimensional versions of problem (2) in Chapter 4. Specifically, given the random sample $x_1, \ldots, x_n$ consider the following constrained optimization problems.

(13) Maximize $L_h^0(y_0, \ldots, y_{m-1}) = \prod_{i=1}^{n} s_0(x_i) \exp\left[-\alpha h^{-1} \sum_{i=1}^{m} (y_i - y_{i-1})^2\right];$

subject to

$$h \sum_{i=0}^{m-1} y_i = 1 \quad \text{and} \quad y_i \ge 0, \qquad i = 0, \ldots, m-1,$$

where s_0 is given by (2) and $y_m = 0$.

(14) Maximize $L_h^1(y_1, \ldots, y_{m-1}) = \prod_{i=1}^{n} s_1(x_i) \exp\left[-\alpha h^{-1} \sum_{i=1}^{m} (y_i - y_{i-1})^2\right];$

subject to

$$h \sum_{i=1}^{m} y_i = 1 \qquad \text{and} \qquad y_i \geq 0, \qquad i = 1, \ldots, m-1$$

where s_1 is given by (6) and $y_0 = y_m = 0$.
The constant spline (member of $S_0(t_0, \ldots, t_m)$) corresponding to the solution of problem (13) and the linear spline (member of $S_1(t_0, \ldots, t_m)$) corresponding to the solution of problem (14) are called *discrete maximum penalized likelihood estimates* (DMPLE).

Theorem 1. The DMPLE exist and are unique.

Proof. It is a straightforward matter to show that (12) is actually an inner product on $S_i(t_0, \ldots, t_m)$, $i = 0, 1$. Moreover, since these inner product spaces are finite dimensional they must be reproducing kernel Hilbert spaces and integration must be continuous (see Appendix I). The theorem now follows from Theorem 1 of Chapter 4. ∎

5.2. Consistency Properties of the DMPLE

The following theorem is actually very satisfying, since it shows that, in contrast to the histogram described in Chapter 3, the DMPLE are dimensionally stable. From this point of view, we should think of DMPLE as "stable histograms." The proof of the theorem is quite lengthy and will not be given. The interested reader is referred to Scott, Tapia, and Thompson [7], where a detailed account can be found.

Theorem 2. Let h be the size of the mesh used to obtain the DMPLE guaranteed by Theorem 1. Then the constant spline DMPLE converges in sup norm as $h \to 0$ to the quadratic monospline MPLE guaranteed by Theorem 2 of Chapter 4 for $H_0^1(a, b)$. Furthermore, the linear spline DMPLE converges in $H_0^1(a, b)$ norm as $h \to 0$ to this same monospline MPLE.

We now turn to the very important question of consistency of the DMPLE. Let s_h denote the constant spline DMPLE given by problem (13) for a particular value of h. Recall that $\bar{f}$ is given by (1) and $x_1, \ldots, x_n$ is our random sample.

Theorem 3. Consider the constant spline DMPLE, with the number of mesh partitions given by $m = [Cn^q]$ (so $h = 0(n^{-q})$) where $C > 0$, $0 <$

$q < \frac{1}{4}$ and $[d]$ denotes the integer closest to d. Suppose that f is continuous on (a, b). Then for $x \epsilon (a, b)$

$$\lim_{n \to \infty} s_h(x) = \bar{f}(x) \text{ almost surely (a.s.).}$$

Proof. For the sample $x_1, \ldots, x_n$, let

$$\begin{aligned} v_0 &= \text{number of samples in } (-\infty, a) \\ v_k &= \text{number of samples in } [t_{k-1}, t_k), \qquad k = 1, \ldots, m \\ v_{m+1} &= \text{number of samples in } [b, \infty). \end{aligned}$$

Then $\sum_{k=0}^{m+1} v_k = n$ and we define $n' = \sum_{k=1}^{m} v_k$. For this particular situation problem (13) is equivalent to

$$\text{(15)} \quad \text{maximize } \log(L_h^0) = \sum_1^m v_k \log(y_k) - \alpha h^{-1} \sum_0^{m-1} (y_{k+1} - y_k)^2$$

subject to

$$h \sum_0^{m-1} y_k = 1 \qquad \text{and} \qquad y_k \geq 0, \qquad k = 0, \ldots, m-1.$$

From the theory of Lagrange multipliers for equality and inequality constraints there exist multipliers $\lambda, \mu_0, \ldots, \mu_{m-1}$ such that for the solution of (15) we must have

$$\text{(16)} \quad \frac{\partial[\log(L_h^0)]}{\partial y_i} + \frac{\lambda \partial \left[h \sum_0^{m-1} y_k - 1 \right]}{\partial y_i} + \mu_i = 0, \qquad i = 0, \ldots, m-1,$$

and the complementarity condition

$$\text{(17)} \qquad \mu_i y_i = 0, \qquad i = 0, \ldots, m-1.$$

Condition (16) becomes

$$\text{(18)} \qquad \mu_i + \frac{v_i}{y_i} + \frac{2\alpha}{h} \delta^2 y_i + \lambda = 0, \qquad i = 1, \ldots, m-1,$$

where

$$\text{(19)} \qquad \delta^2 y_i = y_{i+1} - 2y_i + y_{i-1}.$$

Multiplying (18) by y_i, summing over i, using (17), recalling the definition of n' and the first constraint in problem (15) we obtain the following expression for the multiplier λ

$$\lambda = -n'h - 2\alpha \sum_{i=1}^{m-1} y_i\, \delta^2 y_i. \tag{20}$$

Substituting (20) into (18) and dividing by nh our necessary conditions become

$$\frac{\mu_i}{nh} + \frac{v_i}{nhy_i} + \frac{2\alpha}{nh^2}\delta^2 y_i - \frac{n'}{n} - \frac{2\alpha}{nh}\sum_{j=1}^{m-1} y_j\, \delta^2 y_j = 0. \tag{21}$$

From the constraints of problem (15) we know that $0 \le y_j \le \frac{1}{h}$ so that

$$\delta^2 y_j = 0\left(\frac{1}{h}\right) \tag{22}$$

and

$$\sum_{j=1}^{m-1} y_j\, \delta^2 y_j = 0\left(\frac{1}{h^3}\right). \tag{23}$$

Using (22) and (23) in (21), we obtain for our necessary condition

$$\frac{v_i}{nhy_i} - \frac{n'}{n} + 0\left(\frac{1}{nh^4}\right) = 0, \qquad i = 0, \ldots, m-1. \tag{24}$$

We are really interested in the interval containing x, and in this context the subscript i is misleading, since it depends on x and n. Consequently, let us introduce the notation $[x:n]$ for the mesh interval containing x for a particular value of n and let $v[x:n]$ denote the number of samples in $[x:n]$. Observe that $s_h(x) = y_i$ in this context, so that (24) implies

$$\frac{v[x:n]}{nhs_h(x)} - \frac{n'}{n} + 0\left(\frac{1}{nh^4}\right) = 0. \tag{25}$$

The behavior of the first quantity in (25) is not at all obvious, and we will need the following lemma to complete the proof of the theorem.

Lemma 1. Under the conditions of Theorem 3

$$\lim_{n\to\infty} \frac{v[x:n]}{nh} = f(x) \qquad \text{a.s.}$$

Proof. We shall employ an argument motivated by Section 5.1 of [3]. In order to emphasize the dependence on n, we write $m(n)$ for the number of partitions. In this case the mesh spacing will be $h_n = (b-a)/m(n)$. Given the random sample $\{x_1, \ldots, x_n\}$, define the triangular array of random variables

$$Y_{nj} = \frac{I[x:n](x_j) - p[x:n]}{h_n},$$

where $I[x:n]$ denotes the indicator function of the interval $[x:n]$ and

$$p[x:n] = \int_{[x:n]} f(t)\, dt.$$

Now $\{Y_{nj}: j = 1, \ldots, n\}$ forms an independent sequence for each n, each random variable having zero mean. Also $I[x:n](x_j)$ is a binomial random variable with expectation given by $p[x:n]$. Let

$$S_n = \sum_{j=1}^{n} Y_{nj}.$$

To prove the lemma, we wish to show that $S_n/n \to 0$ almost surely. We have

$$\text{Var}(S_n) = np[x:n](1 - p[x:n])/h_n^2.$$

Hence, by Chebyshev's inequality for any $\epsilon > 0$,

$$(26) \qquad P\{|S_n| > n\epsilon\} < \frac{(1 - p[x:n])p[x:n]}{nh_n^2\epsilon^2}.$$

From the Borel–Cantelli lemma, a sufficient condition that S_n/n converge almost surely to zero is that

$$\sum_{n=1}^{\infty} P\{|S_n| > n\epsilon\} < +\infty.$$

From the Mean Value Theorem $p[x:n] = h_n f(\xi_n)$, where $\xi_n \in [x:n]$. Combining this fact with (26) leads to

$$(27) \qquad \sum_{n=1}^{\infty} P\{|S_n| > n\epsilon\} \le \sum_{n=1}^{\infty} \frac{[1 - h_n f(\xi_n)]f(\xi_n)}{nh_n\epsilon^2}.$$

Since the denominator of the summand on the right-hand side of (27) is $0(n^{1-q})$, this latter series diverges. However, if we consider the subsequence $\{n^2\}$, then $h_{n^2}n^2 = 0(n^{2-2q})$ with $0 < q < \frac{1}{4}$. Hence the majorizing series in (27) is convergent, and thus, by the Borel–Cantelli lemma

$$\frac{S_{n^2}}{n^2} \to 0 \qquad \text{a.s.}$$

Next, let

$$D_n = \max\{|S_k - S_{n^2}| : n^2 \le k < (n+1)^2\}; \quad n = 1, 2, \ldots .$$

We have for $n^2 \le k < (n+1)^2$

$$E[(S_k - S_{n^2})^2] = \sum_{j=1}^{n^2} (E[Y_{kj}^2] + E[Y_{n^2j}^2] - 2E[Y_{kj}Y_{n^2}]) + \sum_{j=n^2+1}^{k} E[Y_{kj}^2]. \tag{28}$$

In order to deal with (28), we make the following important observations. A straightforward application of the Mean Value Theorem to the function $g(x) = x^{2q}$ shows that

$$Cn^{2q} - C(n+1)^{2q} = 0(n^{2q-1}).$$

However, since $m(n) = [Cn^q]$ is integer valued, we see that, with probability $0(n^{2q-1})$, $m(n^2) \ne m(k)$ for $n^2 \le k < (n+1)^2$ and it follows that, with probability $0(n^{2q-1})$, $[x:n^2] \ne [x:k]$ for k in this range.

A somewhat lengthy calculation using the Mean Value Theorem and the definitions shows that for $n^2 \le k < (n+1)^2$

$$E[Y_{kj^2}] = \frac{f(x)}{h_k} + 0(n^{2q}),$$

$$E[Y_{n^2j}^2] = \frac{f(x)}{h_{n^2}} + 0(n^{2q}),$$

and

$$E[Y_{kj}Y_{n^2}] = \frac{f(x)h_{kn^2}}{h_k h_{n^2}} + 0(n^{2q}),$$

where h_{kn^2} denotes the width of the interval $[x:k] \cap [x:n^2]$. Since $h_{n^2} \ne h_k$, with probability $0(n^{2q-1})$, it follows that (28) can be written

$$E[(S_k - S_{n^2})^2] = \sum_{j=1}^{n^2} 0(n^{2q})0(n^{2q-1}) + \sum_{j=n^2+1}^{k} \left[\frac{f(x)}{h_k} + 0(n^{2q})\right] = 0(n^{4q+1}). \tag{29}$$

Now

$$E[D_n^2] \le \sum_{j=n^2}^{(n+1)^2} E[(S_k - S_{n^2})^2] = 0(n^{4q+2}).$$

So from Chebyshev's inequality, we have

$$P[D_n > n^2\epsilon] \le 0(n^{-2+2q}).$$

Hence, $\sum_{n=1}^{\infty} P[D_n > n^2\epsilon] < \infty$, since $0 < q < \frac{1}{4}$.

But, then by the Borel–Cantelli lemma

$$\frac{D^n}{n^2} \to 0 \qquad \text{a.s.} \tag{30}$$

Using the fact that for $n^2 \le k < (n+1)^2$

$$\left|\frac{S_k}{k}\right| \le \frac{|S_{n^2}| + D_n}{n^2},$$

we see that

$$\frac{S_n}{n} \equiv \frac{v[x:n]}{nh_n} - \frac{p[x:n]}{h_n} \to 0 \qquad \text{a.s.}$$

By the continuity of f we see that

$$\lim_{n\to\infty} \frac{p[x:n]}{h_n} = f(x),$$

so that, finally

$$\frac{v[x:n]}{nh_n} \to f(x) \qquad \text{a.s.} \tag{31}$$

This establishes the lemma. ■

We now return to the proof of the theorem. The strong law of large numbers also implies that

$$\lim_{n\to\infty} \frac{n'}{n} = \int_a^b f(t)\,dt \qquad \text{a.s.} \tag{32}$$

Observing that $nh^4 \to \infty$ as $n \to \infty$, using (31) and (32) and letting $n \to \infty$ in (25) we establish the theorem. ■

Remark. If in Theorem 3 we considered a discrete k-th order derivative in the penalty term of our criterion functional, then the analogous consistency result would require $0 < q < (2k+2)^{-1}$.

Remark. Numerical experience indicates that the requirement $0 < q < (2k+2)^{-1}$ is an artifact of the method of proof and is not needed for consistency.

5.3. Numerical Implementation and Monte Carlo Simulation

The numerical examples in this section were calculated using the linear spline DMPLE which is obtained from the penalized likelihood, using a discrete second derivative in the penalty term. We made this choice mainly because the linear spline gives a continuous estimate, and the use of the second derivative in the penalty term seems to give "fuller" estimates. Specifically, given a random sample $x_1, \ldots, x_n$, an interval (a, b) a positive scalar α and a positive integer m we let

$$h = (b - a)/m, \tag{33}$$

$$t_i = a + ih, \qquad i = 0, \ldots, m, \tag{34}$$

$$p_0 = p_m = 0 \tag{35}$$

and solve the $m - 1$ dimensional constrained optimization problem

$$\text{maximize}\, L(p_1, \ldots, p_{m-1}) = \sum_{i=1}^{n} \log p(x_i) - \frac{\alpha}{h} \sum_{k=1}^{m-1} [p_{k+1} - 2p_k + p_{k-1}]^2 \tag{36}$$

subject to

$$h \sum_{k=1}^{m-1} p_k = 1 \qquad \text{and} \qquad p_k \geq 0, \qquad k = 1, \ldots, m-1,$$

where

$$p(t) = \begin{cases} p_k + \dfrac{p_{k+1} - p_k}{h}(t - t_k) & t \in [t_k, t_{k+1}) \\ 0 & t \notin (t_0, t_m). \end{cases} \tag{37}$$

A computer program has been written to solve problem (36) and has been incorporated in the IMSL program library [4]. This program uses the modification of Newton's method due to Tapia [9], [10] and is described in Appendix II.1. The nonnegativity constraints are handled as in Appendix II.2. This algorithm takes advantage of the special banded structure of the Hessian matrix of the Lagrangian functional for problem (36). Thus the amount of work turns out to be $O(m)$ instead of the expected $O(m^3)$ per iteration. Notice that the sample size n enters only in evaluating the gradient and Hessian and not in the matrix inversions. No initial guesses for Newton's method are required. A bootstrap algorithm is employed. First the problem is solved for $m = 7$, an easy problem to solve. This estimate provides initial guesses for $m = 11$, then for $m = 19$, and so on. The bootstrap algorithm takes advantage of the robustness of the DMPLE with respect to h.

The choice of α is very important and more difficult than the selection of the mesh spacing h. Asymptotically, of course, any positive value gives consistent results. For finite sample sizes; however, the choice is critical. The α parameter is analogous to the kernel scaling parameter h (for a discussion on kernel estimators see Section 2.5 of Chapter 2). Unfortunately (as is customary) we have also denoted the mesh spacing by h. Hopefully, this will not lead to confusion in the sequel where we will be discussing the robustness of the penalty parameter α and the kernel scaling parameter h. For a fixed sample $x_1, \ldots, x_n$ there are values of α and h which give the "best" approximations for the DMPLE and the kernel estimator respectively. For values smaller than "best," the corresponding estimates peak sharply at the sample points. On the other hand, values larger than "best" correspond to depressed and oversmoothed estimates. For the kernel estimator, asymptotically optimal choices for h can be derived; however, knowledge of the true (unknown) density is required. Scott, Tapia, and Thompson [8] have developed an iterative approach for estimating the optimal h. This approach was described in part in Section 2.5 of Chapter 2. Wahba [11] has also given considerable insight into the problem. In practice, an interactive approach is often used, where the smallest value of α (or h for the kernel estimator) is chosen that reveals fine structure without "too much" oscillatory behavior (consistent with prior knowledge). However, we note that the DMPLE is robust with respect to the choice of α for a choice of mesh. It is bounded from above and below and has finite support (the kernel estimator can approximate Dirac spikes and, on the other extreme, a very diffuse uniform density).

We next give several examples of density estimation problems which were solved using the DMPLE given by the solution of problem (36). Following these examples we present a Monte Carlo simulation to show that the DMPLE compares quite favorably with kernel estimators and is somewhat more robust.

Let us first consider the problem of estimating the bimodal density

$$f(x) = .75\frac{1}{\sqrt{2\pi}}\exp\left[-\frac{1}{2}(x+1.5)^2\right] + .25\left(\frac{3}{\sqrt{2\pi}}\exp\left[-\frac{9}{2}(x-1.5)^2\right]\right). \tag{38}$$

In Figures 5.1 through 5.6, we show the effect of varying α over 5 $\log_{10}$ units for the DMPLE solution to problem (36) when $n = 300$. We note that discrete maximum penalized likelihood estimation lends itself particularly well to an evolutionary approach. We start with an oversmoothed α of 1,000 in Figure 5.1. Since the graph in Figure 5.2, using $\alpha = 100$, is significantly different, and no high frequency wiggles have yet appeared, we reduce α to 10 in Figure 5.3. Still no high frequency wiggles are in evidence. In Figure 5.4,

```
BIMODAL NORMAL WITH MEANS = -+  1.50 VARIANCE OF LEFT = 1 WITH WEIGHT AND VARIANCE OF RIGHT =  0.2500  0.1111
        SAMPLE SIZE =  300 WITH   75 SAMPLES ON RIGHT     VS.
        DISCRETIZED MAXIMUM LIKELIHOOD PENALIZED ESTIMATE
        WITH WEIGHING PARAMETER ALPHA          0.1000000000D+04
        41 MESH POINTS FROM     -5.20000 TO    2.80000
        DISCRETE MESH INTERVAL    0.20000
        INTEGRATED MEAN SQUARE ERROR           0.1686135855D-01
        INTEGRATED SQUARE ERROR                0.9732630984D-01
        MAXIMUM ABSOLUTE DIFFERENCE            0.2832363062D+00
        LOG LIKELIHOOD TERM                   -0.7290920059D+03
        LOG PENALTY TERM                      -0.1759336143D+03
```

ABSCISSA	ORDINATE *	ORDINATE O	
-5.00D+00	6.55D-04	0.0	
-4.80D+00	1.29D-03	9.57D-03	
-4.60D+00	2.45D-03	2.71D-02	
-4.40D+00	4.46D-03	5.09D-02	
-4.20D+00	7.82D-03	7.95D-02	1.
-4.00D+00	1.31D-02	1.11D-01	2.
-3.80D+00	2.12D-02	1.45D-01	2.
-3.60D+00	3.30D-02	1.79D-01	2.
-3.40D+00	4.92D-02	2.12D-01	1.
-3.20D+00	7.05D-02	2.43D-01	6.
-3.00D+00	9.71D-02	2.71D-01	2.
-2.80D+00	1.29D-01	2.95D-01	9.
-2.60D+00	1.63D-01	3.13D-01	9.
-2.40D+00	2.00D-01	3.26D-01	9.
-2.20D+00	2.34D-01	3.32D-01	9.
-2.00D+00	2.64D-01	3.32D-01	9.
-1.80D+00	2.86D-01	3.24D-01	9.
-1.60D+00	2.98D-01	3.09D-01	9.
-1.40D+00	2.98D-01	2.88D-01	9.
-1.20D+00	2.86D-01	2.60D-01	9.
-1.00D+00	2.64D-01	2.28D-01	9.
-8.00D-01	2.34D-01	1.92D-01	9.
-6.00D-01	2.00D-01	1.53D-01	9.
-4.00D-01	1.63D-01	1.13D-01	7.
-2.00D-01	1.29D-01	7.54D-02	9.
0.0	9.72D-02	4.18D-02	5.
2.00D-01	7.07D-02	1.54D-02	4.
4.00D-01	5.05D-02	3.64D-19	20
6.00D-01	4.08D-02	8.73D-04	10
8.00D-01	5.42D-02	1.73D-03	70
1.00D+00	1.10D-01	1.03D-17	80
1.20D+00	2.07D-01	9.31D-03	9.
1.40D+00	2.91D-01	1.79D-02	9.
1.60D+00	2.88D-01	2.07D-02	9.
1.80D+00	2.01D-01	1.66D-02	6.
2.00D+00	9.78D-02	8.11D-03	5.
2.20D+00	3.33D-02	7.37D-19	10
2.40D+00	7.96D-03	6.64D-04	10
2.60D+00	1.36D-03	0.0	
2.80D+00	1.78D-04	0.0	
3.00D+00	2.40D-05	0.0	
3.20D+00	5.45D-06	0.0	
3.40D+00	1.86D-06	0.0	
3.60D+00	6.74D-07	0.0	
3.80D+00	2.38D-07	0.0	
4.00D+00	8.C8D-08	0.0	
4.20D+00	2.64D-08	0.0	
4.40D+00	8.26D-09	0.0	
4.60D+00	2.49D-09	0.0	
4.80D+00	7.20D-10	0.0	
5.00D+00	2.00D-10	0.0	

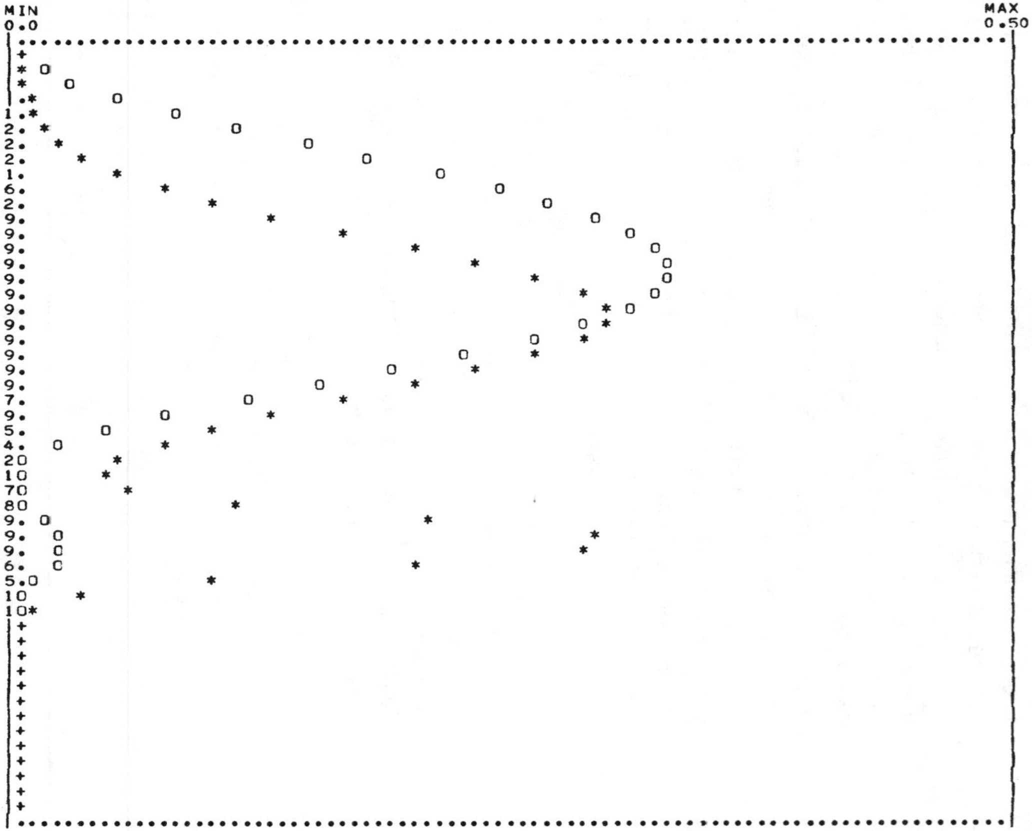

END OF PLOT.

```
SAMPLE SIZE =  300 WITH   75 SAMPLES ON RIGHT      VS.
DISCRETIZED MAXIMUM LIKELIHOOD PENALIZED ESTIMATE
WITH WEIGHING PARAMETER ALPHA          0.1000000000D+03
41 MESH POINTS FROM      -5.20000  TO  2.80000
DISCRETE MESH INTERVAL     0.20000
INTEGRATED MEAN SQUARE ERROR           0.1059614190D-01
INTEGRATED SQUARE ERROR                0.4773349040D-01
MAXIMUM ABSOLUTE DIFFERENCE            0.2316315958D+00
LOG LIKELIHOOD TERM                   -0.5875622343D+03
LOG PENALTY TERM                      -0.5877577822D+02
```

ABSCISSA	ORDINATE *	ORDINATE O
-5.00D+00	6.55D-04	0.0
-4.80D+00	1.29D-03	3.46D-26
-4.60D+00	2.45D-03	7.51D-04
-4.40D+00	4.46D-03	5.30D-03
-4.20D+00	7.82D-03	1.52D-02
-4.00D+00	1.31D-02	3.07D-02
-3.80D+00	2.12D-02	5.12D-02
-3.60D+00	3.30D-02	7.64D-02
-3.40D+00	4.92D-02	1.06D-01
-3.20D+00	7.05D-02	1.41D-01
-3.00D+00	9.71D-02	1.79D-01
-2.80D+00	1.29D-01	2.18D-01
-2.60D+00	1.63D-01	2.57D-01
-2.40D+00	2.00D-01	2.93D-01
-2.20D+00	2.34D-01	3.23D-01
-2.00D+00	2.64D-01	3.45D-01
-1.80D+00	2.86D-01	3.57D-01
-1.60D+00	2.98D-01	3.60D-01
-1.40D+00	2.98D-01	3.51D-01
-1.20D+00	2.86D-01	3.32D-01
-1.00D+00	2.64D-01	3.03D-01
-8.00D-01	2.34D-01	2.65D-01
-6.00D-01	2.00D-01	2.19D-01
-4.00D-01	1.63D-01	1.69D-01
-2.00D-01	1.29D-01	1.17D-01
0.0	9.72D-02	6.81D-02
2.00D-01	7.07D-02	2.65D-02
4.00D-01	5.05D-02	1.69D-18
6.00D-01	4.08D-02	4.05D-03
8.00D-01	5.42D-02	5.69D-03
1.00D+00	1.10D-01	3.32D-17
1.20D+00	2.07D-01	2.99D-02
1.40D+00	2.91D-01	6.16D-02
1.60D+00	2.88D-01	8.02D-02
1.80D+00	2.01D-01	8.11D-02
2.00D+00	9.78D-02	6.67D-02
2.20D+00	3.33D-02	4.29D-02
2.40D+00	7.96D-03	1.77D-02
2.60D+00	1.36D-03	0.0
2.80D+00	1.78D-04	0.0
3.00D+00	2.40D-05	0.0
3.20D+00	5.45D-06	0.0
3.40D+00	1.86D-06	0.0
3.60D+00	6.74D-07	0.0
3.80D+00	2.38D-07	0.0
4.00D+00	8.08D-08	0.0
4.20D+00	2.64D-08	0.0
4.40D+00	8.26D-09	0.0
4.60D+00	2.49D-09	0.0
4.80D+00	7.20D-10	0.0
5.00D+00	2.00D-10	0.0

END OF PLOT.

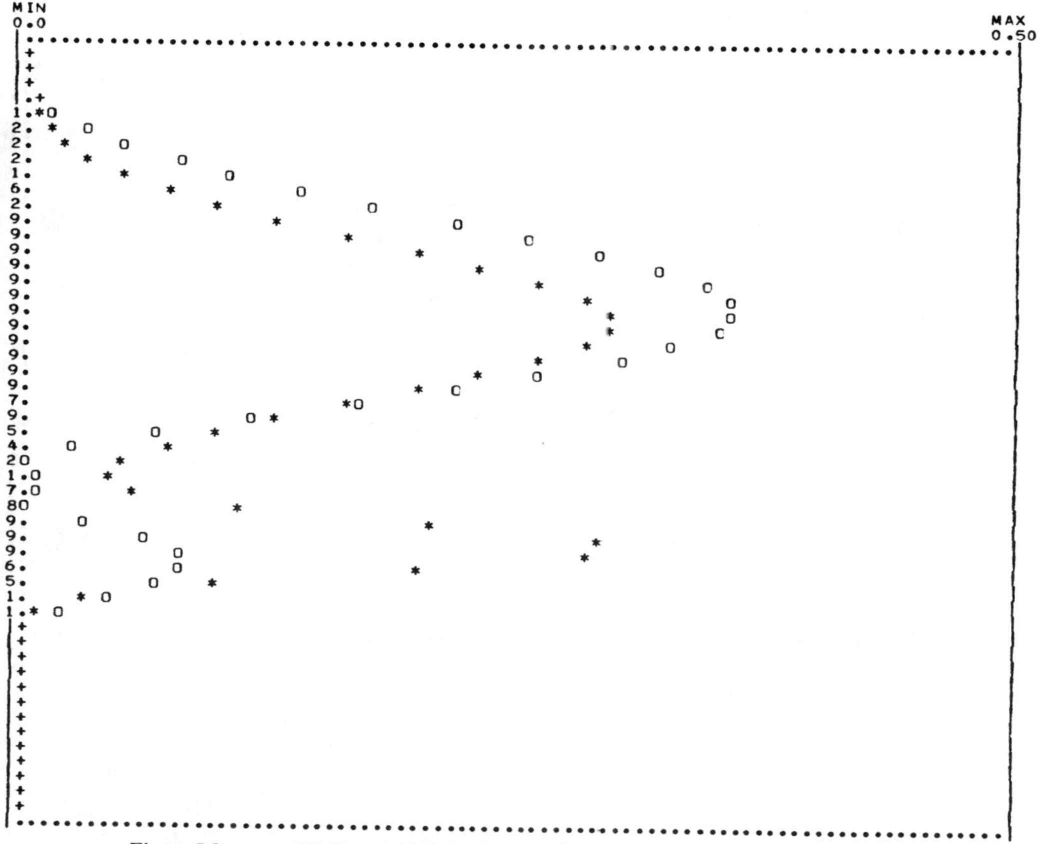

Figure 5.2. $n = 300$ Bimodal DMPLE $\alpha = 10^2$.

```
BIMODAL NORMAL WITH MEANS = -+  1.50 VARIANCE OF LEFT = 1 WITH WEIGHT AND VARIANCE OF RIGHT =   0.2500   0.1111
        SAMPLE SIZE =  300 WITH   75 SAMPLES ON RIGHT     VS.
        DISCRETIZED MAXIMUM LIKELIHOOD PENALIZED ESTIMATE
        WITH WEIGHING PARAMETER ALPHA                0.1000000000D+02
        41 MESH POINTS FROM      -5.20000  TO        2.80000
        DISCRETE MESH INTERVAL    0.20000
        INTEGRATED MEAN SQUARE ERROR                 0.1256887348D-02
        INTEGRATED SQUARE ERROR                      0.5194723592D-02
        MAXIMUM ABSOLUTE DIFFERENCE                  0.1038874681D+00
        LOG LIKELIHOOD TERM                         -0.5007092481D+03
        LOG PENALTY TERM                            -0.7674434095D+01
```

ABSCISSA	ORDINATE *	O	MIN 0.0
-5.00D+00	6.55D-04	0.0	
-4.80D+00	1.29D-03	2.13D-34	
-4.60D+00	2.45D-03	3.86D-26	
-4.40D+00	4.46D-03	8.89D-20	
-4.20D+00	7.82D-03	6.70D-03	1.
-4.00D+00	1.31D-02	1.90D-02	2.
-3.80D+00	2.12D-02	2.97D-02	2.
-3.60D+00	3.30D-02	3.33D-02	2.
-3.40D+00	4.92D-02	4.01D-02	1.
-3.20D+00	7.05D-02	5.70D-02	6.
-3.00D+00	9.71D-02	8.44D-02	2.
-2.80D+00	1.29D-01	1.25D-01	9.
-2.60D+00	1.63D-01	1.72D-01	9.
-2.40D+00	2.00D-01	2.17D-01	9.
-2.20D+00	2.34D-01	2.56D-01	9.
-2.00D+00	2.64D-01	2.87D-01	9.
-1.80D+00	2.86D-01	3.04D-01	9.
-1.60D+00	2.98D-01	3.09D-01	9.
-1.40D+00	2.98D-01	3.03D-01	9.
-1.20D+00	2.86D-01	2.90D-01	9.
-1.00D+00	2.64D-01	2.71D-01	9.
-8.00D-01	2.34D-01	2.42D-01	9.
-6.00D-01	2.00D-01	2.07D-01	9.
-4.00D-01	1.63D-01	1.71D-01	7.
-2.00D-01	1.29D-01	1.35D-01	9.
0.0	9.72D-02	1.02D-01	5.
2.00D-01	7.07D-02	7.42D-02	4.
4.00D-01	5.05D-02	5.67D-02	2.
6.00D-01	4.08D-02	5.95D-02	1.
8.00D-01	5.42D-02	8.65D-02	7.
1.00D+00	1.10D-01	1.32D-01	8.
1.20D+00	2.07D-01	1.77D-01	9.
1.40D+00	2.91D-01	2.01D-01	9.
1.60D+00	2.88D-01	1.96D-01	9.
1.80D+00	2.01D-01	1.62D-01	6.
2.00D+00	9.78D-02	1.12D-01	5.
2.20D+00	3.33D-02	6.07D-02	1.
2.40D+00	7.96D-03	2.16D-02	1.
2.60D+00	1.36D-03	0.0	
2.80D+00	1.78D-04	0.0	
3.00D+00	2.40D-05	0.0	
3.20D+00	5.45D-06	0.0	
3.40D+00	1.86D-06	0.0	
3.60D+00	6.74D-07	0.0	
3.80D+00	2.38D-07	0.0	
4.00D+00	8.08D-08	0.0	
4.20D+00	2.64D-08	0.0	
4.40D+00	8.26D-09	0.0	
4.60D+00	2.49D-09	0.0	
4.80D+00	7.20D-10	0.0	
5.00D+00	2.00D-10	0.0	

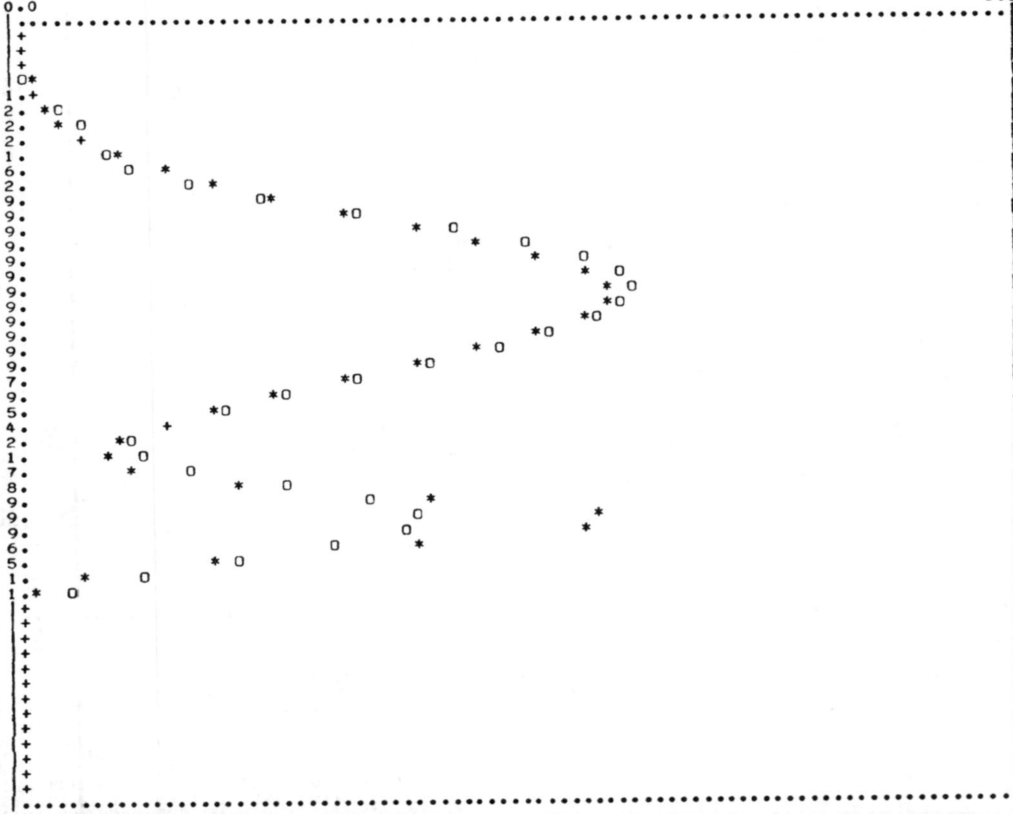

END OF PLOT.

```
SAMPLE SIZE =  300 WITH   75 SAMPLES ON RIGHT      VS.
DISCRETIZED MAXIMUM LIKELIHOOD PENALIZED ESTIMATE
WITH WEIGHING PARAMETER ALPHA          0.1000000000D+01
41 MESH POINTS FROM      -5.20000  TO   2.80000
DISCRETE MESH INTERVAL    0.20000
INTEGRATED MEAN SQUARE ERROR           0.4046088973D-03
INTEGRATED SQUARE ERROR                0.2044613751D-02
MAXIMUM ABSOLUTE DIFFERENCE            0.4449957615D-01
LOG LIKELIHOOD TERM                   -0.4940368135D+03
LOG PENALTY TERM                      -0.3057903913D+01
```

ABSCISSA	ORDINATE *	ORDINATE O
-5.00D+00	6.55D-04	0.0
-4.80D+00	1.29D-03	3.25D-66
-4.60D+00	2.45D-03	8.23D-50
-4.40D+00	4.46D-03	1.93D-39
-4.20D+00	7.82D-03	5.50D-04
-4.00D+00	1.31D-02	2.31D-02
-3.80D+00	2.12D-02	4.10D-02
-3.60D+00	3.30D-02	2.89D-02
-3.40D+00	4.92D-02	3.09D-02
-3.20D+00	7.05D-02	4.95D-02
-3.00D+00	9.71D-02	6.34D-02
-2.80D+00	1.29D-01	1.09D-01
-2.60D+00	1.63D-01	1.79D-01
-2.40D+00	2.00D-01	2.21D-01
-2.20D+00	2.34D-01	2.59D-01
-2.00D+00	2.64D-01	3.00D-01
-1.80D+00	2.86D-01	3.12D-01
-1.60D+00	2.98D-01	3.07D-01
-1.40D+00	2.98D-01	2.85D-01
-1.20D+00	2.86D-01	2.70D-01
-1.00D+00	2.64D-01	2.72D-01
-8.00D-01	2.34D-01	2.43D-01
-6.00D-01	2.00D-01	1.99D-01
-4.00D-01	1.63D-01	1.66D-01
-2.00D-01	1.29D-01	1.37D-01
0.0	9.72D-02	1.08D-01
2.00D-01	7.07D-02	8.06D-02
4.00D-01	5.05D-02	4.28D-02
6.00D-01	4.08D-02	2.59D-02
8.00D-01	5.42D-02	4.15D-02
1.00D+00	1.10D-01	1.25D-01
1.20D+00	2.07D-01	2.18D-01
1.40D+00	2.91D-01	2.65D-01
1.60D+00	2.88D-01	2.51D-01
1.80D+00	2.01D-01	1.81D-01
2.00D+00	9.78D-02	1.04D-01
2.20D+00	3.33D-02	4.35D-02
2.40D+00	7.96D-03	1.73D-02
2.60D+00	1.36D-03	0.0
2.80D+00	1.78D-04	0.0
3.00D+00	2.40D-05	0.0
3.20D+00	5.45D-06	0.0
3.40D+00	1.86D-06	0.0
3.60D+00	6.74D-07	0.0
3.80D+00	2.38D-07	0.0
4.00D+00	8.08D-08	0.0
4.20D+00	2.64D-08	0.0
4.40D+00	8.26D-09	0.0
4.60D+00	2.49D-09	0.0
4.80D+00	7.20D-10	0.0
5.00D+00	2.00D-10	0.0

END OF PLOT.

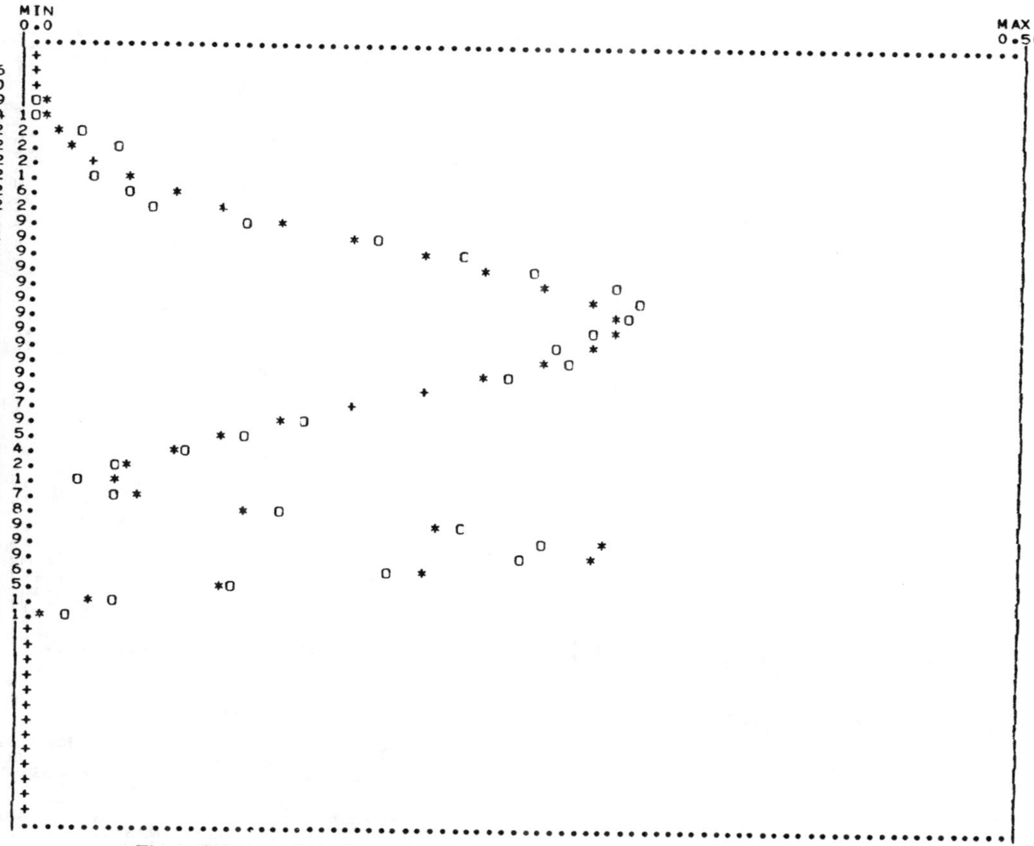

Figure 5.4. $n = 300$ Bimodal DMPLE $\alpha = 1$.

```
BIMODAL NORMAL WITH MEANS = -+  1.50 VARIANCE OF LEFT = 1 WITH WEIGHT AND VARIANCE OF RIGHT =   0.2500  0.1111
        SAMPLE SIZE =  300 WITH   75 SAMPLES ON RIGHT        VS.
        DISCRETIZED MAXIMUM LIKELIHOOD PENALIZED ESTIMATE
        WITH WEIGHING PARAMETER ALPHA              0.1000000000D+00
        41 MESH POINTS FROM       -5.20000  TO     2.80000
        DISCRETE MESH INTERVAL     0.20000
        INTEGRATED MEAN SQUARE ERROR               0.8118148898D-03
        INTEGRATED SQUARE ERROR                    0.4299430667D-02
        MAXIMUM ABSOLUTE DIFFERENCE                0.6367941241D-01
        LOG LIKELIHOOD TERM                       -0.4895295963D+03
        LOG PENALTY TERM                          -0.2220424538D+01
```

ABSCISSA	ORDINATE *	ORDINATE 0
-5.00D+00	6.55D-04	0.0
-4.80D+00	1.29D-03	3.73D-52
-4.60D+00	2.45D-03	2.53D-50
-4.40D+00	4.46D-03	1.49D-45
-4.20D+00	7.82D-03	1.59D-02
-4.00D+00	1.31D-02	1.42D-33
-3.80D+00	2.12D-02	5.57D-02
-3.60D+00	3.30D-02	2.11D-02
-3.40D+00	4.92D-02	2.75D-02
-3.20D+00	7.05D-02	6.49D-02
-3.00D+00	9.71D-02	4.50D-02
-2.80D+00	1.29D-01	8.90D-02
-2.60D+00	1.63D-01	2.15D-01
-2.40D+00	2.00D-01	2.13D-01
-2.20D+00	2.34D-01	2.22D-01
-2.00D+00	2.64D-01	3.24D-01
-1.80D+00	2.86D-01	3.15D-01
-1.60D+00	2.98D-01	3.24D-01
-1.40D+00	2.98D-01	2.61D-01
-1.20D+00	2.86D-01	2.22D-01
-1.00D+00	2.64D-01	3.10D-01
-8.00D-01	2.34D-01	2.56D-01
-6.00D-01	2.00D-01	1.74D-01
-4.00D-01	1.63D-01	1.62D-01
-2.00D-01	1.29D-01	1.41D-01
0.0	9.72D-02	1.03D-01
2.00D-01	7.07D-02	9.32D-02
4.00D-01	5.05D-02	3.73D-02
6.00D-01	4.08D-02	2.70D-02
8.00D-01	5.42D-02	1.17D-16
1.00D+00	1.10D-01	1.41D-01
1.20D+00	2.07D-01	2.49D-01
1.40D+00	2.91D-01	2.93D-01
1.60D+00	2.88D-01	2.82D-01
1.80D+00	2.01D-01	1.62D-01
2.00D+00	9.78D-02	1.02D-01
2.20D+00	3.33D-02	3.23D-02
2.40D+00	7.96D-03	2.02D-02
2.60D+00	1.36D-03	0.0
2.80D+00	1.78D-04	0.0
3.00D+00	2.40D-05	0.0
3.20D+00	5.45D-06	0.0
3.40D+00	1.86D-06	0.0
3.60D+00	6.74D-07	0.0
3.80D+00	2.38D-07	0.0
4.00D+00	8.08D-08	0.0
4.20D+00	2.64D-08	0.0
4.40D+00	8.26D-09	0.0
4.60D+00	2.49D-09	0.0
4.80D+00	7.20D-10	0.0
5.00D+00	2.00D-10	0.0

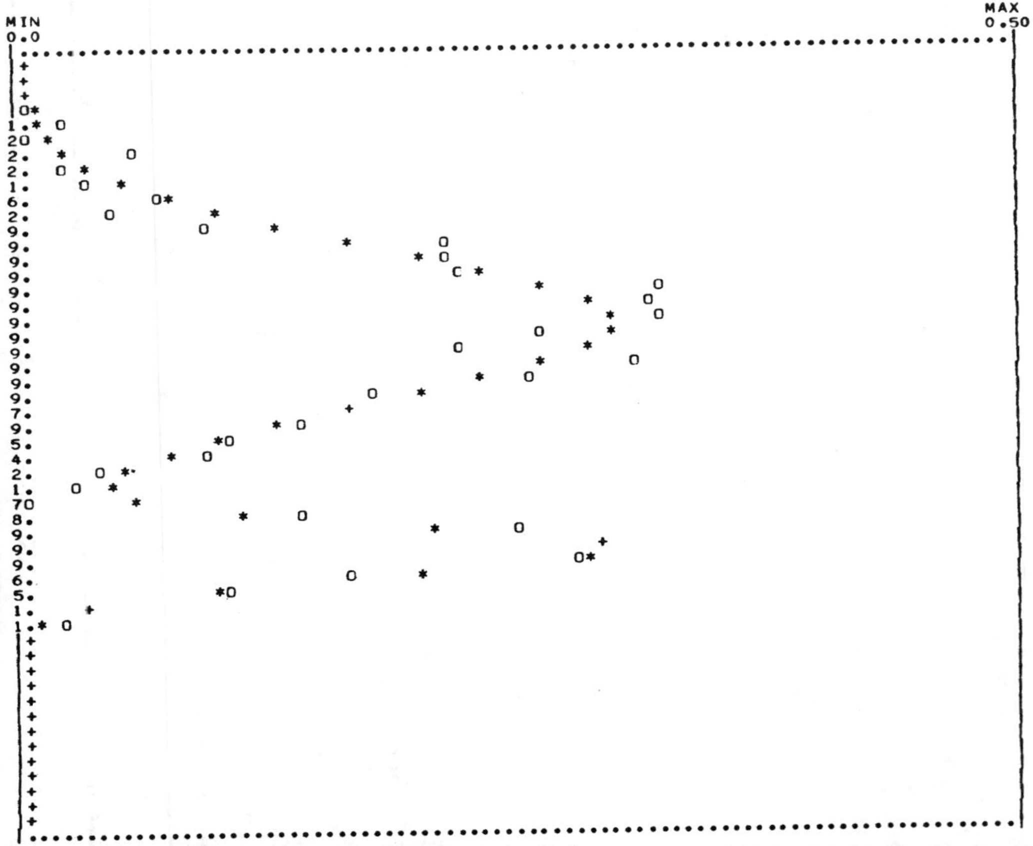

END OF PLOT.

FIG. 5.5. n = 300 Bimodal DMPLE $\alpha = 10^{-1}$

```
BIMODAL NORMAL WITH MEANS = -+  1.50 VARIANCE OF LEFT = 1 WITH WEIGHT AND VARIANCE OF RIGHT =   0.2500   0.1111
         SAMPLE SIZE =  300 WITH   75 SAMPLES ON RIGHT    VS.
         DISCRETIZED MAXIMUM LIKELIHOOD PENALIZED ESTIMATE
         WITH WEIGHING PARAMETER ALPHA           0.1000000000D-01
         41 MESH POINTS FROM       -5.20000  TO   2.80000
         DISCRETE MESH INTERVAL     0.20000
         INTEGRATED MEAN SQUARE ERROR            0.1900463957D-02
         INTEGRATED SQUARE ERROR                 0.9000829638D-02
         MAXIMUM ABSOLUTE DIFFERENCE             0.1182738981D+00
         LOG LIKELIHOOD TERM                    -0.4871999786D+03
         LOG PENALTY TERM                       -0.8969259766D+00
```

ABSCISSA	ORDINATE *	O
-5.00D+00	6.55D-04	0.0
-4.80D+00	1.29D-03	4.81D-45
-4.60D+00	2.45D-03	1.05D-43
-4.40D+00	4.46D-03	3.26D-42
-4.20D+00	7.82D-03	1.67D-02
-4.00D+00	1.31D-02	8.82D-16
-3.80D+00	2.12D-02	6.21D-02
-3.60D+00	3.30D-02	1.87D-02
-3.40D+00	4.92D-02	2.60D-02
-3.20D+00	7.05D-02	7.26D-02
-3.00D+00	9.71D-02	4.02D-02
-2.80D+00	1.29D-01	7.42D-02
-2.60D+00	1.63D-01	2.49D-01
-2.40D+00	2.00D-01	2.16D-01
-2.20D+00	2.34D-01	1.71D-01
-2.00D+00	2.64D-01	3.76D-01
-1.80D+00	2.86D-01	2.67D-01
-1.60D+00	2.98D-01	3.57D-01
-1.40D+00	2.98D-01	2.64D-01
-1.20D+00	2.86D-01	1.68D-01
-1.00D+00	2.64D-01	3.67D-01
-8.00D-01	2.34D-01	2.47D-01
-6.00D-01	2.00D-01	1.52D-01
-4.00D-01	1.63D-01	1.72D-01
-2.00D-01	1.29D-01	1.44D-01
0.0	9.72D-02	9.31D-02
2.00D-01	7.07D-02	1.03D-01
4.00D-01	5.05D-02	2.86D-02
6.00D-01	4.08D-02	3.20D-02
8.00D-01	5.42D-02	1.24D-16
1.00D+00	1.10D-01	1.49D-01
1.20D+00	2.07D-01	2.53D-01
1.40D+00	2.91D-01	2.74D-01
1.60D+00	2.88D-01	3.23D-01
1.80D+00	2.01D-01	1.20D-01
2.00D+00	9.78D-02	1.18D-01
2.20D+00	3.33D-02	2.52D-02
2.40D+00	7.96D-03	2.17D-02
2.60D+00	1.36D-03	0.0
2.80D+00	1.78D-04	0.0
3.00D+00	2.40D-05	0.0
3.20D+00	5.45D-06	0.0
3.40D+00	1.86D-06	0.0
3.60D+00	6.74D-07	0.0
3.80D+00	2.38D-07	0.0
4.00D+00	8.08D-08	0.0
4.20D+00	2.64D-08	0.0
4.40D+00	8.26D-09	0.0
4.60D+00	2.49D-09	0.0
4.80D+00	7.20D-10	0.0
5.00D+00	2.00D-10	0.0

END OF PLOT.

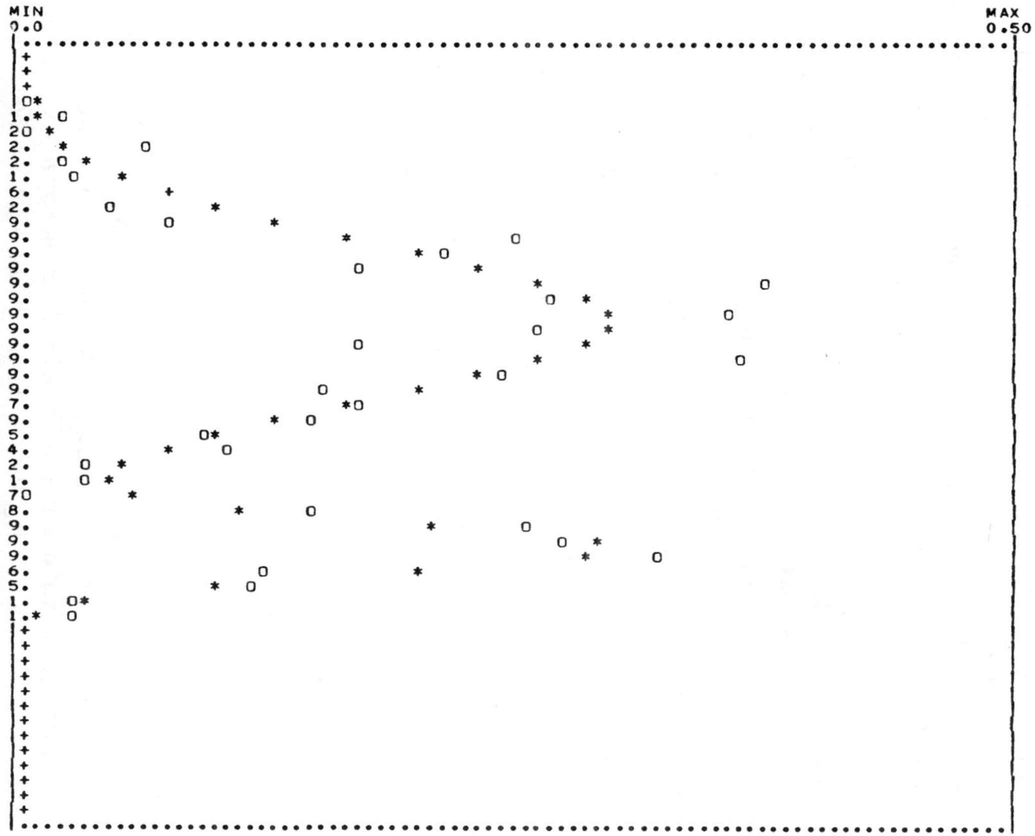

Figure 5.6. $n = 300$ Bimodal DMPLE $\alpha = 10^{-2}$.

when $\alpha = 1$, we see the hint of the beginning of unstable behavior. This is still more in evidence in Figure 5.5, when $\alpha = .1$. Figure 5.6, with $\alpha = .01$, shows a distinct beginning of Dirac degeneracy. Probably, a user, reviewing the effects of the six α choices (of course, without the benefit of the asterisks of the true density) would opt for an α of 10 or 1, since one should seek the highest resolution consistent with stability.

Now, in order to compare the approximation properties of the DMPLE with those of kernel estimators and the robustness of the choice of penalty parameter α in DMPLE with that of the scaling parameter h in kernel estimation, we performed a Monte Carlo simulation. A "reasonable" value for α was chosen for each density (e.g., $\alpha = 30$ for the bimodal Gaussian). For comparison purposes we weighted the simulation study in favor of the kernel estimator by using the optimal choice of h (since in this case the solution is known). The highly popular Gaussian kernel

$$K(x) = \frac{1}{\sqrt{2\pi}} \exp\left(-\frac{x^2}{2}\right) \tag{39}$$

was used, although kernels with finite support enjoy computational savings. The optimal choice for the scaling parameter h as a function of n in this case is

$$h(n) = \left\{ \frac{\int K(x)^2\,dx}{\left[\left[\int x^2 K(x)\,dx\right]^2 \int [f''(x)]^2\,dx\right]} \right\}^{1/5} n^{-1/5}. \tag{40}$$

Random samples were generated on the computer and the integrated mean square error (IMSE) evaluated numerically. The Monte Carlo technique is to report the mean and standard deviation of the IMSE of 25 generated samples from a fixed distribution for a fixed sample size n. We also calculated the kernel estimate (using the "optimal" choice of h) for the same random samples and evaluated the IMSE numerically in the same manner. These results are given in Table 5.1.

We now consider the sensitivity or robustness of these estimators with respect to the parameter α for the DMPLE and the parameter $h(n)$ for the Gaussian kernel estimator. These results are given in Table 5.2.

In Table 5.2, the same normal samples generated for Table 5.1 were used with values of α and $h(n)$ perturbed by a factor of 2. Clearly, as n increases, the DMPLE estimates are insensitive to values of α in this range; however,

Table 5.1 Monte Carlo Estimation of Integrated Mean Square Error of DMPLE and Gaussian Kernel Estimator*

Sampling density	Sample size	DMPLE	Gaussian kernel
$N(0, 1)$	$n = 25$	.010 (.008)	.016 (.012)
$N(0, 1)$	$n = 100$	.0037 (.0021)	.0050 (.0027)
$N(0, 1)$	$n = 400$	.0015 (.0008)	.0020 (.0009)
bimodal	$n = 25$	.010 (.003)	.009 (.007)
bimodal	$n = 100$	.0036 (.0007)	.0036 (.0020)

* Each row represents the mean of the IMSE for 25 trials of the DMPLE and the Gaussian kernel estimator based on 25 random samples from the density in question for fixed n, the standard error of these 25 IMSE is given in parentheses, $\alpha = 10$ was used for the $N(0, 1)$, $\alpha = 30$ was used for the bimodal and the bimodal density is the mixture $.5N(-1.5, 1) + .5N(1.5, 1)$.

the integrated mean square error of the kernel estimates varies by a factor of 2. The sensitivity of the kernel estimator can also be derived from asymptotic arguments. If we rewrite equation (40) as $h(n) = Cn^{-1/5}$, then from (140) of Chapter 2 we see that

$$\text{(41)} \qquad \text{IMSE} \approx \left(\frac{\theta}{C} + \mu C^4\right)n^{-4/5},$$

Table 5.2. Monte Carlo Sensitivity Analysis of α and $h(n)$*

Sample size	DMPLE			Gaussian kernel		
	$\alpha = 5$	$\alpha = 10$	$\alpha = 20$	$.5h(N)$	$h(N)$	$2h(N)$
$n = 25$	.011	.010	.014	.030	.016	.031
$n = 100$	.0043	.0037	.0039	.0103	.0050	.0136
$n = 400$	.0016	.0015	.0014	.0032	.0021	.0067

* The samples used in this table are the same random samples used in Table 5.1 from the sampling density $N(0, 1)$. The table values, as in Table 5.1, represent the means of the IMSE obtained from these 25 trials.

where θ and μ are constants depending on K and f, thus perturbing $h(n)$ (and hence C) by a factor β results in a change in IMSE of β^{-1} or β^4, depending on the relative magnitude of θ and μ. In general, the design parameter of discrete maximum penalized likelihood estimation appears more insensitive to changes of sample size than the kernel width required for kernel density estimation.

Next we use Monte Carlo methods to estimate the rate of convergence of the DMPLE as a function of n, using Gaussian random samples. When plotted on log–log paper, the values in Table 5.3 fall on a straight line with slope $-.773$. The actual regression analysis gave

$$\log_{10}(\text{IMSE}) = -.773 \log_{10} n - .873, \tag{42}$$

with a sample correlation of $r = -.996$. Thus in this case the IMSE $\approx O(n^{-.773})$, which is about the same as that for kernel estimators, namely, $O(n^{-4/5})$ as was discussed in Section 2.5 of Chapter 2.

Table 5.3 Asymptotic Rate of Convergence for DMPLE Based on the $\mathbf{N}(0, 1)$ Sampling Density

Sample size	Number of samples	Mean IMSE
25	50	.0110
100	100	.00347
400	50	.00151
800	50	.000843
1,000	50	.000545
2,000	84	.000360

In Figure 5.7, we show the DMPLE (hollow squares) for the density of annual snowfall in Buffalo, New York. Although the data was merely preprocessed by a scale and translation to lie between -3.3 and 3.3 and a "rough and ready" α of 1 was used, the fit agrees very well with that obtained by an interactive mode using older techniques.

There are a number of practical procedures for selecting α. For example, we might rescale and translate the data so that the 5 percentile lies at -3, the 95 percentile lies at $+3$, and use $\alpha = 1$. The important fact is that numerical experimentation has led us to conjecture that α selection is not as fragile a design consideration as parameter selection in state-of-the-art procedures.

```
SAMPLE SIZE =    63
DISCRETIZED MAXIMUM LIKELIHOOD PENALIZED ESTIMATE
WITH WEIGHING PARAMETER ALPHA              0.1000000000D+01
67 MESH POINTS FROM       -3.30000 TO     3.30000
DISCRETE MESH INTERVAL     0.10000
LOG LIKELIHOOD TERM                       -0.1039526466D+03
LOG PENALTY TERM                          -0.1899760704D+01

ABSCISSA  ORDINATE MIN                                                       MAX
             O      0.0                                                      0.4
                    ............................................................
-3.65D+00 0.00D+00 IO
-3.45D+00 0.00D+00 IO
-3.25D+00 0.00D+00 IO
-3.05D+00 1.48D-02 1.  O
-2.85D+00 1.16D-02 I.  O
-2.65D+00 5.31D-03 I. O
-2.45D+00 1.07D-02 I. O
-2.25D+00 5.02D-02 3.       O
-2.05D+00 9.48D-02 I.             O
-1.85D+00 1.24D-01 1.                O
-1.65D+00 1.58D-01 2.                     O
-1.45D+00 1.83D-01 3.                        O
-1.25D+00 1.88D-01 3.                         O
-1.05D+00 1.76D-01 2.                       O
-8.50D-01 1.68D-01 1.                      O
-6.50D-01 1.82D-01 2.                        O
-4.50D-01 2.16D-01 1.                             O
-2.50D-01 2.63D-01 5.                                   O
-5.00D-02 3.05D-01 2.                                        O
 1.50D-01 3.34D-01 7.                                            O
 3.50D-01 3.37D-01 5.                                             O
 5.50D-01 3.09D-01 3.                                         O
 7.50D-01 2.58D-01 5.                                  O
 9.50D-01 2.00D-01 I.                           O
 1.15D+00 1.58D-01 1.                     O
 1.35D+00 1.48D-01 1.                    O
 1.55D+00 1.57D-01 3.                     O
 1.75D+00 1.67D-01 2.                      O
 1.95D+00 1.71D-01 3.                       O
 2.15D+00 1.68D-01 2.                      O
 2.35D+00 1.53D-01 1.                    O
 2.55D+00 1.29D-01 2.                 O
 2.75D+00 9.80D-02 1.             O
 2.95D+00 5.59D-02 1.       O
 3.15D+00 8.60D-03 I. O
 3.35D+00 0.00D+00 IO
 3.55D+00 0.00D+00 IO
END OF PLOT.        ............................................................
```

Figure 5.7. Snowfall Densities in Buffalo, New York, over a Period of 67 Years.

Finally, we consider an extension of discrete maximum penalized likelihood to the estimation of multivariate densities, using a pseudo-independence algorithm given in [1] and [2]. Let $\boldsymbol{x}$ be a p dimensional random variable with unknown density f. Suppose we have a random sample $\{\boldsymbol{x}_1, \boldsymbol{x}_2, \ldots, \boldsymbol{x}_n\}$ with sample mean $\bar{\boldsymbol{x}}$ and positive definite sample covariance matrix $\sum$. Let Λ denote the $p \times p$ diagonal matrix, where the diagonal consists of the eigenvalues of $\sum$, and let E denote the corresponding $p \times p$ matrix whose

Figure 5.8. Pseudo-Independent Discrete Estimate of Reflective Intensities of Corn Data in the .40–.44 $\mu m \times$ .72–.80 μm Bands.

columns are the normalized eigenvectors. We take

$$R = \Lambda^{-1/2} E. \tag{43}$$

Then,

$$z = R(\boldsymbol{x} - \bar{\boldsymbol{x}}) \tag{44}$$

has mean 0 and identity covariance matrix. We then use our one dimensional DMPLE algorithm on the quasi-independent components of z. The resulting density may then be transformed back into $\boldsymbol{x}$ space.

This procedure is only theoretically justified for Gaussian data. Interestingly, we have not yet found a real data set which works as a counterexample to the algorithm, although hypothetical counterexamples may be

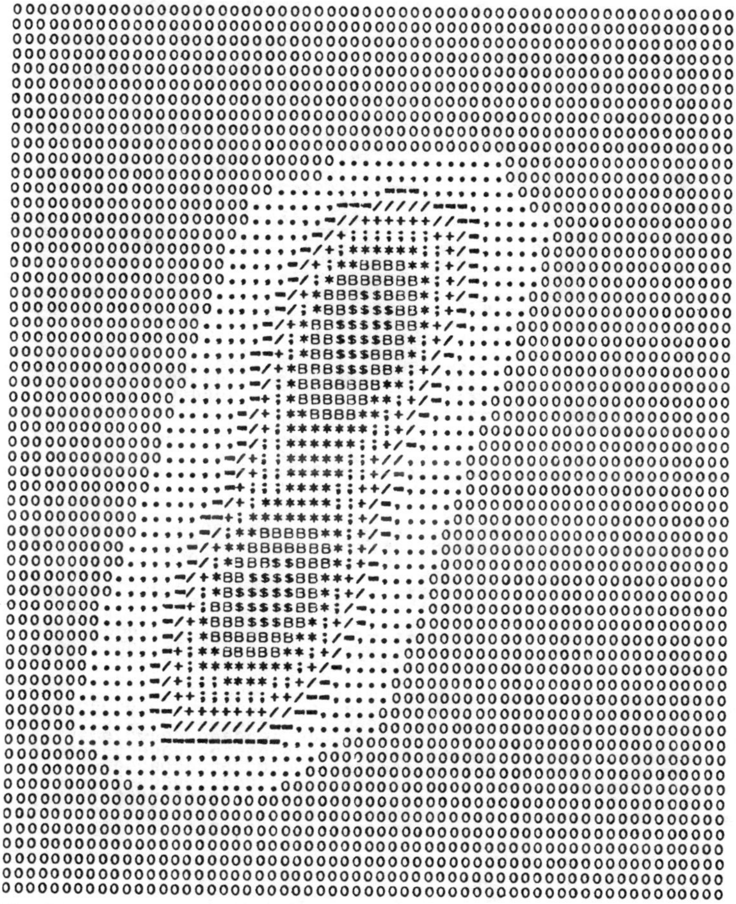

Figure 5.9. Pseudo-Independent Discrete Estimate of Reflective Intensities of Corn Data in the .66–.72 $\mu m \times$ .72–.80 μm Bands.

constructed quite easily. In Figure 5.8 we see the result of applying the quasi-independence algorithm to two channels of a multispectral scanner viewing the reflective light intensity from a corn field in the .40–.44 μm (abscissa) $\times$ .72–.80 μm bands. The DMPLE fit is quite satisfactory, but we might have done as well using the standard NASA LARYS algorithm which assumes Gaussianity. In Figures 5.9 through 5.11 we see the DMPLE fit for intensities in .66–.72 μm $\times$ .72–.80 μm, .66–.72 μm $\times$ .80 $\times$ 1.00 μm, and .72–.80 μm $\times$ .80–1.00 μm bands respectively. For these far more typical (of crop remote sensing data) data sets, the DMPLE procedure easily exhibits the multimodal density structure, unlike the standard parametric procedure.

Figure 5.10. Pseudo-Independent Discrete Estimate of Reflective Intensities of Corn Data in the .66–.72 μm × .80–1.00 μm Bands.

References

[1] Bennett, J.O. (1974). "Estimation of a multivariate probability density function using B-splines." Doctoral dissertation at Rice University, Houston, Texas.

[2] Bennet, J.O. de Figueiredo, R.J.P., and Thompson, J.R. (1974). "Classification by means of B-spline potential functions with appications to remote sensing." *The Proceedings of the Sixth Southwestern Symposium on System Theory*, FA3.

[3] Chung, K.L. (1968). *A Course in Probability Theory.* New York; Harcourt Brace and World.

[4] Subroutine NDMPLE. IMSL Libraries, Houston, Texas.

[5] Scott, D.W. (1976). "Nonparametric probability density estimation by optimization theoretic techniques." Doctoral dissertation at Rice University, Houston, Texas.

[6] Scott, D.W., Tapia, R.A., and Thompson, J.R. (1976). "An algorithm for density estimation." *Computer Science and Statistics: Ninth Annual Symposium on the Interface*, Harvard University, Cambridge, Massachusetts.

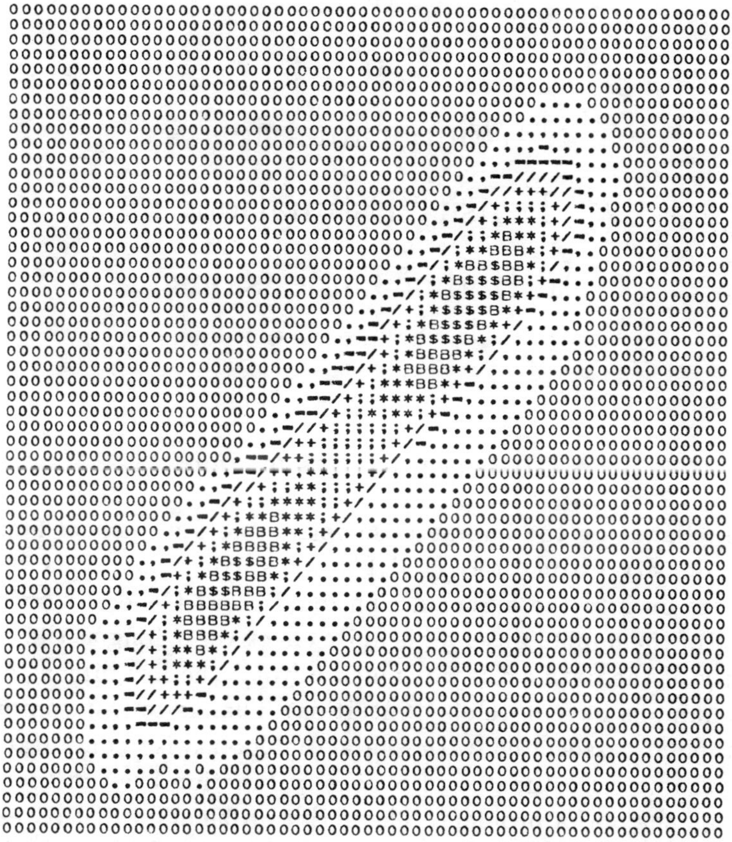

Figure 5.11. Pseudo-Independent Discrete Estimates of Reflective Intensities of Corn Data in the .72–.80 $\mu m \times$ .80–1.00 μm Bands.

[7]——(1980) "Nonparametric probability density estimation by discrete maximum penalized likelihood criteria," *Annals of Statistics* 8: 820-832.

[8]——(1977) "Kernel density estimation revisited." *Nonlinear Analysis* 1:339-72.

[9] Tapia, R.A. (1974). "A stable approach to Newton's method to general mathematical programming problems in $\Re^n$. *Journal of Optimization Theory and Application* 14: 453-76.

[10]—— (1977). "Diagonalized multiplier methods and quasi-Newton methods for constrained optimization." *Journal of Optimization Theory and Application* 22: 135-94.

[11] Wahba, G. (1976). "A survey of some smoothing problems and the method of generalized cross-validation for solving them." TR # 457, Department of Statistics, University of Wisconsin.

CHAPTER

6 Nonparametric Density Estimation In Higher Dimensions

6.1. Graphical Considerations

The reliance on the human observer in one dimensional nonparametric density estimation (NDE) is quite common. This reliance continues on into the higher dimensional case. Generally speaking, this use of the human observer is quite effective, as long as the dimensionality of the data set is not much beyond the low dimensionality where humans have developed intuition. Thanks to fast computing, the problems associated with the actual pointwise computation of a nonparametric estimator of a density function are not, generally speaking, the major problem. The major problem for density estimation in higher dimensions is knowing where to compute that density function. Furthermore, for most problems of data analysis, we will be able to use the philosophy of nonparametric density estimation as guidance, without actually having to come up with a density estimator suitable for graphical display.

Suppose we have a sample of size 300 from a five dimensional density. An investigator who estimated the density using fixed mesh (say 20 intervals per dimension) would be required to estimate the density f at 3.2×10^6 points. Since we cannot visualize all five dimensions of the density in a three dimensional plot, we must first attempt to decide which of these grid points are likely to produce the highest (and therefore most interesting) densities.

To demonstrate the qualitative difference between low dimensional and higher dimensional NDE problems, let us consider the choice between two different quanta of information:

A: a sample of size 100 from an unknown density,

B: exact knowledge of the probability density function on an equispaced mesh of size 100 between the 1 and 99 percentiles.

For one dimensional data most investigators will select quantum B. For four dimensional data, however, the mesh in quantum B would give us only slightly more than three mesh points per dimension. We might find that we had our 100 precise values of the density function evaluated at 100 points where the density was effectively zero. Here we see the high price we pay for an equispaced Cartesian mesh in higher dimensions. If we insist on using it, we will be spending most of our time searching about in empty space.

But note that quantum A remains useful in four dimensional space, for it gives 100 points which will tend to come from regions where the density is relatively high. Thus they provide anchor points from which we can examine, in spherical search fashion, the fine structure of the density.

Unfortunately most investigators in nonparametric density estimation generally try to transform information of type A into information of type B. Clearly, in the case of nonparametric density estimation, as in many other areas of applied statistics, it is a mistake to believe that we can get from the one dimensional problem to problems of higher dimensionality by a simple wave of the hand. Practically speaking, almost anything reasonable works for one dimensional data. We still know very little about what works for the higher dimensional problems. Representational problems are dominant. The difficulty is not so much being able to estimate a density function at a particular point, but knowing where to look. We can, if we are not careful, spend an inordinate amount of time coming up with excellent estimates of zero.

John W. Tukey of Princeton University founded some 25 years ago a new school of data analysis called exploratory data analysis (EDA). This highly graphics based analysis purports, as does nonparametric density estimation (NDE), to relax the distributional assumptions (e.g., normality) of classical statistics and "to let the data set speak for itself." The main characteristic of EDA is the combination of the speed and precision of the digital computer with the visual intuition of a human observer. A very important set of techniques in EDA involves "cloud analysis." The PRIM-9 mainframe software of Tukey shows three dimensional displays of three dimensional data points. By spinning this "data cloud" about one axis or another, the human observer can hope to pick up the continuity of the underlying density. The downsizing of the PRIM-9 program to the MacIntosh computer by D^2 Software in its *MacSpin* program (and more recently by the SAS Institute in its *JMP* program) brings the cloud analysis technique of EDA within the capabilities of most users of microprocessors. In Figures 6.1 and 6.2, we show two views of R.A. Fisher's well known iris data base.

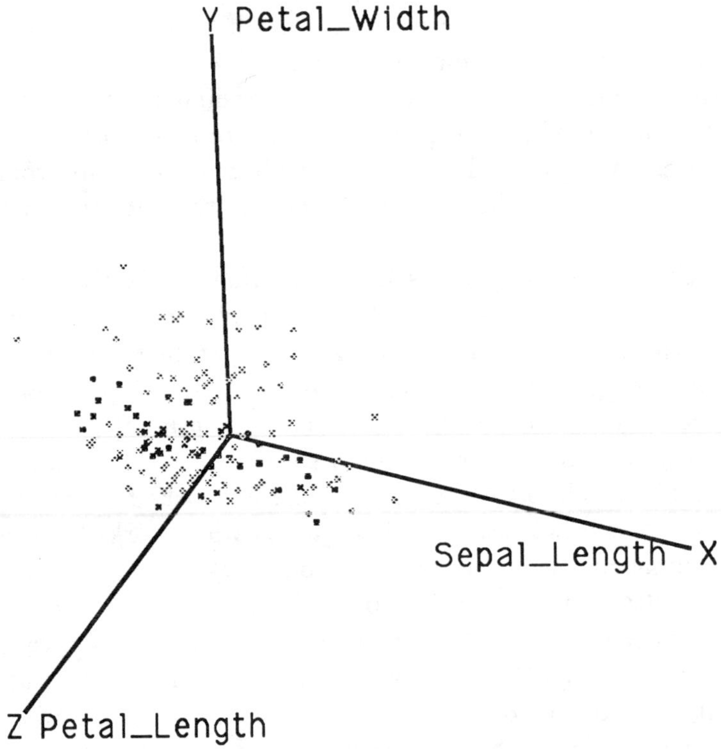

Figure 6.1.

By flailing around a bit, we can go from the confusing, apparently one cluster view shown in Figure 6.1 to a picture where the three clusters (corresponding to three varieties of iris) might be argued to be seen (well, sort of) as in Figure 6.2. We note that the usage of the cloud analysis program is bound to the three dimensions of the human visual system. As a matter of fact, Fisher's iris data actually has a fourth variable, sepal width. By showing only values of this fourth variable within a particular interval, we can begin to bring in the influence of the fourth variable. But, in so doing, we will not have treated the fourth variable in the same fashion as the other three. Graphically, data sets of dimensions four and higher can only be dealt with three dimensions at a time. The human observer is very much stuck in the loop. The computer is used to carry out rapid transformation of the coordinates. All the analysis is carried out, intuitively by the human observer.

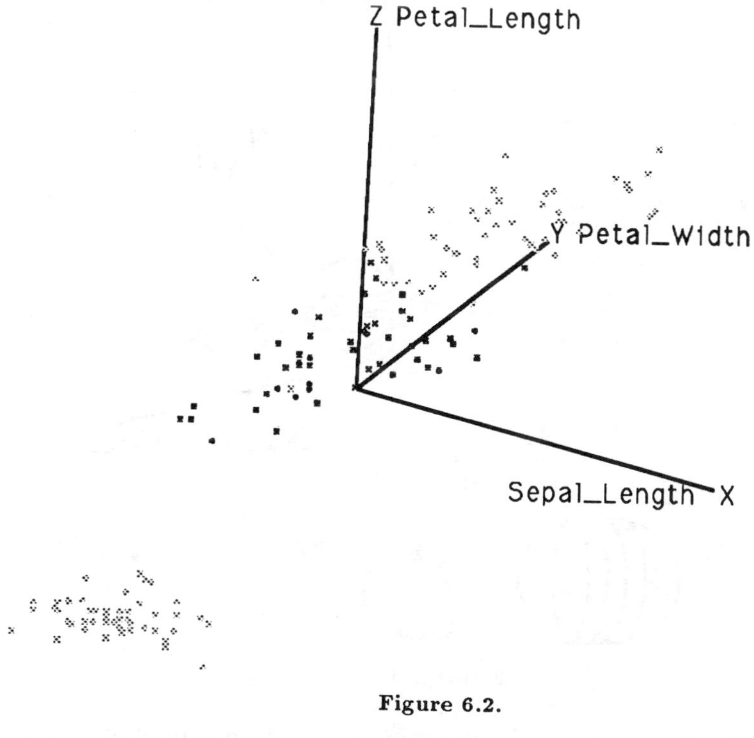

Figure 6.2.

Unquestionably, a major difference between NDE and EDA is that EDA stays with data points instead of attempting to graph an estimate of the underlying density which gave rise to the data. In other words, NDE exploits the continuity of the underlying density rather more explicitly than does EDA. Skipping forward a bit, let us observe (following Scott and Thompson [8] and Scott [9]) the kind of insights which occur immediately when we compute a nonparametric estimator of the density of the iris data and simply display the contours of the regions which have an estimated density value greater than or equal to 20% of the maximum of the density as shown in Figure 6.3. The three varieties of iris are readily seen. Generally speaking, nonparametric density estimation approaches tend to perform better than the cloud analysis of exploratory data analysis for many tasks. Asking the human eye to use persistence of vision in order to pick up on the continuity of a cluster of points as we rotate it through space is really pushing things a bit. If we can show contours representing relative density of the data points, it probably is wise to do so.

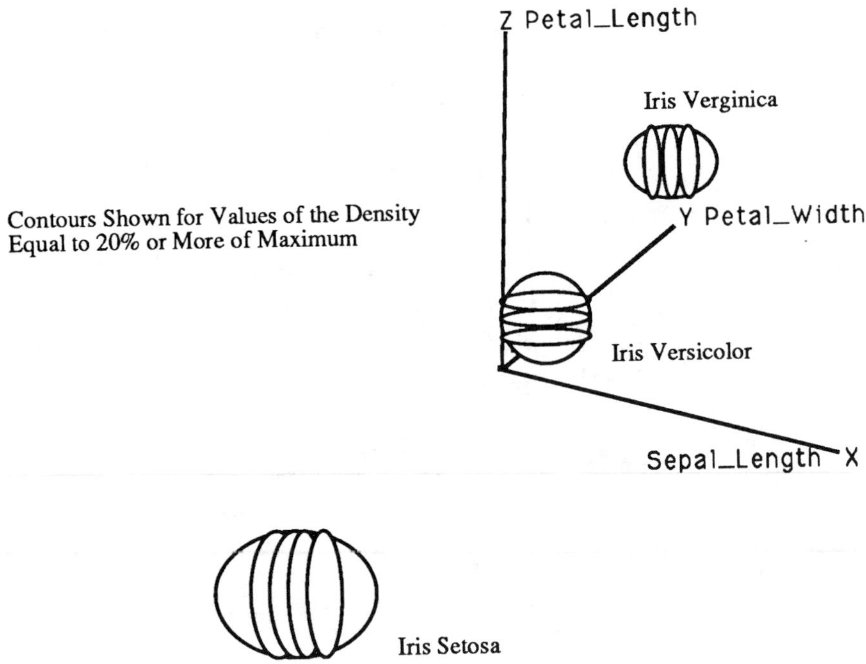

Figure 6.3.

The preceding example indicates a fairly standard property both of cloud analysis and of NDE, namely, the ultimate endproduct desired is likely to be something other than a rotated cloud of points or a nonparametric density estimate. In the case of the iris data, the task is to see whether three varieties can be recovered from the data. In other cases, the problem might involve identification of wheat from satellite data or distinguishing incoming warheads from decoys. It is clear that in most repetitive tasks (particularly the one dealing with warheads) it is probably desirable to automate the process, to take the human observer out of the loop.

Unfortunately, most investigators in NDE follow the EDA protocol of leaving the human observer in the loop. There is a tendency to play the graphics/gestalt game indefinitely. The human observer is relied on to intervene in real time to select bandwidths, to rotate coordinate systems, to color-code grey levels, etc.

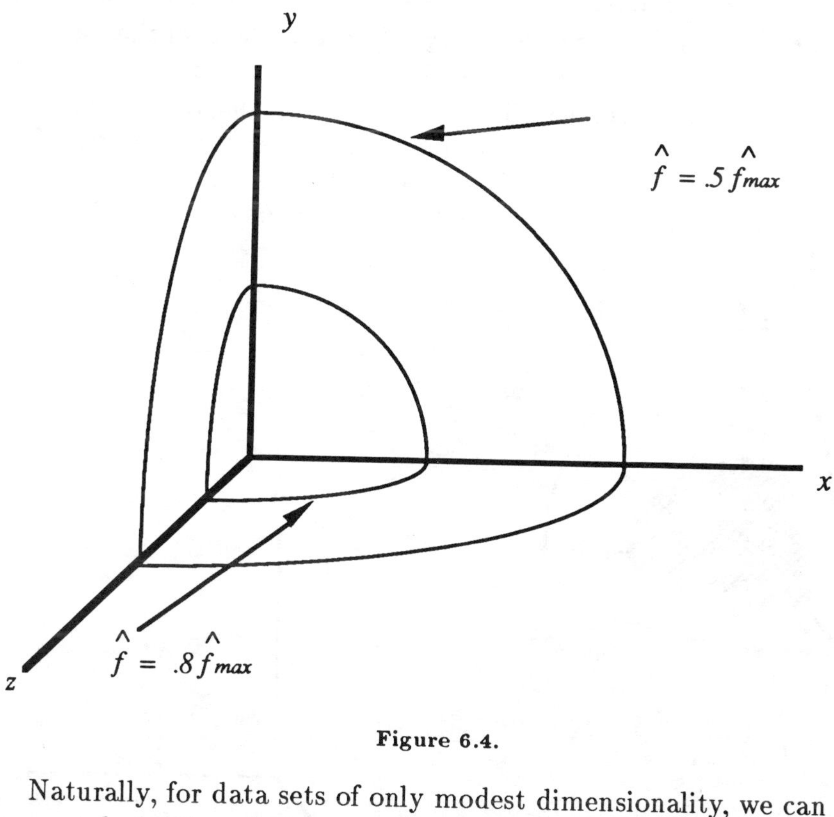

Figure 6.4.

Naturally, for data sets of only modest dimensionality, we can get a great deal of benefit from our three dimensional perception system. We function in a three dimensional world, and we can therefore see a great deal as long as the dimensionality of the data set is not so great as to drive us beyond three dimensions. What is the dimensionality beyond which we exceed our natural 3D perceptions? The answer is, in a sense, two rather than three for the NDE enthusiast, for we need the third dimension for the density function itself. The answer is not so clear-cut, however, as we note using the *onion peel* algorithm developed by Fwu , Tapia and Thompson in 1981 [4]. If we assume that we have a good estimator for the density function $\hat{f}(x, y, z)$, then we can draw contours where $\hat{f}$ is equal to various fractions of its maximum. Thus, we can obtain a graphical representation such as that in Figure 6.4.

The generalization of the *onion peel* algorithm to higher dimensions is apparently quite straightforward. In Figure 6.5, we demonstrate such a generalization for the case of six dimensions. Here, we have divided the six variables into two sets: x, y, z and u, v, w. We use a standard *onion peel* approach for x, y, z. Level surfaces of the value of the nonparametric density estimator $\hat{f}$ are plotted in terms of x, y, z in the first, *onion peel* , box. The second box, the one in u, v, w , may be looked upon as a *control box.* As we move the cursor around for particular u, v, w values, the shape of the level surfaces will change accordingly. This approach sounds more satisfactory than it is in actuality.

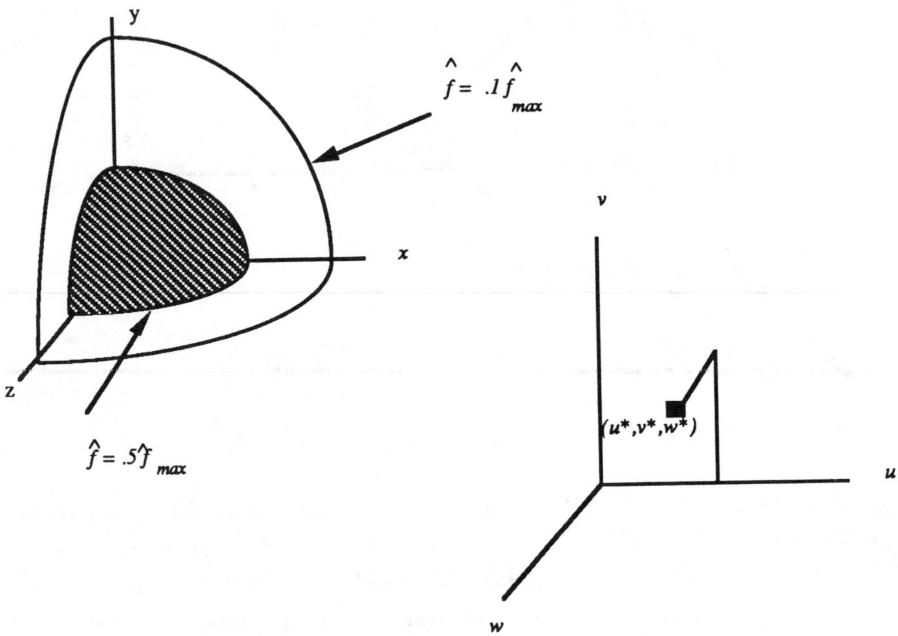

Figure 6.5.

Note, for example, that u, v, w are handled very differently from x, y, z. Our level curves of the density are in x, y, z for a particular point u, v, w. Thus, our feel for the u, v, w values is generally much poorer than that for x, y, z. Of course, once we have, for example, computed values on some mesh in 6-space, it is quite possible to see the density picture change in real time, using interpolations from the mesh points. Persistence of vision will give us some benefit from using time, more or less, as a surrogate variable which will assist in the analysis of the higher dimensional data. Still, any notion that the six dimensional problem is just like the three dimensional problem is quite incorrect. Moreover, attempts to use other variables in the display screen, such as color, in order to squeeze out higher dimensional information has not proved successful in our experience. Color provides contrast and ease of going between level sets of density function values, but as a device for giving greater dimensional perception, it seems of little use beyond what one can obtain from grey levels. This is due, perhaps, to the fact that whereas grey levels are naturally ordered insofar as human perceptions are concerned, the same is not true of colors. In general, the only variables which appear to be useful to buy readily perceptible dimensional information are the three spatial variables and time. Such variables as grey levels, sound intensity, etc., seem to buy very little as surrogates for dimensions of our data. There is no doubt that additional enhancements could be obtained if an involved learning process were developed to train the observer better to use variables not normally used as surrogates for spatial variables. Based on our experience at Rice, it seems unlikely that one can readily handle data graphically for dimension six, where three of the variables are handled asymmetrically from the other three using the control box strategy in Figure 6.5.

This leaves us with the option of developing algebraic algorithms for dealing with the higher dimensional situations. Naturally, this has been done for some time in detection theory, where investigators distinguish between types of crops, say, using light intensities in, say, 12 narrow bands of the light spectrum. The usual procedure for developing such algebraic algorithms involves the assumption of a particular form of the density function of light intensities, e.g., Gaussian. In our experience, the assumption of Gaussianity tends to get more and more dubious as the dimensionality of the data increases.

6.2. Resampling Algorithms Based on Some Nonparametric Density Estimators

It is seldom the case that a graphical display of a nonparametric density estimator is the final end product needed. This is fortunate, since, as we have observed, such a display becomes difficult for dimensions of three. Past a dimensionality of four, such a display is seldom practical.

There are cases, however, where the "curse of dimensionality" can be readily overcome. One of these is that of the creation of a pseudo-data set with properties very much like those of a real data set.

To motivate the SIMDAT algorithm [11], let us first consider the Dirac-comb density estimator associated with a one dimensional data set $\{x_i\}_{i=1}^n$. The Dirac-comb density estimator is given by

$$\hat{f}_\delta(x) = \frac{1}{n}\sum_{i=1}^{n} \delta(x - x_i). \tag{1}$$

We recall that $\delta(x)$ can be defined as

$$\delta(x) = \lim_{\tau \to 0} \frac{1}{\sqrt{2\pi}\tau} e^{-\frac{x^2}{2\tau^2}}. \tag{2}$$

Such a density is a "function" which is zero everywhere except at the data points. At each of these points, the density becomes a line stretching to infinity and with mass $1/n$. As a nonparametric density estimator, $\hat{f}_\delta(x)$ would appear to be terrible. It says that all that can happen in any other experiment is a repeat of the data points already observed, each occurring with probability $1/n$.

Yet, for many purposes $\hat{f}_\delta(x)$ is quite satisfactory. For example, the mean of $\hat{f}_\delta(x)$ is simply

$$\mu_\delta = \int_{-\infty}^{\infty} x\hat{f}_\delta(x)dx = \frac{1}{n}\sum_{i=1}^{n} x_i = \bar{x}. \tag{3}$$

Similarly,

$$\sigma_\delta^2 = \int_{-\infty}^{\infty} (x - \bar{x})^2 \hat{f}_\delta(x)dx = \frac{1}{n}\sum_{i=1}^{n}(x - \bar{x})^2 = s^2. \tag{4}$$

The Dirac-comb density estimator may clearly be extended to higher dimensions. For example, if we have a sample of size n from a density of dimension p, $\hat{f}_\delta(x)$ becomes

$$\hat{f}_\delta(X) = \frac{1}{n}\sum_{i=1}^{n} \delta(X - X_i) \tag{5}$$

where

$$\delta(X) = \lim_{\tau\to 0}\left(\frac{1}{\sqrt{2\pi}\tau}\right)^p \exp\left(-\frac{\sum_{j=1}^{p} x_j^2}{2\tau^2}\right) \tag{6}$$

with x_j being the $j'th$ component of X.

We could, for example, for the two dimensional case, develop a 95% confidence interval for the correlation coefficient,

$$\rho = \frac{Cov(x,y)}{\sigma_x\sigma_y}. \tag{7}$$

Suppose that we have a sample of size n: $\{x_i, y_i\}_{i=1}^n$. Then we construct

$$\hat{f}_\delta(x,y) = \frac{1}{n}\sum_{i=1}^{n} \delta((x,y) - (x_i,y_i)). \tag{8}$$

Using this estimator, we construct 10,000 resamplings of size n, i.e., for each of the 10,000 resamplings we draw samples from the n data points (with replacement) of size n. For each of the resamplings, we compute the sample correlation

$$r_j = \frac{\sum_{i=1}^{n}(x_{ji} - \bar{x}_j)(y_{ji} - \bar{y}_j)}{\sqrt{\sum_{i=1}^{n}(x_{ji} - \bar{x}_j)^2 \sum_{i=1}^{n}(y_{ji} - \bar{y}_j)^2}} \tag{9}$$

Then, we rank the sample correlations from smallest to largest. A 95% confidence estimate is given by

$$r_{(250)} < \rho < r_{(9,750)}. \tag{10}$$

Such a resampling procedure has been popularized by Efron [3], who calls it the "bootstrap." Although it is clear that such a procedure may have some use for estimating the lower moments of some interesting parameters, we should never lose sight of the fact that it is, after all, based on the profoundly discontinuous Dirac-comb estimator $\hat{f}_\delta$.

It is quite easy to find an example where Dirac-comb nonparametric density estimator based techniques (i.e., the bootstrap) produce disastrous results. For example, suppose we have a sample of size 100 of firings at a bullseye of radius 5 centimeters. If the distribution of the shots is circular normal with mean the center of the bullseye and deviation one meter, then with a probability in excess of .88, none of the shots will hit inside the bullseye. Then any Dirac-comb resampling procedure will tell us the bullseye is a safe place, if we get a base sample (as we probably will) with no shots in the bullseye.

Clearly, it would usually be preferred to use resampling schemes based on better nonparametric density estimators. One such estimator would be

$$\hat{f}_\delta(X) = \frac{1}{n}\sum_{i=1}^{n} K(X - X_i, \Sigma_i), \tag{11}$$

where $K(X)$ is a normal distribution centered at zero with locally estimated covariance matrix Σ_i.

Although a simulation algorithm employing such an estimator has much to recommend it, it would appear extremely difficult to execute. It is at this point, that we should recall what it is we seek: not a good nonparametric density estimator, but a random sample from such an estimator. One is tempted to try a scheme which goes directly from the actual sample to the pseudo-sample. Of course, this is precisely what the bootstrap estimator does, with the very bad properties associated with a Dirac-comb. It is possible, however, to go from the sample directly to the pseudo-sample in such a way that the resulting estimator behaves very much like that of the normal kernel approach above. This is the SIMDAT algorithm of Taylor and Thompson [11].

We assume that we have a data set of size n from a p-dimensional variable X, $\{X_i\}_{i=1}^n$. First of all, we shall assume that we have already rescaled our data set so that the marginal sample variances in each vector component are the same. For a given integer m, we find, for each of the n data points, the $m-1$ nearest neighbors. These will be stored in an array of size $n \times (m-1)$.

Let us suppose that we wish to generate a pseudo-sample of size N. Of course, there is no reason to suppose that n and N will, in general, be the same (as is the case generally with the bootstrap). To start the algorithm, we sample one of the n data points with probability $1/n$ (just as with the bootstrap). Starting with this point, we recall it and its $m-1$ nearest neighbors from memory, and compute the mean of the resulting set of points:

$$\bar{X} = \frac{1}{m}\sum_{i=1}^{m} X_i. \tag{12}$$

Next, we code each of the m data points about $\bar{X}$:

$$\{X'_j\} = \{X_j - \bar{X}\}_{j=1}^{m}. \tag{13}$$

Clearly, although we go through the computations of sample means and coding about them here as though they were a part of the simulation process, the operation will be done once only, just as with the determination of the $m-1$ nearest neighbors of each data point. The $\{X'_j\}$ values as well as the $\bar{X}$ values will be stored in an array of dimension $n \times (m+1)$.

Next, we shall generate a random sample of size m from the one dimensional uniform distribution:

$$U\left(\frac{1}{m} - \sqrt{\frac{3(m-1)}{m^2}}, \frac{1}{m} + \sqrt{\frac{3(m-1)}{m^2}}\right). \tag{14}$$

We now generate our centered pseudo-data point X', via

$$X' = \sum_{l=1}^{m} u_l X'_l. \tag{15}$$

Finally, we add back on $\bar{X}$ to obtain our pseudo-data point X:

$$X = X' + \bar{X}. \tag{16}$$

The procedure above, as m and n get large, becomes very like that of the normal kernel approach mentioned earlier. To see why this is so, we consider the sampled vector X_l and its $m-1$ nearest neighbors:

$$\{X_l\}_{l=1}^{m} = \begin{bmatrix} x_{1l} \\ x_{2l} \\ \vdots \\ x_{pl} \end{bmatrix}_{l=1,\ldots,m} \tag{17}$$

For a moment, let us treat this collection of m points from a truncated distribution with mean vector μ and covariance matrix Σ. Now, if $\{u_l\}_{l=1}^{m}$ is an independent sample from the uniform distribution in (14), then

$$E(u_l) = \frac{1}{m};\ Var(u_l) = \frac{m-1}{m^2};\ Cov(u_i, u_j) = 0, \text{ for } i \neq j. \tag{18}$$

Next, we form the linear combination:

$$Z = \sum_{l=1}^{m} u_l X_l. \tag{19}$$

For the $r'th$ component of the vector Z, $z_r = u_1 x_{r1} + u_2 x_{r2} + \ldots + u_m x_{rm}$, we observe the following relationships:

$$E(z_r) = \mu_r, \tag{20}$$

$$Var(z_r) = \sigma_r^2 + \frac{m-1}{m}\mu_r^2, \tag{21}$$

and

$$Cov(z_r, z_s) = \sigma_{rs} + \frac{m-1}{m}\mu_r \mu_s. \tag{22}$$

Note that if the mean vector of X were $(0, 0, \ldots, 0)$, then the mean vector and covariance matrix of Z would be the same as that of X, i.e., $E(z_r) = 0$, $Var(z_r) = \sigma_r^2$, and $Cov(z_r, z_s) = \sigma_{rs}$. Naturally, by translation to the local sample mean of the nearest neighbor cloud, we will not quite have achieved this result. But we will come very close to the generation of an observation from the truncated distribution which generated the points in the nearest neighbor cloud.

Clearly, for m moderately large, by the central limit theorem, SIMDAT comes close to sampling from n normal distributions with the mean and covariance matrices corresponding to those of the n, m

nearest neighbor clouds. If we were seeking rules for consistency of the nonparametric density estimator corresponding to SIMDAT, we could use the formula of Mack and Rosenblatt [7] for nearest neighbor nonparametric density estimators:

$$m = Cn^{4/(p+4)}. \tag{23}$$

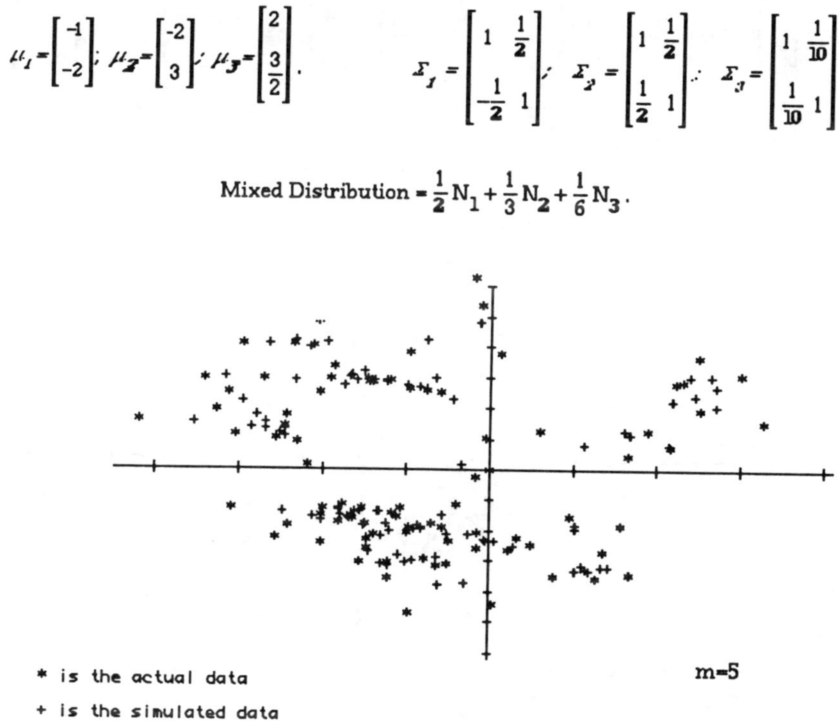

Figure 6.6.

Of course, people who carry out nonparametric density estimation realize that such formulae have little practical relevance, since C is usually not available. Beyond this, we ought to remember that our goal is not to obtain a nonparametric density estimator, but rather to generate a data set which appears like that of the data set before us. Let us suppose that we err on the side of making m far too small, namely, $m = 1$. That would yield simply the bootstrap. Suppose we err on the side of making m far too large, namely, $m = n$. That would yield an estimator which roughly sampled from a multivariate normal distribution with the mean vector and covariance matrix computed from the data.

In Figure 6.6, we see a sample of size 85 from a mixture of three normal distributions with the weights indicated, and a pseudo-data set of size 85 generated by SIMDAT with $m = 5$.

$$\mu_1 = \begin{bmatrix} -1 \\ -2 \end{bmatrix};\ \mu_2 = \begin{bmatrix} -2 \\ 3 \end{bmatrix};\ \mu_3 = \begin{bmatrix} 2 \\ \frac{3}{2} \end{bmatrix}. \qquad \Sigma_1 = \begin{bmatrix} 1 & \frac{1}{2} \\ -\frac{1}{2} & 1 \end{bmatrix};\ \Sigma_2 = \begin{bmatrix} 1 & \frac{1}{2} \\ \frac{1}{2} & 1 \end{bmatrix};\ \Sigma_3 = \begin{bmatrix} 1 & \frac{1}{10} \\ \frac{1}{10} & 1 \end{bmatrix}.$$

$$\text{Mixed Distribution} = \frac{1}{2}N_1 + \frac{1}{3}N_2 + \frac{1}{6}N_3.$$

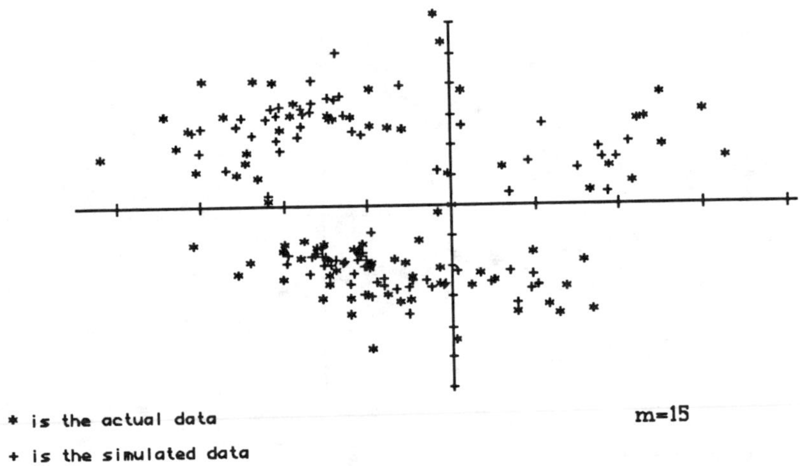

Figure 6.7.

We note that the emulation of the data is reasonably good. In Figure 6.7, we go through the same exercise, but with $m = 15$. There effects of a modest oversmoothing are noted. In general, if the data set is very large, say of size 1,000 or greater, good results are generally obtained with $m \approx .02n$. For smaller values of n, m values in the $.05n$ range appear to work well. A version of SIMDAT in the S language, written by E. Neely Atkinson, is available under the name "GENDAT" from the S Library available through e-mail from Bell Labs. The new edition of the IMSL Library will contain a SIMDAT subroutine entitled "RNDAT."

In this section, we have observed the usefulness of noting what we really seek rather than using graphically displayed nonparametric density estimators as our end product. The cases where we really must worry about obtaining the graphical representation of a density are, happily, rare. But, as with the case of SIMDAT, the context of nonparametric density estimation is very useful in many problems where the explicit graphical representation of a nonparametric density estimator is not required.

6.3. The Search for Modes in High Dimensions

Let us start with the relatively simple problem of estimating the mean of a multivariate normal distribution $N(\mu, \sigma^2 I)$ based on a sample of size one from X. We will use as our criterion function, for a shrinkage estimator $\hat{\mu}$, the risk (mean square error of the estimator divided by that for the usual estimator, X):

$$R(\hat{\mu}(X)) = \frac{1}{p\sigma^2} E(\hat{\mu}(X) - \mu)^T (\hat{\mu}(X) - \mu). \tag{24}$$

In 1961, James and Stein [5] established the fact that there were estimators, provided $p \geq 3$, which gave a lower value of expected quadratic loss than the usual estimator X. One such shrinkage estimator is that first suggested by Kendall and Stuart [6] and examined by Thompson [12] for the one dimensional case, and by Alam and Thompson [1] for the multidimensional case:

$$\hat{\mu}(X) = \frac{||X||^2}{||X||^2 + p\sigma^2} X. \tag{25}$$

In the case where we are estimating the mean of a spherical multivariate normal density using summed quadratic error, for shrinkage estimators such as that of Alam and Thompson, the allocation between the means dimension by dimension is unimportant, for fixed

$$||\mu||^2 = \sum_{i=1}^{p} \mu_i^2. \tag{26}$$

In Figure 6.8, we note how the shrinkage estimator has MSE uniformly less than that of the customary estimator X. We observe that this estimator has arbitrarily shrunk X toward the origin. Any other point for a shrinkage focus would have done as well.

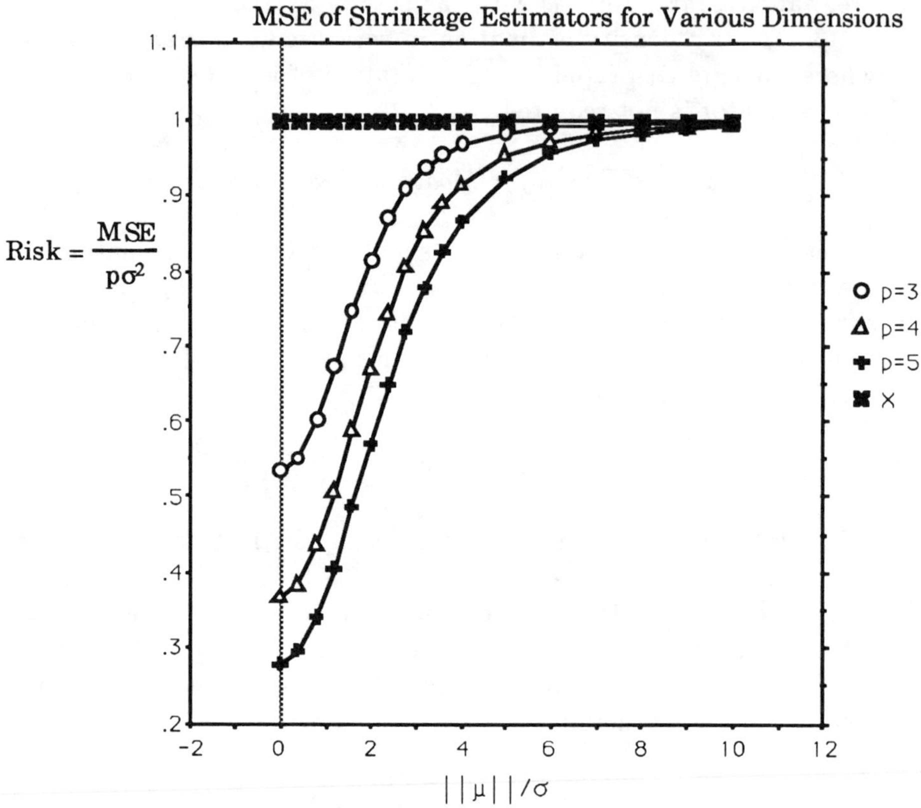

Figure 6.8.

To get some feel for this phenomenon, we consider the one-dimensional case. Suppose we wish to estimate the mean on the basis of one observation X. We will restrict ourselves to estimators of the form

$$\mu_0 = aX. \tag{27}$$

Let us pick a in such a way as to minimize

$$R(aX) = E[(aX - \mu)^2] = a^2\sigma^2 + \mu^2(1-a)^2. \tag{28}$$

Taking the derivative with respect to a and setting it equal to 0, we find the a of choice to be

$$a = \frac{\mu^2}{\mu^2 + \sigma^2}. \tag{29}$$

But then, we see that the risk of aX is always less than that of X, i.e.,

$$R(aX) = \frac{\mu^2}{\mu^2 + \sigma^2}\sigma^2 < \sigma^2. \tag{30}$$

Since, in practice, we know neither μ nor σ^2, the idealized shrinkage estimator is not available. Still, it is a matter of interest to ask why it is that a, if it were available to us, would enable us to beat X as an estimator. To understand better, we rewrite aX as:

$$aX = \frac{X/\mu}{1 + (\sigma^2/\mu^2)}\mu. \tag{31}$$

So then, aX gives us the truth—μ—degraded by a multiplier which, should μ be small relative to σ^2, would automatically discount large values of X as outliers, and give as the estimator essentially zero. On the other hand, if μ is large relative to σ, then we will be left essentially the usual estimator X. It is interesting to observe that we have not used the assumption of normality to arrive at our result.

If we somehow knew σ^2 but not μ, we could substitute X for μ:

$$\hat{\mu}(X) = \frac{X^2}{X^2 + \sigma^2}X. \tag{32}$$

The degradation in quality by replacing μ with X in the shrinkage factor would be such that for small (relative to σ) values of μ we would do rather better with $\hat{\mu}(X)$ than with X. For larger values of μ, we would do worse than with X as we show in Figure 6.9. The shrinkage estimator also is not uniformly better than X for $p = 2$ when the criterion used is mean square error. However, as we have observed in Figure 6.8, $\hat{\mu}(X)$ dominates X for $p \geq 3$.

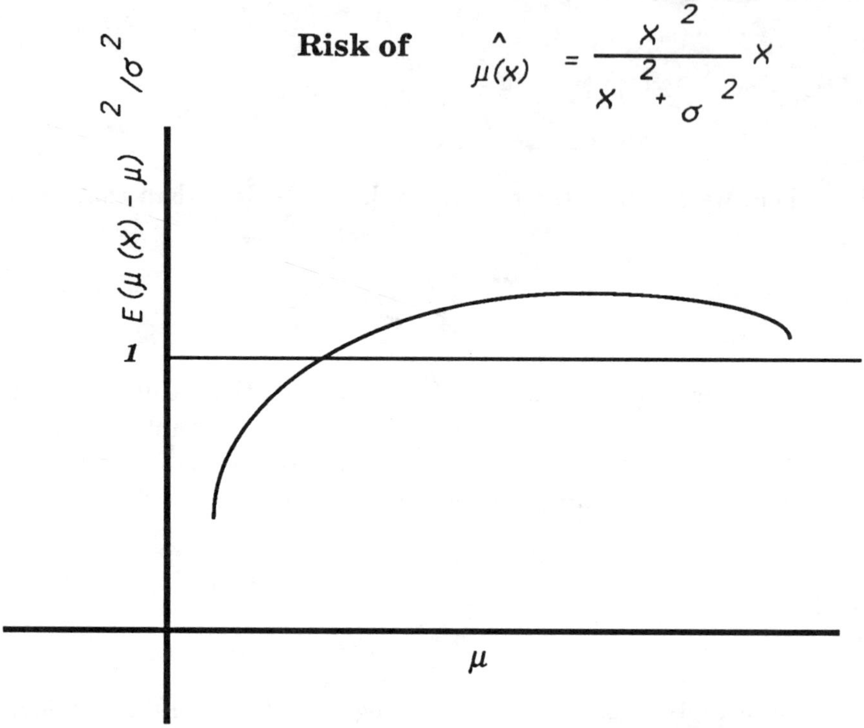

Figure 6.9.

To see better what is going on, let us suppose that we have one observation X_1 to use for the estimation of μ. We have additional $p-1$ observations $X_2, \ldots, X_p$ to be used in an a of the form

$$a = \frac{\sum_{i=1}^{p} X_i^2}{\sum_{i=1}^{p} X_i^2 + p\sigma^2}. \tag{33}$$

Since the expectation of the square of a random variable X with finite mean and variance is given by

$$E(X^2) = \mu^2 + \sigma^2, \tag{34}$$

and the average of samples of such a random variable converges almost surely to its expectation, we have:

$$\sum_{i=1}^{p} \frac{X_i^2}{p} \to \mu^2 + \sigma^2, \text{ almost surely in } p\ . \tag{35}$$

Hence,

$$a \to \frac{\mu^2 + \sigma^2}{\mu^2 + 2\sigma^2} \text{ for } p \text{ large.} \tag{36}$$

This gives, for large p, a and X_1 being asymptotically independent,

$$E[(aX_1 - \mu)^2] \approx \frac{\mu^4 + 3\sigma^2\mu^2 + \sigma^4}{\mu^4 + 4\sigma^2\mu^2 + 4\sigma^4}\sigma^2 < \sigma^2. \tag{37}$$

The estimation of the mean of a spherical normal distribution reduces immediately to the univariate problem considered above. Apparently, for the normal distribution, the asymptotic result starts impacting for $p = 3$. By the above argument, we note that it is the assumption of a loss function of a particular form which gives the Stein improvement. It should be noted that the improvement is available even when the weights are unequal across the dimensions *provided the weights are known.*

Philosophically, there is much to object to about the way its advocates have proposed using Stein shrinkage (see Thompson [13], [14]). Naturally, shrinkage to a point which is believed *a priori* to be close to the mean is an idea acceptable to Bayesians and non-Bayesians alike. But the Stein result assumes no such prior information. Consequently, one might be tempted to deal with the joint estimation of p completely unrelated quantities by shrinking their p-dimensional estimate to the origin or any other arbitrary point. As we have seen, the gain in efficiency of shrinkage estimators is due to considerations having to do with the Euclidean metric, i.e., squared deviation, when combined with the added consideration of relatively high dimensionality. It is revealing, however artificial Stein estimators would generally be in practice, to note that we have observed a case where things appear to get better rather than worse as the dimensionality increases.

There are many other matters of interest concerning the amelioration of data analysis as the dimensionality increases. For example, let us again consider the p-dimensional random vector X which is spherically normally distributed with mean the origin and covariance matrix $\sigma^2 I$. If we consider the squared distance of X from the origin, we have

$$D^2 = \frac{1}{p}\sum_{j=1}^{p} x_j^2. \tag{38}$$

We have further

$$E[D^2] = \sigma^2 \text{ and } Var[D^2] = \frac{2\sigma^4}{p}. \tag{39}$$

The implications are that as p becomes large, X lies roughly on the surface of a p-sphere centered at the origin and having radius $\sqrt{p}\sigma$.

As we have observed, the holistic perception of a probability density function by nonparametric means becomes exceedingly difficult once we get past dimension three or four. If we insist on graphical perceptions of a density of dimension eight, say, we are unlikely to be very satisfied with our end product. Certainly, any notion of using such graphical analysis to produce rapid analysis with the observer out of the loop, appears contradictory.

On the other hand, we know that, very frequently, we are simply concerned about neighborhoods of points where the density function is relatively high. A remote sensing investigator attempting to look at the reflective light intensities from crops of a particular type in twelve narrow bands of the spectrum is most likely to be interested in "lines," i.e., combinations where the intensities are unusually high. Such a collection of combinations will be referred to as the "signature" of the crop. Knowing the "signatures" of all ground features likely to be observed from a satellite will enable the remote sensing scientist to build automated algorithms which give almost instantaneous readout of the ground cover of the region of the earth below the satellite. Here we have an example of a fairly common phenomenon facilitating the analysis of higher dimensional data: it is frequently the case that we can extract most of the information relevant to our task from the locations of some of the modes of the underlying probability density function.

One might well ask if it is not reasonable to suppose that the task of mode estimation does not experience the same curse of dimensionality as that of density estimation. An intuitive example will indicate that this may not be the case. Suppose we wish to locate the angle from an observer to the highest peak in a range on the horizon. The observer standing in one spot would essentially have at his disposal something very like a one dimensional density estimator. But if the observer were given, in addition, photographs taken from a helicopter flying over the mountain range, he would be much better able to pick out the tallest peak. Clearly, sketching the one dimensional density estimator would be easier than sketching the two dimensional density estimator. But the peak could be the more easily found with the two dimensional photograph. The search for peaks, then, is not quite the same task as that of holistic density estimation.

Boswell [2] has demonstrated that the task of the estimation of local modes is much easier than that of density estimation for densities of dimension three or higher. The simplest of the algorithms studied by Boswell, the Mean Update Algorithm, was first investigated by Fwu, Tapia and Thompson [4] for dimensions of three and below. Let us suppose we have rescaled our data so that the marginal variances for each variable are equal. Then the Mean Update Algorithm for finding a mode is given by

Mean Update Algorithm

Let μ_1 be the initial guess

Let m be a fixed parameter;

$i = 1$;

Repeat until $\mu_{i+1} = \mu_i$;

Begin

Find the sample points $\{X_1, X_2, \ldots, X_m\}$ which are closest to μ_i;

Let $\mu_{i+1} = \frac{1}{m}\sum_{j=1}^{m} X_j$;

$i = i + 1$;

end.

Let us consider some intuitive reasons for the improved behavior of this estimator with increasing dimension. In Figure 6.10, we note observations from a two dimensional distribution with a single mode at the origin. We note also the projections of these observations on the x−axis. For $m = 2$, starting from the point labelled "0," we note that the mean update algorithm in one dimension stalls after only one step. On the other hand, working with the two dimensional data and starting again from the same point, we note that the algorithm does not stop until after 3 steps. The customary measure of performance is the Euclidean distance of the stopping point from the mean of the distribution.

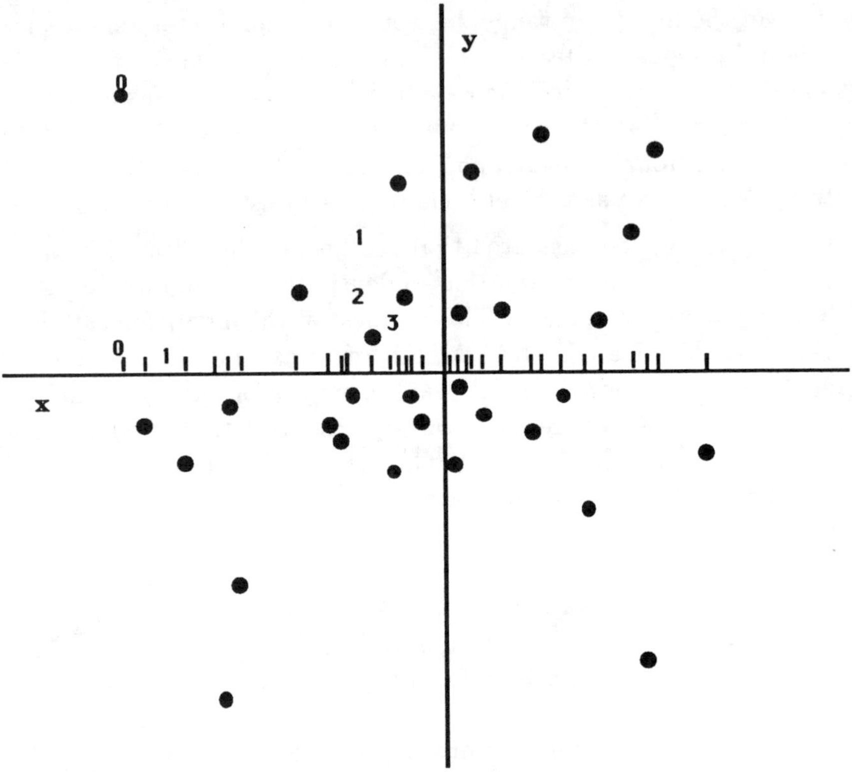

Figure 6.10.

Note that even if we were to use the two dimensional algorithm to give us a value whose performance were judged solely by the projection of the two dimensional stopping point on the x−axis, we would do better, on average, than using the mean update algorithm on the data projected on the x−axis. A glance at Figure 6.10 indicates the reason for the generally improved performance of the algorithm in two rather than one dimension. Points actually far away from each other may project onto a one dimensional space in such a way as to appear rather close together, hence causing premature stopping.

Next let us consider the use of the mean update algorithm for the task of estimating the mean of a p-variate normal distribution with covariance matrix $\sigma^2 I$. Clearly, for this case, the mean update algorithm is not as efficient as $\bar{X}$. Returning again to the one dimensional projection in Figure 6.9, we note that since the m nearest neighbors are highly correlated with the starting observation, for m small, the mean update algorithm has efficiency very nearly as bad as the use of the starting observation itself.

As the dimension, p, becomes large, however, a small number of nearest neighbors will tend to surround the population mean as effectively as a random subsample of size m. Thus, for p large, the efficiency of the mean update algorithm is very close to m/n, i.e., the same as a random subsample of size m from a random sample of size n. In Figure 6.11, we show Boswell's results for a variety of dimensions (p) and nearest neighbors (m).

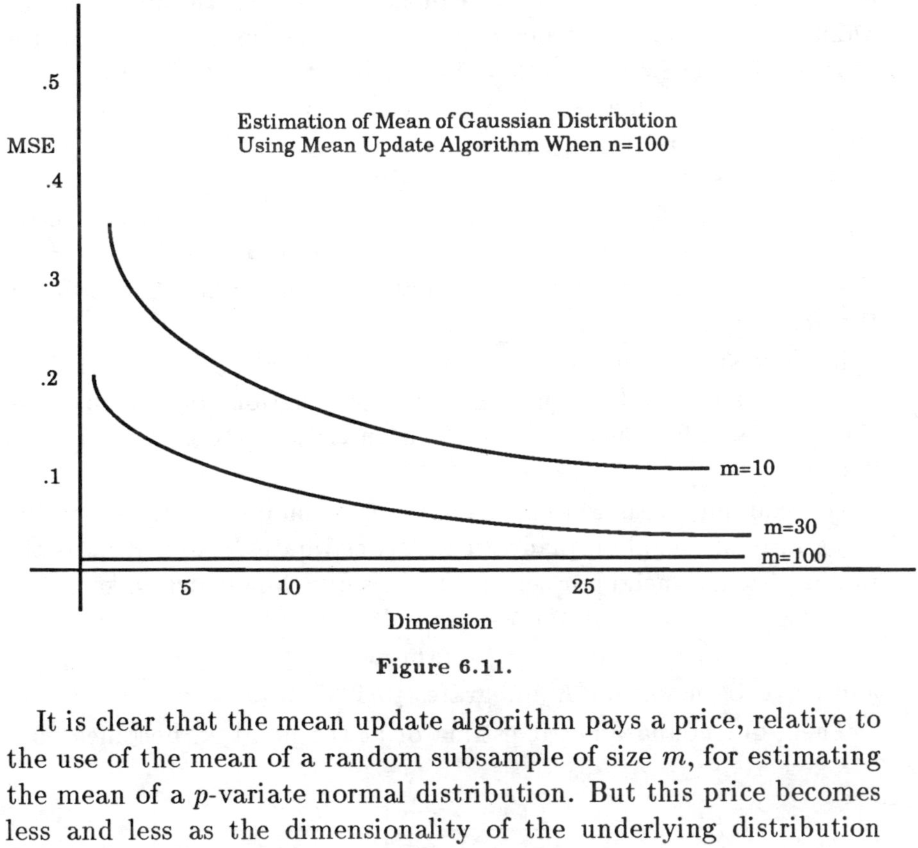

Figure 6.11.

It is clear that the mean update algorithm pays a price, relative to the use of the mean of a random subsample of size m, for estimating the mean of a p-variate normal distribution. But this price becomes less and less as the dimensionality of the underlying distribution increases.

The mean update algorithm is, naturally, intended for the estimation of modes from an underlying distribution with several modes. A prototype of such a distribution is the Gaussian mixture

$$f = \sum_{i=1}^{s} \alpha_i N(\mu_i, \Sigma_i), \tag{40}$$

where the α_i are all positive and sum to 1.

Naturally, for $s > 1$, the mean update algorithm is generally preferred to $\bar{X}$. When we use the mean update algorithm, we must be willing to sacrifice some efficiency for robustness against the possibility of multiple modes. The point is that it appears that, unlike the nonparametric density estimator, graphically displayed, the mean update algorithm does not become unmanageable for high dimensionality. Therefore, if we can make do with the modes of an unknown density, then it makes a great deal of sense to carry out the estimation of the modes rather than attempting to deal with graphical displays of estimates of the density function. Even in those cases where there is interest in evaluation of the density function over that part of the support where it is relatively large, it probably makes sense to carry out the mode estimation as the first part of the investigation. Then, using the location of these modes as centers of interest, we can carry out the estimation of the unknown density in the neighborhoods of these modes.

If one wished to employ a mode finding algorithm similar to those common in the evaluation of maxima of functions of known form, then we need, formally, to base them on estimators of the unknown density.

For example, we might choose to base our mode finding estimation using nearest neighbor quadratic kernel estimates. Thus, if r_m is the radius of the m-neighbor set of X_c, consider the quadratic kernel

$$\begin{aligned} K(X) &= c(1 - \tfrac{1}{2}\|X\|^2), \quad \text{if } \|X\|^2 < 1 \\ &= 0, \qquad\qquad\qquad \text{otherwise} \end{aligned}$$

and c is chosen so that K integrates to 1.

Then, our nonparametric estimator at the point X, becomes

$$\hat{f}_n(X;m) = \frac{c'}{n(r_m(X_c))^p} \sum_{i=1}^{m} \left[1 - \frac{1}{2}\frac{\|X_i - X\|^2}{(r_m(X_c))^2}\right]. \tag{41}$$

Furthermore, the partial derivatives of the quadratic kernel nearest neighbor estimators are given by

$$\frac{\partial}{\partial x_j}\hat{f}_n(X;m) = \frac{c'}{n(r_m(X_c))^{p+2}} \sum_{i=1}^{m}[x_{ij} - x_j] \tag{42}$$

and,

$$\nabla^2 \hat{f}_n(X;m) = \frac{c'm}{n(r_m(X_c))^{p+2}} I, \tag{43}$$

where I is the identity matrix of dimension p.

We observe that, if the smoothing parameter, $r_m(X_c)$ is held fixed, once initialized, then $\nabla^2 \hat{f}_n(X;k)$ is negative definite and constant for all X. Regardless of the value of $r_m(X_0)$, as long as the neighbor set is held fixed, the unique value satisfying $\nabla \hat{f}_n(X;m) = 0$ is given by the mean of the neighbor set. Hence, the mean update is the maximizer of the local quadratic model of the estimator of the density function. Thus, we have shown that the mean update algorithm is equivalent to a steepest ascent algorithm based on a nearest neighbor nonparametric density estimator using a particular quadratic kernel.

Boswell observes that the use of the quadratic kernel (and hence of the Mean Update Algorithm) can stall prematurely before a mode is found. The density estimator is, of course, discontinuous when the nearest neighbor set changes. As an alternative, he suggests an algorithm which is based on Gaussian kernels for the nonparametric density estimator. We note that a fixed bandwidth kernel density estimator with bandwidth h is given by

$$\hat{f}_n(X;h) = -\frac{c}{nh^p}\sum_{i=1}^{n}\left[\prod_{j=1}^{p} K\left(\frac{x_{ij}-x_j}{h}\right)\right]. \tag{44}$$

A necessary condition for $\hat{f}_n$ to have a mode at X is that

$$\nabla \hat{f}_n(X;h) = 0. \tag{45}$$

The $k'th$ partial derivative of this density estimator is given by

$$\frac{\partial}{\partial x_k}\hat{f}_n(X;k) = \frac{c}{nh^{p+1}} \sum_{i=1}^{n}\left[K'\left(\frac{x_{ik}-x_k}{h}\right)\prod_{j=1,j\neq k}^{p} K\left(\frac{x_{ij}-x_j}{h}\right)\right]. \tag{46}$$

Let us use the univariate Gaussian kernel

$$K(t) = \frac{1}{\sqrt{2\pi}}e^{-\frac{1}{2}t^2}. \tag{47}$$

Using the fact that for the Gaussian kernel, $K'(t) = -tK(t)$, we have

$$(48)\quad \frac{\partial}{\partial x_k}\hat{f}_n(X;h) = \frac{c}{nh^{p+2}}\sum_{i=1}^{n}\left[(x_{ik}-x_k)\prod_{j=1}^{p}K\left(\frac{x_{ij}-x_j}{h}\right)\right].$$

The kernel product

$$(49)\qquad \pi_i(X;h) = \frac{c}{nh^p}\left[\prod_{j=1}^{p}K\left(\frac{x_{ij}-x_j}{h}\right)\right]$$

depends upon the location X, the smoothing parameter h, and upon the $i'th$ sample point. However, it is independent of the coordinate with respect to which differentiation is performed. Thus, the critical points of $\hat{f}_n$ are those $\tilde{X}$ for which

$$(50)\qquad \tilde{X} = \frac{\sum_{i=1}^{n}\pi_i(\tilde{X};h)X_i}{\sum_{i=1}^{n}\pi_i(\tilde{X};h)}.$$

Writing this in slightly different fashion, we have

$$(51)\qquad \tilde{X} = \sum_{i=1}^{n} w_i(\tilde{X};h)X_i$$

where

$$(52)\qquad w_i(X;h) = \frac{\pi_i(X;h)}{\sum_{i=1}^{n}\pi_i(X;h)} = \frac{\pi_i(X;h)}{\hat{f}_n(X;h)}.$$

This leads us immediately to Boswell's Weighted Mean Update Algorithm:

Weighted Mean Update Algorithm

Assume as input an initial guess X_c and a smoothing parameter h

Repeat until (stopping criteria are met)

$X_c = \sum_{i=1}^{n} w_i(X_c;h)X_i$

End repeat

Return X_c

As above, the weights are given by

$$w_i(X;h) = \frac{\pi_i(X;h)}{\sum_{i=1}^n \pi_i(X;h)} \tag{53}$$

and

$$\pi_i(X;h) = \exp\left[-\frac{1}{2h^2}(\|X_i - X\|^2)\right]. \tag{54}$$

Boswell also proposes a slight amendment to the Weighted Mean Update Algorithm in the form of the Truncated Weighted Mean Algorithm shown below.

Truncated Weighted Mean Update Algorithm

Select a truncation value k_T

Proceed as in the Mean Update Algorithm but at each step

Set $w_{k_T+1} = w_{k_T+2} = \ldots = w_n = 0$

Rescale so that the weights sum to 1

We recall that for the one dimensional kernel density estimator very little improvement was available going from one kernel to the next. Boswell demonstrates dramatic improvement for the use of the Weighted Mean Update Algorithm over the Mean Update Algorithm for the estimation of the major mode in the case of relatively low dimensionality (say $p < 5$). Stalling due to a local concentration of points away from that mode is less of a problem when we use a weighting procedure which reaches out past the local concentration. On the other hand, the more global nature of the Weighted Mean Algorithm would appear to put it at a disadvantage for the estimation of the location of minor modes.

For analysis of comparative performances in the location of minor modes, Boswell considers the density

$$f(X) = .3N(X; -.0526\underline{1}, I) + .7N(X; 2.4474\underline{1}, I) \tag{55}$$

where $N(X;\nu;\Sigma)$ is the normal density with mean ν and covariance matrix Σ, and $\underline{1}$ is the vector of all 1's. Samples of size 100 were generated, and various m values investigated.

Table 6.1

Average Mean Squared Error Based on 25 Simulations

p	m	Mean Update	Weighted Mean	Truncated Weighted Mean
1	5	.2318	.2774	.2404
1	10	.3637	.7611	.3138
1	20	.6371	2.3160	.3902
1	30	1.3600	4.7360	.9151
1	40	3.2570	5.1390	2.1310
2	5	.2678	.5030	.4375
2	10	.2245	1.5030	.2141
2	20	.3706	4.7450	.4799
2	30	1.5370	6.1820	1.4100
2	40	4.1920	5.9830	1.6200
3	10	.1964	.9155	.3275
3	20	.2856	3.6430	.4054
3	30	.7158	5.5920	.3179
3	40	2.4960	6.0070	.6051
4	10	.1622	.5605	.3175
4	20	.0791	5.2840	.1754
4	30	.1590	6.0550	.0788
4	40	2.441	5.996	.0468
5	10	.1133	.5124	.2422
5	20	.0582	3.3530	.1509
5	30	.0735	5.7500	.0654
5	40	1.338	5.969	.0417
10	10	.1149	.0498	.2910
10	20	.0612	2.1700	.1460
10	30	.05735	5.7570	.0612
10	40	.9219	5.9770	.0414
15	10	.1091	.04317	.2928
15	20	.0520	2.9560	.1302
15	30	.07142	5.8250	.0500
15	40	.7748	6.0430	.0323
20	10	.1001	.2776	.3423
20	20	.0594	2.6540	.1397
20	30	.1024	4.540	.0607
20	40	1.1990	5.7460	.0401

The starting value throughout was $.5\underline{1}$, selected since it would tend to favor motion toward the minor mode near the origin. For both the Weighted Mean Update and the Truncated Weighted Mean Update, the h value of $.5r_m(X)$ was used. For the Truncated Weighted Mean Update, truncation occured after m, i.e., $k_T = m$.

In Table 6.1, we compare these three algorithms for various dimensions, based on a (rather skimpy) Monte Carlo simulation of size 25 for each p and k value shown. Throughout, the criterion used is the mean squared error, namely

$$MSE(\hat{X}) = Ave\left\{\frac{1}{p}\sum_{i=1}^{p}(\hat{x}_i - x_i^*)^2\right\}. \tag{56}$$

Here $\hat{X}$ is the estimator of the local mode, and X^* is the mode.

The situation concerning the various update strategies is ambiguous. Clearly, a great deal of further investigation needs to be carried out. However, the observation in the one dimensional case to the effect that kernel shape does not appear to be particularly important seems to be born out in the higher dimensional cases as well. Thus, the simple-minded Mean Update Algorithm, equivalent to a steepest ascent approach using a particular quadratic kernel, appears to work rather well.

Very significant, however, is the improvement of the algorithms from $p = 1$ through roughly $p = 5$, with no apparent deterioration for dimensions thereafter. Unlike the nonparametric estimation of probability density functions, the estimation of local modes does not become unmanageable for high dimensions.

Let us suppose, for a moment, that our underlying density to be estimated nonparametrically is very close to a mixture of normal densities with close to diagonal covariance structure for each component, and that the dimensionality is reasonably high (say 5 or more). Then it appears that the simple Mean Update Algorithm should work rather well. There will be a tendency, once we have started with a point from one of the component normal densities, to step to other points from that density (see discussion following (39)) if we have not selected the number of nearest neighbors to be too large (certainly, no larger than the fraction of the data which corresponds to the particular normal density). And, again, the high dimensionality will tend to put our stepwise averages close to the mean of the component density.

If the structure of the underlying probability density is sufficiently favorable, then dimensionality higher than 3 (a value of p beyond which nonparametric density estimation by graphical analysis is difficult) can actually be a blessing rather than a curse for examining the locations of the local modes. As to what sort of structure would confuse the Mean Update Algorithm, the most obvious is simply an underlying density which, instead of having distinct modes, has the multidimensional version of the curved manifold structure shown for two dimensions in Figure 6.12.

What sorts of densities do we realistically expect to see in higher dimensional problems? Certainly, in many situations, the ubiquity of the Central Limit Theorem should cause us to see something close to a single Gaussian density. Again, the problem of contamination should cause us rather frequently to see essentially mixtures of Gaussian densities. Beyond that, we can only speculate. As yet, the investigation of higher dimensional data structures is unfortunately disdained by all but a few.

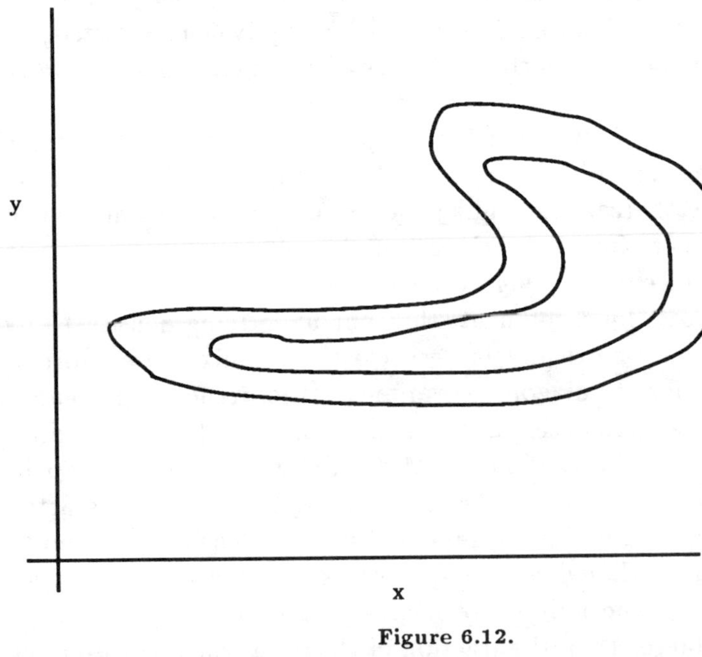

Figure 6.12.

6.4. Appropriate Density Estimators in Higher Dimensions

It would be convenient if the global nature of the maximum penalized likelihood estimator explicated in Chapter 4 extended conveniently to the higher dimensional situation. In the next chapter, we will note that such procedures are the method of choice for such problems as the computation of hazard functions. Unfortunately, for the estimation of density functions, unpublished work by de Montricher would seem to indicate that such methods become unfeasible for $p > 3$. The reason is that the penalty terms consisting of the integral of the squares of partial derivatives must include all partial derivates up to the $[p/2]+1$. Thus, for $p = 4$, we would require numerical computation of derivatives up to the third order. Such differences are notoriously unstable.

Consequently, we are driven back to the very local kernel density estimation methods. It is customary to transform the data set so that the transformed data has unit covariance function. Then, we might use, based on a sample of size n,

$$\hat{f}(X) = \frac{1}{nh^p}\sum_{j=1}^{n} K\left(\frac{X - X_j}{h}\right) \tag{57}$$

where K is a symmetric p-variate density function. Typically, the kernel used will be a Gaussian density with mean zero and unit covariance matrix:

$$K(X) = \left[\frac{1}{(2\pi)^{p/2}}\right] \exp\left[-\frac{\|X\|^2}{2}\right]. \tag{58}$$

Because of the infinite support of the normal density function, it may be computationally convenient to use an approximating quadratic kernel, such as

$$\begin{aligned} K(X) &= c(1 - \tfrac{1}{2}\|X\|^2), \quad \text{if } \|X\|^2 < 1 \\ &= 0, \quad \text{otherwise.} \end{aligned}$$

Silverman [10] presents the following insight on the selection of the bandwidth (smoothing parameter) h. We assume that the p-dimensional kernel K is a radially symmetric probability density function and that the unknown density f has continuous and bounded derivatives. Let us define the constants α and β by

$$\alpha = \int t_1^2 K(\mathbf{t})dt, \tag{59}$$

and

$$\beta = \int K^2(\mathbf{t})dt. \tag{60}$$

Then, using Taylor's Theorem, we have the approximations

$$\text{Bias}_h[\hat{f}(X)] \approx \frac{1}{2}h^2\alpha \nabla^2 f(X), \tag{61}$$

and

$$\text{Var}_h[\hat{f}(X)] \approx \frac{1}{nh^p}\beta f(X). \tag{62}$$

This gives approximate mean integrated square error

$$IMSE \approx \frac{1}{4}h^4\alpha^2 \int [\nabla^2 f(X)]^2 dX + \frac{1}{nh^p}\beta. \tag{63}$$

Thus, the optimal window width is given approximately by

$$h_{opt}^{p+4} \approx \frac{p\beta}{n\alpha^2 \int (\nabla^2 f)^2}, \tag{64}$$

where the Hessian $\nabla^2 f(X)$ is given by

$$\nabla^2 f(X) = \frac{\partial^2 f(X)}{\partial x_i \partial x_j};\ i, j = 1, \ldots, p. \tag{65}$$

But then the mean integrated square error converges to zero like $n^{-1/(p+4)}$. And, of course, the optimal bandwidth depends on the unknown density itself. An iterative procedure such as that given in Chapter 2 is a possibility. Thus if $\hat{f}_i$ is the estimate based on h_i, we could attempt to obtain a revised h via

$$\hat{h}_{i+1}^{p+4} = \frac{p\beta}{n\alpha^2 \int (\nabla^2 \hat{f}_i)^2}. \tag{66}$$

Such a procedure is seldom practical, due to the instability of numerical estimators for the Hessian.

In general, it is not reasonable to suppose that a general bandwidth estimator should be used for all values of X in the p-dimensional space. Accordingly, Silverman suggests the possibility of an adaptive algorithm for modifying the bandwidth at each of the data points. By this procedure, we start with a pilot density estimate $\tilde{f}$. We then determine local bandwidth factors λ_i, by

$$\lambda_i = \left[\frac{g}{\tilde{f}(X_i)}\right]^{\alpha}, \tag{67}$$

where

$$\log(g) = \frac{1}{n}\sum_{i=1}^{n}\log(\tilde{f}(X_i)) \tag{68}$$

and α is a design parameter between 0 and 1.

Then the adaptive kernel estimate is defined by

$$\hat{f}(X) = \frac{1}{n}\sum_{i=1}^{n}\frac{K\left(\frac{(X-X_i)}{(h_i\lambda_i)}\right)}{(h\lambda_i)^p}. \tag{69}$$

This algorithm could, of course, be iterated again, using the new density estimate as a new pilot estimate.

An attractive procedure from the standpoint of focusing on the local is that in which we build up a kind of histogram estimator using as the "interval width" the value of the Euclidean distance $r_m(X)$ of the $m'th$ nearest data point to the X of interest. Thus, the nearest neighbor density estimator is given by

$$\hat{f}(X) = \frac{m/n}{V(r_m(X))}, \tag{70}$$

where V is the volume of a hypersphere of dimension p and radius $r_m(X)$, i.e.,

$$V(r_m(X)) = \frac{\pi^{p/2}(r_m(X))^p}{(p/2)!}, \tag{71}$$

for p even, and

$$V(r_m(X)) = \frac{\pi^{(p-1)/2}\left(\frac{p-1}{2}\right)!(2r_m(X))^p}{p!} \tag{72}$$

for p odd.

The nearest neighbor density estimator is quite natural for many applications. It does have the difficulty of not integrating to unity. This deficiency can be remedied in a number of ways.

As an easily remembered strategy which embodies the flavor of both the nearest neighbor approach and the adaptive kernel estimate, we suggest the estimator related to the SIMDAT resampling scheme in Section 6.2, namely

$$\hat{f}(X) = \frac{1}{n}\sum_{i=1}^{n} N(X - X_i, \Sigma_i), \tag{73}$$

where $N(X)$ is a normal distribution centered at zero with locally estimated covariance matrix Σ_i, the covariance of the $m-1$ nearest neighbors of data point X_i and X_i itself. We recommend values of m approximately equal to $.02n$ work for n large. Naturally, prior information may change this selection.

In general, the problem of estimating a density function in the higher dimensional situation should be carried out in the light of what is really being sought. Usually, the problem can be solved much more simply than trying to obtain a density function throughout $\Re_p$. If the problem is exploratory in nature (i.e., we really do not quite know what we want), then we strongly recommend going through the seeking out of centers of high density as we dealt with in Section 6.3 first. Then, localizing the problem to a detailed investigation of regions around these centers can provide local nonparametric density estimators. We will show in the next section another example of how a data analysis problem can be solved without obtaining a nonparametric density estimator, but using the philosophy of nonparametric density estimation.

6.5. A Test for Equality of Density Functions

Frequently, we want to make a judgement as to whether two data sets represent random observations from the same density function. Once again, it is not necessary that we obtain nonparametric density estimators carefully crafted for graphical display. Indeed, we can readily develop tests which do not require real time intervention for fine tuning.

Let us suppose we have two samples of sizes n_1 and n_2, respectively. We assume, without loss of generality, that $n_1 \leq n_2$. That is, let us suppose we are considering

$$\{X_i\}_{i=1}^{n_1} \text{ and } \{Y_j\}_{j=1}^{n_2}. \tag{74}$$

We would like to make a judgement as to whether they were generated by the same density function.

Template Test. If the two data sets were generated by normal distributions, then our task would be rather straightforward. We start by transforming the first data set $\{X_i\}_{i=1}^{n_1}$ by a linear transformation such that for the transformed data set $\{U_i\}_{i=1}^{n_1}$ the mean vector becomes zero and the covariance matrix becomes I.

$$U = AX + b. \tag{75}$$

Next, we apply the same transformation to the second data set $\{Y_j\}_{j=1}^{n_2}$. We compute the sample mean $\overline{Y}$ and covariance matrix Σ of the transformed data set. Assuming that both data sets come from the same density, the likelihood ratio Λ should be close to unity, where,

$$\Lambda = \frac{\prod_{i=1}^{n_2} \exp[-\frac{1}{2}(u_{1i}^2 + \ldots + u_{pi}^2)]}{\prod_{i=1}^{n_2} \sqrt{|\Sigma^{-1}|} \exp[-\frac{1}{2}\sum_{j,l=1}^{p} \sigma^{jl}(u_{ji} - \overline{Y}_j)(u_{li} - \overline{Y}_l)]}, \tag{76}$$

where σ^{jl} is the $j, l'th$ element of the inverse of Σ. In the case where both generating densities are Gaussian, then if they are both the same and if our sample sizes are large, we know that $-2\log(\Lambda)$ is, if both underlying densities have the same mean and covariance matrix, approximately distributed as chi-square with one degree of freedom.

In the more realistic situation, we may not have good reason to suppose that the generating densities are normal. In such a case, if the two densities are well behaved (i.e., unimodal or nearly so and without heavy tails), we may still use the likelihood ratio as a a kind of template criterion function. However, we lose the neat asymptotic form of the distribution of Λ under the null hypothesis. We can, however, use a resampling procedure, such as the SIMDAT algorithm of Section 6.2, to obtain appropriate significance levels.

For example, let us apply the linear transformation which takes the first data set to standard form to each of 1,000 quasi-data sets of size n_2 generated by the SIMDAT algorithm applied to the first data set. Each of the transformed quasi-data points will be denoted by V. We then compute

$$\Lambda = \frac{\prod_{i=1}^{n_2} \exp[-\frac{1}{2}(u_{1i}^2 + \ldots + u_{pi}^2)]}{\prod_{i=1}^{n_2} \exp[-\frac{1}{2}(v_{1i}^2 + \ldots + v_{pi}^2)]} \tag{77}$$

If 50 or more times the likelihood ratio is greater than or equal to unity, then we fail to reject the hypothesis (at the 5% level) that both distributions are the same.

The above Gaussian template test has the advantage of being simple and quick. Frequently, it can serve as a successful screening procedure. However, we note that it is strongly oriented to measuring differences between the sample means of the two populations. Thus, it can easily fail when one or both of the two populations is multimodal. Other scenarios can be devised to fool the algorithm. It is interesting to note, however, how successful such empirical template methods can be for many non-Gaussian situations.

Nearest Neighbor Test. A test can be built on the observation that if the two data sets have the same distribution, then the nearest neigbor to a point in either should come from either data set in proportion to the numbers of data points in each. As a first step, we will linearly transform (using the same linear transformation on each) each of the two data sets such that the pooled data set has diagonal covariance matrix I and mean zero. For each point in the first data set, we compute the Euclidean distances between each of the points $\{\rho(X_i, X_j)\}$. Then we compute the distance from each of the points in the first data set to each of the points in the second data set $\{\rho(X_i, Y_j)\}$. For each point in the first data set, we determine whether the closest data point is in the first set or the second. If the closest point is in the first set, we increase the score function, S, by one. The probability the score will be increased for a given point if the underlying densities are the same is given by

$$p = \frac{n_1 - 1}{n_1 + n_2 - 1}. \tag{78}$$

Then our test statistic becomes

$$Z = \frac{S - n_1 p}{\sqrt{n_1 p(1-p)}}. \tag{79}$$

For $Z \leq 1.96$, we fail to reject the null hypothesis that the underlying densities are the same at the 5% level.

It should be noted that going from the Template Test to the Nearest Neighbor Test requires the increase of an order of magnitude in computation time. The number of distances to be computed with the Nearest Neighbor Test is

$$\frac{(n_1+1)n_1}{2} + n_1 n_2. \tag{80}$$

As $\min(n_1, n_2)$ reaches 1,000, we are reaching running times in the many minutes range on personal computers. As this figure reaches 10,000, we are immobilizing a large parallel processor. The modern data analyst is frequently confronted with data sets of this size. When embarrassed by such riches of data, we may find it necessary actually to employ random selection of a subset of data from each data set before employing our test.

One way to confuse the Nearest Neighbor Test is to take two different one dimensional densities f_1 and f_2 and multiply each by the same $(p-1)$ dimensional f_{p-1}. As p increases, for constant sample sizes, data generated by the two resulting densities will tend, speciously, to pass the test. Consequently, one may find it appropriate to apply the test, first for the one dimensional marginals, then for the two dimensional marginals, and so on.

Nonparametric Density Estimator Based Test. Moving to an approach based on a nonparametric density estimator, we obtain a nonparametric density estimator using, for example, (73) for the first data set. Let us call the nonparametric density estimator $\hat{f}_1$. Using the SIMDAT algorithm, we generate N quasi-data sets of size $n = n_1$ from the first data set. This will give us the N resampled data sets $\{X_{i,j}\}$. Similarly, using the second data set, we obtain N resampled data sets $\{Y_{i,j}\}$. We compute for each of the $2N$ resampled data sets the score functions

$$S_1(j) = \sum_{i=1}^{n} \log \hat{f}_1(X_{i,j}) \text{ and } S_2(j) = \sum_{i=1}^{n} \log \hat{f}_1(Y_{i,j}). \tag{81}$$

If the two underlying density functions are the same, then the two sets of score functions should be similarly distributed, and a Wilcoxon-Mann-Whitney test can be employed. Pooling the $2N$ sets together, we find the sum of the ranks T in the pooled set of the S_1. Under the null hypothesis that the two sets are the same, we have

$$E(T) = \frac{N(N+1)}{2} \tag{82}$$

and

$$\text{Var}(T) = \frac{N^2(N+1)}{12}. \tag{83}$$

Then, for N large (and typically we can use $N = 1,000$), we can use a one-sided normal test. That is, we will fail to reject the hypothesis that the two underlying density functions are the same at the 5% level if

$$\frac{T - E(T)}{\sqrt{\text{Var}(T)}} \leq 1.96. \tag{84}$$

Clearly, the power of the computer will enable us to come up with a vast variety of resampling tests which can be employed automatically, without human intervention and fine tuning of bandwidths. In the parametric case, the estimation of parameters is easier and more robust than tests for equality of these parameters. For the situation where we do not know the functional form of the underlying densities, it is much easier to develop tests for equality of two densities than it is to gain good graphical estimates for either. We have yet another example of the value of knowing what it is we need before blindly assuming that a nonparametric density estimator is an essential step in the process. Happily, this is seldom the case.

References

[1] Alam, Khursheed and Thompson, J.R. (1968). "Estimation of the mean of multivariate normal distribution." Indiana University Technical Report.

[2] Boswell, S.B. (1983). *Nonparametric Mode Estimation for Higher Dimensional Densities.* Doctoral Dissertation (Rice University).

[3] Efron, Bradley (1979). "Bootstrap methods—another look at the jacknife." *Annals of Statistics.* 7: 1-26.

[4] Fwu, C., Tapia, R.A, and Thompson, J.R. (1981)."The nonparametric estimation of probability densities in ballistics research." *Proceedings of the Twenty-Sixth Conference of the Design of Experiments in Army Research Development and Testing.* 309-326.

[5] James, W. and Stein, Charles (1961). "Estimation with quadratic loss function." *Proceedings of the Fourth Berkeley Symposium.* Berkeley: University of California Press. 361-370.

[6] Kendall, M.G. and Stuart, David(1961). *The Advanced Theory of Statistics, v. II.* New York: Hafner Publishing Company. 22.

[7] Mack, Y.P. and Rosenblatt, Murray (1979). "Multivariate k-nearest neighbor density estimates. *Journal of Multivariate Analysis.* 9: 1-15.

[8] Scott, D.W. and Thompson, J.R. (1983). "Probability density estimation in higher dimensions." *Computer Science and Statistics.* J.E. Gentle, ed., Amsterdam: North Holland. 173-179.

[9] Scott, D.W. (1987). "Data analysis in 3 and 4 dimensions with nonparametric density estimation." in *Statistical Image Processing.* Wegman and DePriest, eds., New York: Marcel Dekker. 281-305

[10] Silverman, B.W. (1986). *Density Estimation for Statistics and Data Analysis.* London: Chapman and Hall.

[11] Taylor, M.S. and Thompson, J.R. (1986)."A data based algorithm for the generation of random vectors." *Computational Statistics and Data Analysis.* 4: 93-101.

[12] Thompson, J.R. (1968)."Some shrinkage techniques for estimating the mean." *Journal of the American Statistical Association.* 63: 113-122.

[13] Thompson, J.R. (1969). "On the inadmissibility of X as the estimate of the mean of a p-dimensional normal distribution for $p \geq 3$." Indiana University Technical Report.

[14] Thompson, J.R. (1989). *Empirical Model Building.* New York: John Wiley & Sons. 177-183.

CHAPTER

7 Nonparametric Regression and Intensity Function Estimation

7.1. Nonparametric Regression

Regression is perhaps the most popular data based technique for examining the interplay between variables, particularly those relationships loosely described as "cause and effect." Regression may be used for predicting the yield of a chemical reaction for a given temperature, input concentration and reaction time, or it can be employed to predict next year's gross national product given a hundred supposedly relevant variables. In many cases, regression models will be used for optimization purposes, as in the chemical reaction situation where we may wish to select temperature, concentration and reaction time to produce maximum yield.

Generally speaking, regression is used in a more or less ad hoc fashion, without much thought as to the mechanism underlying the system under consideration. The following argument is frequently made for using linear regression. We have n observations of a variable y and of each of p related variables. Suppose we postulate a relationship between y and p variates $(x_1, \ldots, x_p) = X$:

$$y_i = g(x_{1i}, \ldots, x_{pi}) + \text{zero average noise}. \tag{1}$$

We will assume enough differentiability of the unknown function g that we can employ a Maclaurin's expansion to give:

$$(2)\quad y_i = a + b_1 x_{1i} + \ldots + b_p x_{pi} + R_p(X_i) + \text{zero average noise.}$$

So far so good. But then, with a wave of the hand, we assume that

$$(3)\qquad R_p(X) \approx 0.$$

The investigator then uses as a surrogate for reality

$$(4)\qquad y_i = b_0 + b_1 x_{1i} + \ldots + b_p x_{pi} + \epsilon_i$$

where

$$(5)\qquad E(\epsilon_i) = 0,\ E(\epsilon_i^2) = \sigma^2, \text{ and } E(\epsilon_i \epsilon_j) = 0,$$

for all i and $i \neq j$.

Sometimes, this reduction to a linear model is satisfactory. Generally, other than as a crude first approximation, it is not. Attempts are frequently made to overcome the inadequacies of linear approximations to a highly nonlinear reality by adding still more x variables but retaining the linearity of the model. Such approaches only increase instability and are generally ill advised. Econometricians have used such an approach to predict eleven of the last two recessions.

As a practical matter, regression models, linear or nonlinear, should be considered exploratory devices, to be replaced as soon as possible by theory based models supported by consideration of the mechanism underlying the process. In other words, it is not very practical to discard theory for a misplaced confidence in linear regression models. Those areas of quantitative science which are most successful tend to follow the paradigm of replacing the ad hocery of a linear model with a mechanistic one as soon as possible. This is true, for example, of chemistry and physics. There are, however, areas where investigators are pleased to use the linear model indefinitely, even though the fit to reality is well known to be unsatisfactory for the particular application at hand. In some areas, linear models seem to have a life of their own, to be considered the actual reality, the data notwithstanding. As we shall argue in a later section on chaos theory, a model accepted in spite of its total dissonance with the data, is a presocratic throwback with which science can well do without.

For the case of the linear model, estimation of the regression coefficients is generally very quick and easy. Suppose again that the model is given by

$$y_i = b_0 x_{0i} + b_1 x_{1i} + \ldots + b_p x_{pi} + \epsilon_i = BX_i + \epsilon_i, \tag{6}$$

where x_{0i} is always taken to equal 1. Then, suppose we wish to find the regression coefficients which minimize the sum of squares of the miss between the model and the data, i.e., we wish to minimize

$$S(B) = \sum_{i=1}^{n} [y_i - BX]^2. \tag{7}$$

Simply taking the partical derivatives of $S(B)$ with respect to each of the $p+1$ regression coefficients and setting each equal to zero, gives

$$\hat{B} = (X^T X)^{-1} X^T y. \tag{8}$$

We will not take the time here to discuss those cases where the matrix to be inverted is singular, or nearly so. The point to be made is that if the linear model really is sufficiently close to reality as to be used as such, then we are in a fortunate (and rare) situation. Computation is almost instantaneous, and holistic perception of the interrelationship between y and X is readily obtained.

To begin consideration of nonlinear relationships between y and X, let us first return to the simple case where $p = 1$. The only non-trivial functional relationship between y and x which can be readily perceived graphically is that of the straight line. Accordingly, since the time of the beginnings of physical chemistry and chemical engineering in the nineteenth century, scientists have considered trying to understand monotonic relationships by going through a pharmacopoeia of transformations which would produce a linear relationship between the two variables.

For example, let us suppose that, unbeknownst to us but in actuality,

$$y = 7e^{.3x}. \tag{9}$$

In Figure 7.1 below, we note that an untransformed graph of y versus x gives us little indication as to the functional relationship between the two variables.

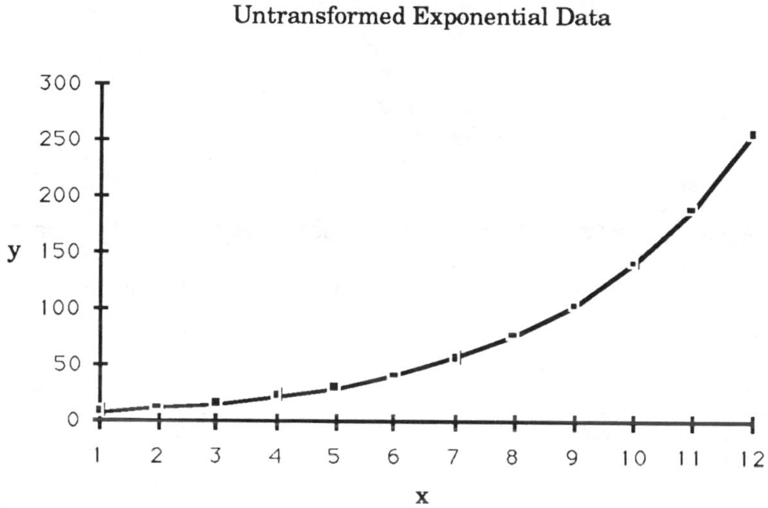

Figure 7.1.

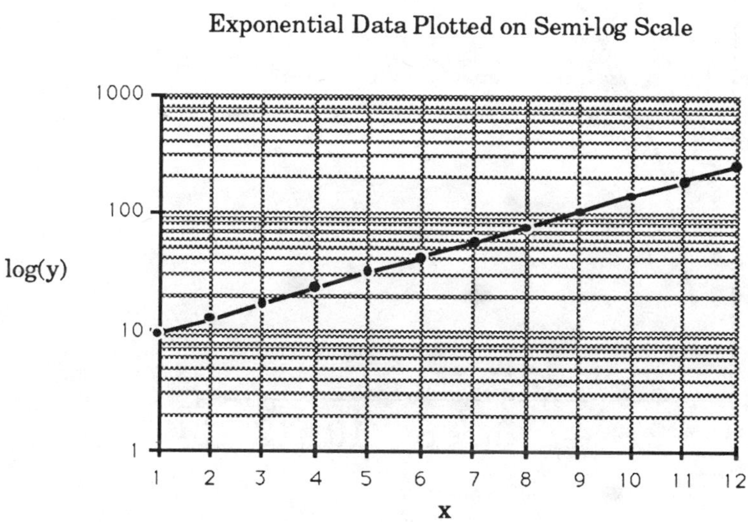

Figure 7.2.

If we plot $\log(y)$ versus x, however, then the picture becomes clearer as we see in Figure 7.2. Naturally, the plot on semi-log paper is no panacea. Consider the case where the relationship is given by

$$y = 4x^{.2}. \tag{10}$$

In Figure 7.3, we note that the semi-log plot does not yield a straight line.

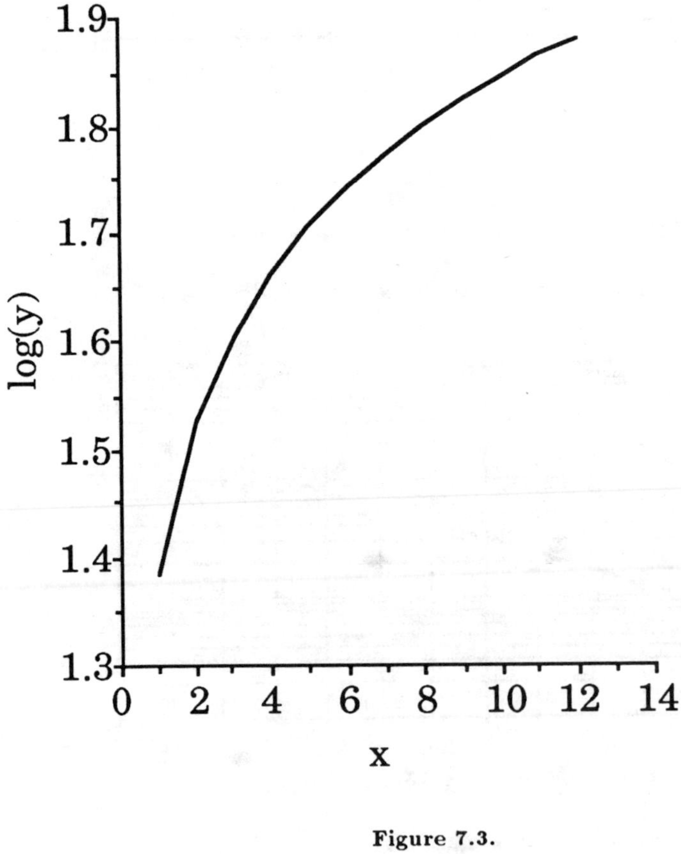

Figure 7.3.

However, in the log-log plot of Figure 7.4, a straight line is obtained.

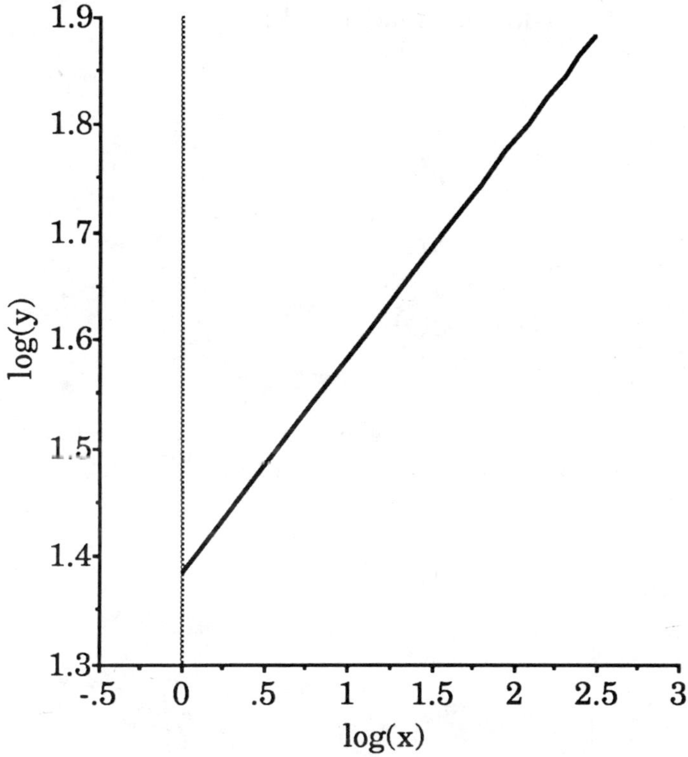

Figure 7.4.

In the situation where x is one dimensional, if we can find a transformation on x and/or y which will reduce the transformed data to a linear relationship, then it is an easy matter to infer the functional relationship between x and y. Naturally, the measure of the set of simple transformations in the set of all possible transformations is zero. However, the simple transformations of power, logarithm, exponential, etc. correspond to phenomena at the level of deep modeling which will frequently correspond to reality. Building on this hundred year old technology, Tukey [4] has suggested a "Transformational Ladder" which, when applied to a monotone relation in the data, frequently yields deep insights into the actual mechanism driving the process.

Transformational Ladder

$$\exp(e^y)$$
$$\exp(y)$$
$$y^4$$
$$y^2$$
$$y$$
$$y^{\frac{1}{2}}$$
$$y^{\frac{1}{4}}$$
$$\log(y)$$
$$\log(\log(y))$$

In Figure 7.5, we note Tukey's simple graphical aid for determining whether to go up or down the Transformational Ladder. Curves of Type A point down the Transformational Ladder. Curves of Type B point up the ladder.

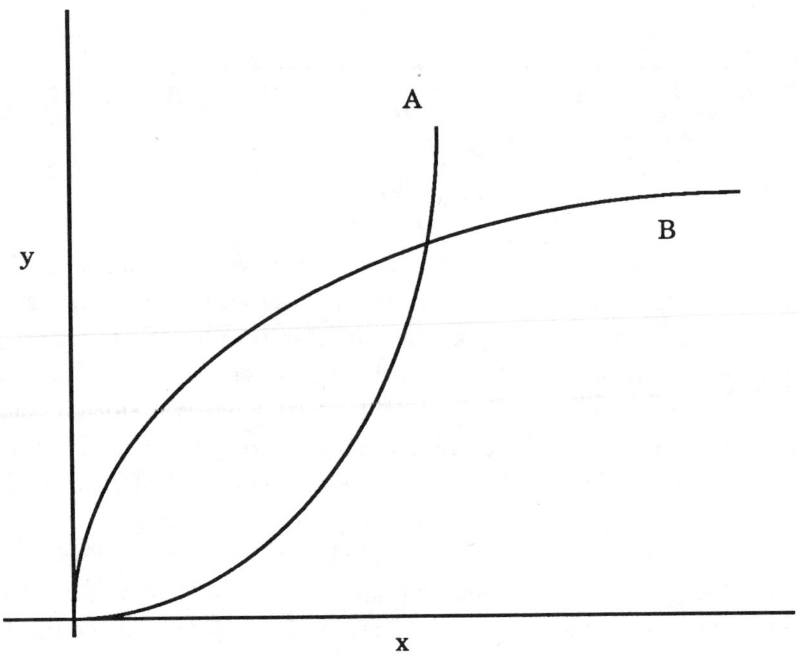

Figure 7.5.

Once we leave monotone relationships, we generally lose our hope of ready insight into the basic mechanism underlying a process via the

Transformational Ladder. Following Tukey, we use a nearest neighbor approach to deal with those situations where the x-y relationship is not monotone.

One ready means for smoothing a regression relationship is the very local procedure of digital filtering. Among the simplest of the digital filters is the hanning window. Using such a window, which generally supposes equal spacing in the x domain, we replace each observation by the average of the observation weighted twice with that of its two adjacent (in the x domain) nearest neighbors. Thus, the hanning filter gives

$$y(x) \leftarrow \frac{y(x-1) + 2y(x) + y(x+1)}{4}. \tag{11}$$

The hanning filter is excellent for "smoothing out rough edges." It is much less satisfactory for discounting the effect of wild observations. Generally speaking, those who rely on hanning for getting rid of wild points must apply it several times, frequently with devastating consequences. We give an example of this unfortunate property in Table 7.1.

Table 7.1

Repeated Application of a Hanning Window

x	$y(x)$	H	HH	HHH
1	1	1	1	1
2	1	1	1	16.61
3	1	1	63.44	94.66
4	1	250.75	250.75	235.14
5	1000	500.50	375.62	313.18
6	1	250.75	250.75	235.14
7	1	1	63.44	94.66
8	1	1	1	16.61
9	1	1	1	1

We note how the attempt to hann the one big outlier (1,000) from the data set contaminates the rest of the data set. Hanning has the unhappy property that when used many times, the data will tend to a straight line. It was Tukey who introduced hanning into the time series field fifty years ago, and it was Tukey who discovered an alternative filter, which would eliminate the major effects of outliers without allowing their effects to leak throughout the smoothed data set. This median smooth algorithm is as simple as it is effective. We simply replace $y(x)$ by the median of itself and the two (x space) nearest neighbors. That is

$$y(x) \leftarrow \text{med}\,[y(x-1), y(x), y(x+1)]. \tag{12}$$

Unlike the hanning algorithm, we do not have to worry about repeated applications obscuring the data. In fact, it is seldom the case that more than two or three applications are necessary until the resulting transformed set is invariant to further applications of the smoother. The median smooth, repeated until it has no further effect, is termed the "3R." The 3R filter can be used to eliminate the effect of wild points, and then followed by a hanning filter to smooth away the rough edges (some of which will typically be induced by the 3R itself). There are many nuances to Tukey's technique for nonparametric regression procedures. However, most of the power of the procedures is illustrated by the use of 3RH on the data set in Table 7.2 consisting of 21 days of returned goods at a large department store.

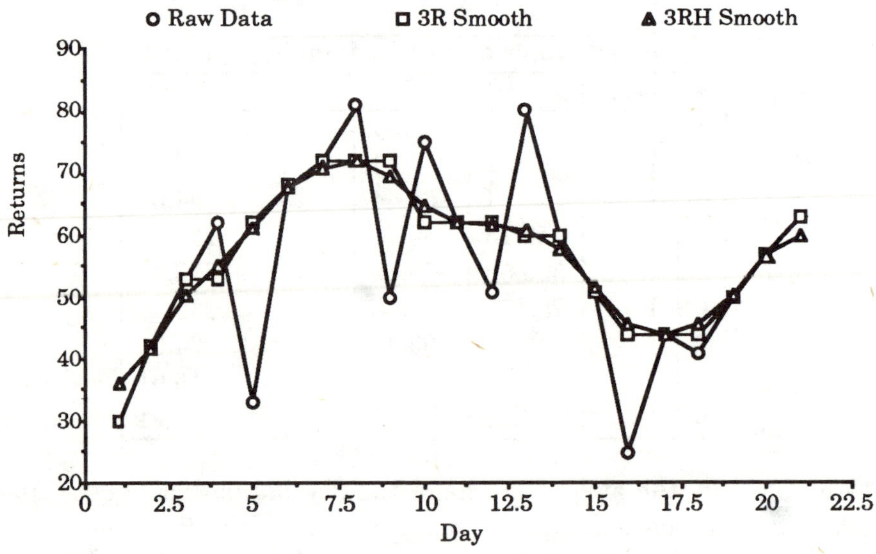

Figure 7.6.

We note that the "3" column uses the median smooth once. The "3R" column shows all the values which change after a second application of the median smooth. No values change for any further application of the median smooth. At the ends, we follow the convention of not changing the values of the endpoints for the 3R filter and averaging the end value with the next to end value for the H filter. We note the smooths in Figure 7.6.

Table 7.2

Smooths of Returned Item Data

Day	Returns	3	3R	3RH
1	30	30		36.00
2	42	42		41.75
3	53	53		50.25
4	62	53		55.25
5	33	62		61.25
6	68	68		67.50
7	72	72		71.00
8	81	72		72.0
9	50	75	72	69.50
10	75	62		64.50
11	62	62		62.00
12	51	62		61.50
13	80	60		60.50
14	60	60		57.75
15	51	51		51.50
16	25	44		45.75
17	44	41	44	44.00
18	41	44		45.50
19	50	50		50.25
20	57	57		56.75
21	63	63		60.00

We note that we are now in a fundamentally different situation than we were when dealing with a y monotone in x. In the monotone situation, we have the possibility of gaining some deep insights into the mechanism generating the data set. And even if we do not find a transformation which appears to relate to a recognizable mechanism (a seventh root transformation, for example, may sometimes nearly transform the data to linearity but is unlikely to be easily identifiable as relating to a fundamental mechanism), we can, with some optimism, extrapolate a monotone curve outside the range of the data. Such is not the case with a data set which we must treat with a 3RH smoother. We cannot readily use the local smoothing process to gain fundamental insights about the underlying process. Furthermore, we should generally be hesitant about extrapolating outside the data base. The use of such nearest neighbor procedures as 3RH is mainly to provide an ad hoc smoother and interpolator.

Leaving the tidy, well ordered domain of the one dimensional x, we can use the insights gained examining 3RH to build reasonable nonparametric regression procedures for higher dimensional X. We note that for $p > 1$, it is seldom the case that the X values will be given over a uniform mesh. As a first step, then, we will apply a linear transformation to X such that the transformed values will be centered at zero with identity covariance matrix.

Suppose our problem is, on the basis of insights gained from a data set of size n, to obtain a reasonable conjecture as to what y will be at a point X_0. For m equal to the appropriate fraction of n, say $.05n$ for samples under 100 and $.02n$ for samples over a thousand, we find the m nearest X neighbors of X_0 and pair them with their y values. Ranking these points according to their distances from X_0, we have $(X_{0,1}, y_{0,1}), \ldots, (X_{0,m}, y_{0,m})$. Then, an equal weight nearest neighbor estimator for y is given by

$$\hat{y}(X_0) = \frac{1}{m}\sum_{i=1}^{m} y_{0,i}. \tag{13}$$

We note that this formula is quite easily automated and can be used more or less blindly. That means, if we should happen to do something as foolish as attempting to estimate $y(X_0)$ for X_0 outside the convex hull of the X_i, the formula will be only too happy to oblige, though the answers obtained would probably be worth very little. Again, for high dimensional problems of modest data size, we may be hoping to estimate $y(X_0)$ essentially in the middle of a data desert, i.e, at a point around which we have little information. In such a case, the formula will give us an answer, but it will not be much use. It is in the nature of regression theory, parametric as well as nonparametric, that it is easy to get numerical answers, but the utility of these answers very much depends on the data set at hand in relation to the X value where we wish to estimate y and the purposes for which the answers will be used. It is the ability of a regression model confidently to present us with answers beyond the justification of the data at hand and without any real understanding of the underlying process that makes regression one of the most dangerous of the techniques of statistics.

Obviously, just as with the hanning window, we may prefer to give higher weight to those (y, X) values where the X's are closer to X_0. This is particularly the case for n small.

In general, the problem of nonparametric regression is much easier than the problem of nonparametric density estimation. To obtain some feel as to why this is the case, let us suppose that $(y, x_1, \ldots, x_p)$ has the joint probability density $f(y, X)$. Then the regression function of y on X_0, $\tilde{y}(X_0)$, is given by

$$(14)\quad \tilde{y}(X_0) = E(y|X_0) == \int yf(y|X_0)dy = \int \frac{yf(y, X_0)dy}{\int f(y*, X_0)dy*}.$$

Thus, $\tilde{y}(X)$ is, in a sense, the averaging of a one dimensional density function. The estimation of $f(y|X)$, on the other hand, would involve a $(p+1)$-dimensional problem.

Once again we have an example of how the estimation and graphical display of a density function can frequently be sidestepped.

Returning to the nearest neighbor nonparametric regression estimator, consistency in mean square error is guaranteed for estimators of the form

$$(15)\qquad \hat{y}(X_0) = \sum_{i=1}^{m} w_i(X_0) y_{0,i}$$

provided we satisfy the appropriate version of the Parzen-Rosenblatt conditions, namely, the weights are non-negative and sum to unity, and $m \to \infty$ and $m/n \to 0$ as $n \to \infty$.

Thus, if the conditions are satisfied,

$$(16)\qquad \lim_{n\to\infty} E[\hat{y}(X) - \tilde{y}(X)]^2 = 0.$$

7.2. Nonparametric Intensity Function Estimation

Let us return to Graunt's Life Table on page 3. We shall attempt to resolve the ambiguities of the age boundaries with the following assumption. Namely, the deaths from 0 to 6 will be understood to mean deaths that occur in the first year through the end of the fifth year. The 6 to 16 interval will be taken to be deaths which occur from the beginning of the sixth year through the end of the 15th year, etc. If we make a rough plot of the probability of death on a per year basis at the center of interval we obtain the graph in Figure 7.7.

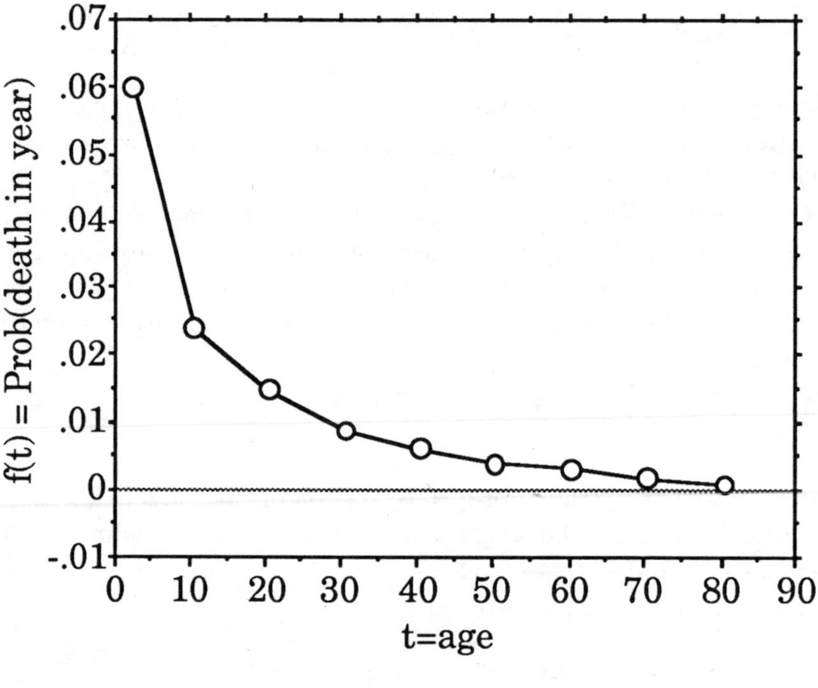

Figure 7.7.

This relationship perhaps looks rather negative exponential, though, as we have observed, the human eye really only can identify a straight line. Plotting $\log(f(t))$ versus t in Figure 7.8, we obtain a graph which is very close to linear.

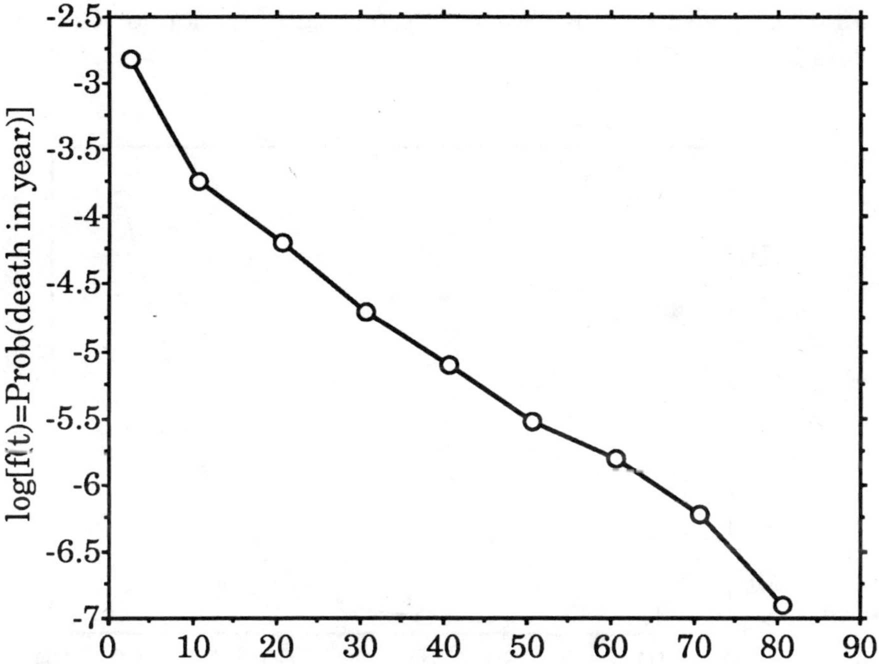

Figure 7.8.

Let us go futher and examine the hazard function, $H(t)$ of Graunt's data where

$$(17)\quad H(t) = \lim_{\Delta t \to 0} \frac{\text{Prob(death in } [t, t+\Delta t])/(\Delta t)}{\text{Prob(no death before } t)} = \frac{f(t)}{1-F(t)}.$$

For the case where

$$(18)\qquad f(t) = \lambda e^{-\lambda t},$$

we have

$$(19)\qquad H(t) = \frac{\lambda e^{-\lambda t}}{1-(1-e^{-\lambda t})} = \lambda.$$

In other words, a density of deaths which is exponential would have the interesting property that there was no deterioration in life expectancy as a result of age. At the beginning of each day, all living persons in the population would have the same expectation of surviving the day. Such a phenomenon would apply in a human population only under special circumstances, such as the presence of a rather virulent plague. We show the hazard function for Graunt's data in Figure 7.9.

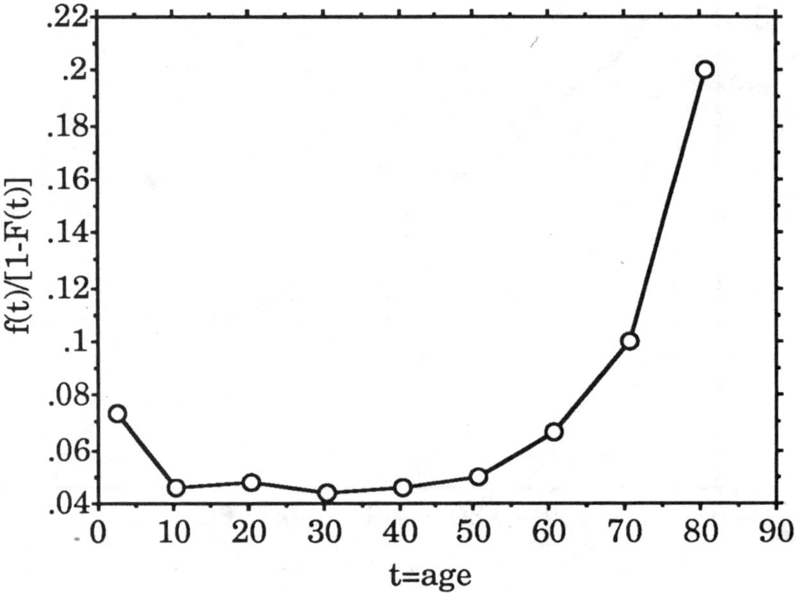

Figure 7.9.

We see that the density of time to death is not really exponential. In fact, the hazard function shows an increased risk of death in the first years of life, followed by a nearly flat hazard until the age of 60. Still, the daily risk of death for a seventy-year-old individual is only roughly twice that of a twenty-year-old individual. The data is indicative of the presence of an epidemic (or epidemics) which tends to minimize the aging process itself as a factor in the force of mortality. This is the kind of information, one supposes, Thomas Cromwell had in mind when commissioning the tabulation of the age at death over a century before Graunt's analysis.

Let us proceed to a convenient generalization of the exponential mortality density. We shall postulate that

$$\text{Prob(death in } [t, t + \Delta t]) = \lambda(t)\Delta t. \tag{20}$$

Then,

$$\text{Prob(0 failures by } t + \Delta t) = P_0(t + \Delta t) = P_0(t)(1 - \lambda(t)\Delta t). \tag{21}$$

So,

$$\frac{P_0(t+\Delta)-P_0(t)}{\Delta t} = -\lambda(t)P_0(t). \tag{22}$$

Thus

$$\frac{dP_0(t)}{dt} = -\lambda(t)P_0(t). \tag{23}$$

Finally,

$$P_0(t) = \exp\left[-\int_0^t \lambda(\tau)d\tau\right]. \tag{24}$$

We can now compute the density of the time of death via

$$f(t) = \frac{d}{dt}[F(t)] = \frac{d}{dt}[1-P_0(t)] = \exp\left[-\int_0^t \lambda(\tau)d\tau\right]\lambda(t). \tag{25}$$

The hazard function is then

$$H(t) = \frac{f(t)}{P_0(t)} = \frac{\exp[-\int_0^t \lambda(\tau)d\tau]\lambda(t)}{\exp[-\int_0^t \lambda(\tau)d\tau]} = \lambda(t). \tag{26}$$

The function $\lambda(t)$ is normally referred to as an *intensity function.* As an empirical device, the intensity function frequently gives insights into the mechanism of complex phenomena indexed by time. Accordingly we shall spend some time developing nonparametric estimators for $\lambda(t)$ in a cancer related process [1].

Let us consider a study consisting of n patients who present with a particular type of solid tumor. At the time when a patient has the tumor excised, we set his or her clock equal to zero. The patient is then followed for a time period of length T_i. During the time period $(0, T_i]$, the patient exhibits metastases at times $0 < t_{i1} < \ldots < t_{im_i}$. If the patient disappears from the study at time T_i for any reason, including death, we will take T_i for the censoring time of that patient.

At least two processes underlie the metastatic process. Firstly, there is the process by which a tumor breaks free from a primary tumor to form a metastasis. Secondly, there is the growth process whereby the metastatic tumor grows to observable size. The estimation of an empirical intensity function is a prudent first step in the modeling process. We shall make the simplifying assumption that each patient has the same intensity function $\lambda(t)$. Thus for a given patient, the probability that no metastasis will be observed from time t_i to time t_j is given by

$$\exp\left[-\int_{t_i}^{t_j}\lambda(\tau)d\tau\right]. \tag{27}$$

Then, the joint likelihood, given the times of metastasis observations, is

$$L(\lambda)=\left[\prod_{i=1}^{n}\prod_{j=1}^{m_i}\lambda(t_{ij})\right]\exp\left[-\sum_{i=1}^{n}\int_0^{T_i}\lambda(\tau)d\tau\right]. \tag{28}$$

The likelihood function, without some stiffening, has no unique maximum. The naive Dirac-comb estimator is given by

$$\lambda_\delta(t)=\sum_{i=1}^{n}\sum_{j=1}^{m_i}\delta(t-t_{ij}), \tag{29}$$

where δ is the Dirac delta function. Generally, such an estimate would be deemed unacceptable for most applications.

Histogram Estimator. The analogue of the histogram density estimator is given by

$$\hat{\lambda}_h(t)=\frac{\psi(t,h(t))}{h(t)\xi(t)}, \tag{30}$$

where

$$\psi(t,h(t)) = \text{metastases counted in}\left(t-\frac{h(t)}{2},t+\frac{h(t)}{2}\right), \tag{31}$$

and

$$\xi(t) = \text{ number of patients still at risk at time } t. \tag{32}$$

Using an argument similar to that on pages 46 and 47, it can be shown that

$$MSE[\hat{\lambda}_h(t)|\xi(t)]=\frac{\lambda(t)}{h\xi(t)}+\frac{h^2\lambda^4(t)}{4}+O\left(\frac{1}{\xi(t)}\right)+O(h^3). \tag{33}$$

A consideration of the two leading terms suggests the use of

$$h(t)=\left(\frac{2}{\xi(t)}\right)^{1/3}\frac{1}{\lambda(t)}. \tag{34}$$

We note the power of 1/3 just as with the histogram estimator of density functions. Naturally, we do not know $\lambda(t)$, but we can use an iterative procedure similar to that for bandwith selection of density estimators as shown on page 67.

$$h_{j+1}(t) = \left(\frac{2}{\xi(t)}\right)^{1/3} \frac{1}{\lambda_{h,j}(t)}. \tag{35}$$

We use the customary stopping criterion that $\hat{\lambda}_{h,j+1}(t)$ be close to $\hat{\lambda}_{h,j}(t)$. In the event the estimator cycles, we can use convergence of the Cesaro sum

$$\sum_{j=0}^{N} \frac{\hat{\lambda}_{h,j}}{N+1} \tag{36}$$

as a stopping criterion, where N may typically be taken to be 20.

Constrained Maximum Likelihood Estimator Without Penalty Term. We need not stiffen (28) if we know that the intensity function is nondecreasing. Let us suppose $\lambda(t) \in M(0,T)$, the class of all nonnegative monotone nondecreasing functions on $(0,T)$. We shall seek to

$$\max_{\lambda \in M(0,T)} \log L(\lambda) = \sum_{i=1}^{n} \sum_{j=1}^{m_i} \log \lambda(t_{ij}) - \sum_{i=1}^{n} \int_0^{T_i} \lambda(\tau) d\tau. \tag{37}$$

It is clear that increases in $\log(L(\lambda))$ can occur only by increases in $\lambda(.)$ at the metastasis points. Since $\lambda(.)$ is nondecreasing, any increase of $\lambda(.)$ away from these points will decrease the log likelihood via the integral terms, with no accompanying increase via the log terms. Thus, it is clear that a solution to (37) must consist of step functions closed on the left with no salti except at some of the metastasis points.

Let us sort the metastasis times from all patients into a single list:

$$0 < \tau_1 < \ldots < \tau_s < \tau_{s+1} = T = \max T_i. \tag{38}$$

Let c_j = number of metastases observed at time τ_j, and let

$$d_j = \sum_{i=1}^{n} \max[0, \min\{T_i - \tau_j, \tau_{j+1} - \tau_j\}]; \text{ for } j = 1, \ldots, s. \tag{39}$$

We can then reduce (37) to

$$\max_{\lambda_i \geq 0} \log L(\lambda_1, \ldots, \lambda_s) = \sum_{i=1}^{s} (c_i \log(\lambda_i) - d_i \lambda_i). \tag{40}$$

To solve (40), we let $z_j = \lambda_j; j = 1, \ldots, s$. If $z_1, \ldots, z_{s-1}$ have already been selected, then z_s is that value of z which is greater than or equal to z_{s-1} and maximizes

$$G_s(z) = c_s \log(z) - d_s z. \tag{41}$$

Since the unique maximum of $G_s(z)$ occurs at $z = c_s/d_s$,

$$\begin{aligned} z_s &= \frac{c_s}{d_s} \text{ if } \frac{c_s}{d_s} > z_{s-1} \\ &= z_{s-1} \text{ if } \frac{c_s}{d_s} \leq z_{s-1}. \end{aligned} \tag{42}$$

Next, we let $D_1(z_{s-1})$ be the maximum of the last term of the sum in (40), given the choice of z_{s-1}, i.e.,

$$\begin{aligned} D_1(z_{s-1}) &= c_s \log(z_{s-1}) - d_s z_{s-1} \text{ if } z_{s-1} \geq \frac{c_s}{d_s} \\ &= c_s \log\left(\frac{c_s}{d_s}\right) - d_s \frac{c_s}{d_s} \text{ if } z_{s-1} < \frac{c_s}{d_s}. \end{aligned} \tag{43}$$

We observe that $D_1(z_{s-1})$ is at first constant and then decreases monotonically to $-\infty$ as z_{s-1} increases.

Next, suppose we have $z_1, \ldots, z_{s-2}$. Then z_{s-1} is that value which is greater than or equal to z_{s-2} and maximizes

$$c_{s-1} \log(z_{s-1}) - d_{s-1} z_{s-1} + D_1(z_{s-1}). \tag{44}$$

Observing that the derivative decreases monotonically from ∞ at 0 and goes to $-\infty$ as $z_{s-1} \to \infty$, the maximum is unique. If $z_{s-1} \leq c_s/d_s$, the derivative is $c_{s-1}/z_{s-1} - d_{s-1}$, and the maximum occurs at c_{s-1}/d_{s-1} if $c_{s-1}/d_{s-1} \leq c_s/d_s$. If $z_{s-1} > c_s/d_s$, the derivative equals $(c_{s-1} + c_s)/z_{s-1} - (d_{s-1} + d_s)$, and the maximum occurs at $(c_{s-1} + c_s)/(d_{s-1} + d_s)$ if $c_{s-1}/d_{s-1} > c_s/d_s$.

Let

$$D_2(z_{s-2}) = \max_{z_{s-1} \geq z_{s-2}} [c_{s-1} \log(z_{s-1}) - d_{s-1} z_{s-1} + D_1(z_{s-1})]. \tag{45}$$

The maximum must occur at z_{s-2} or at one of the points c_{s-1}/d_{s-1}, $(c_{s-1}+c_s)/(d_{s-1}+d_s)$, depending on z_{s-2}. Again, $D_2(z_{s-2})$ is first constant in z_{s-2} and then decreases to $-\infty$.

We next assume $D_k(z_{s-k})$ is already defined and is first constant in z_{s-k} and then decreases to $-\infty$. Choose z_{s-k} which is greater than or equal to z_{s-k-1} and maximizes

$$c_{s-k}\log(z_{s-k}) - d_{s-k}z_{s-k} + D_k(z_{s-k}). \tag{46}$$

Again, the maximum is unique and

$$D_{k+1}(z_{s-k-1}) = \max_{x\geq z_{s-k-1}}(c_{s-k}\log(x) - d_{s-k}x + D_k(x)) \tag{47}$$

is of the same form as D_k. Proceeding in this way, we obtain the optimal value

$$z_1 = \lambda_1 = \lambda(\tau_1). \tag{48}$$

We then proceed to the solution in $[\tau_2, T]$ with fixed z_1, and so on.

Maximum Penalized Likelihood Estimator. The technique of maximum penalized likelihood estimation developed in Chapters 4 and 5 lends itself very nicely to the nonparametric estimation of intensity functions. Again, let us consider a penalized version of (28), namely

$$\max_{\lambda\in S} J(\lambda,\alpha) = \sum_{i=1}^{n}\sum_{j=1}^{m_i}\log(\lambda(t_{ij})) - \sum_{i=1}^{n}\int_0^{T_i}\lambda(\tau)d\tau - \|\lambda\|^2, \tag{49}$$

where S is a closed convex set of the Sobolev space

$$H^s(0,T) = \{\lambda|\lambda^{(j)} \in L^2(0,T)\ ;\ j = 0,\ldots,s \geq 1\}, \tag{50}$$

$$\|\lambda\|^2 = \sum_{j=0}^{s}\alpha_j\int_0^T(\lambda^{(j)}(\tau))^2d\tau, \text{ with } \alpha_j \geq 0, \text{ for } j = 0,\ldots,s, \tag{51}$$

and $T = \max(T_i)$. That $H^s(0,T)$ is a reproducing kernel Hilbert space is proved in [3]. Let $J : H \to R$ be continuous in S, twice Gâteaux differentiable in S with the second Gâteaux variation uniformly negative definite in S. From Theorem 7 of Appendix I, we know that J has a unique maximizer in S.

But $H^s(0,T)\bigcap\{\lambda|\lambda \geq 0\}$ is closed and convex. Since $H^s(0,T)$ is a RKHS, pointwise evaluation is a continuous operation. To establish the continuity of J in $H^s(0,T)\bigcap\{\lambda|\lambda \geq 0\}$, we need only show that for any sequence such that $\lambda_n \to \lambda_0$ in $H^s(0,T)$ norm,

$$\int_0^t \lambda_n(\tau)d\tau \to \int_0^t \lambda_0(\tau)d\tau. \tag{52}$$

But this is obvious because of the $H^s(0,T)$ norm.

Finally, a straightforward computation show that the second Gâteaux variation of J at λ in the η direction is given by

$$J''(\lambda)(\eta,\eta) = -\sum_{i=1}^{n}\sum_{j=1}^{m_i}\frac{\eta^2(t_{ij})}{\lambda^2(t_{ij})} - 2\|\eta\|^2 < -\|\eta\|^2. \tag{53}$$

But then J is uniformly negative definite on $H^s(0,T)\bigcap\{\lambda|\lambda \geq 0\}$.

This establishes in $H^s(0,T)\bigcap\{\lambda|\lambda \geq 0\}$ the existence of a unique solution to (49) . Furthermore, since $H^s(0,T)\bigcap\{\lambda|\lambda \geq 0, \lambda' \geq 0\}$ is closed and convex, there is a unique solution to (49) when the constraint of monotonicity is added. We note that the proof of existence and uniqueness immediately leads us to an easy creation of a numerical algorithm for both the unconstrained and constrained versions of (49). We simply divide the interval $(0,T)$ into N subintervals of equal length and solve (49) for the $\{c_i \geq 0\}$ in

$$\hat{\lambda}(t) = c_i \text{ for the } i'th \text{ subinterval.} \tag{54}$$

The existence and uniqueness result shows that in the limit as N goes to ∞ the step function approximation will not deteriorate into Dirac spikes.

Because of the naturally occurring term $-\int_0^T \lambda(\tau)d\tau$ in $J(\lambda,\alpha)$, we can use $\alpha_0 = 0$. At the practical level, in the discrete implementation no instability results when we let $\alpha_1 = 0$. The method for selection of the smoothing parameter α_2 is similar to that for penalized likelihood techniques in the density estimation case, namely to decrease α_2 until high frequency wiggles appear, and then increase α_2 to the preceding value.

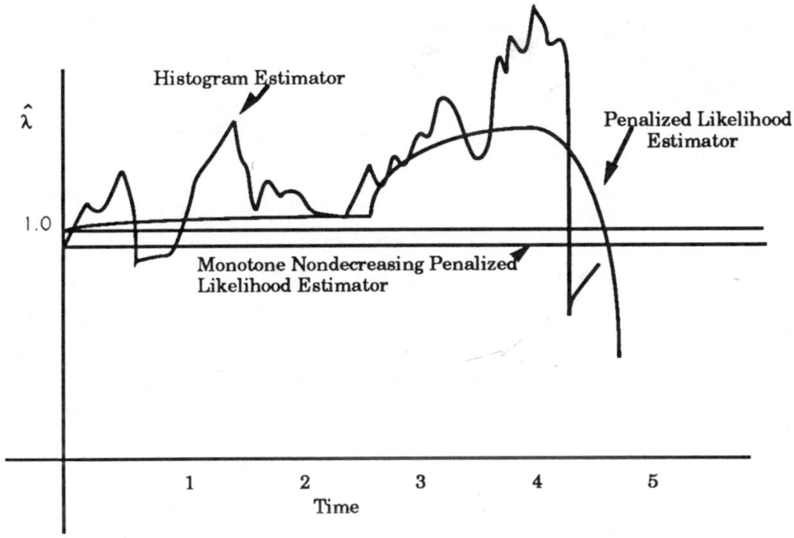

Figure 7.10.

As a practical matter, penalized likelihood should generally be the method of choice for estimating intensity functions. The histogram estimator tends to be rather jagged due to patients coming in or out of a particular interval. The second of the three basic methods considered, the constrained maximum likelihood estimator without penalty term, namely the solution to (37) has the advantage of being programmable for the hand held calculator, but for the user with a modest personal computer, the penalized likelihood estimator is computable in a few seconds. Moreover, the constraint of monotonicity is frequently unjustified.

In Figure 7.10, we show the results of a histogram estimator, a penalized likelihood estimator, and a penalized likelihood estimator constrained to be nondecreasing. The simulated data set of size 500 has constant λ on the interval $(0, 5)$. We note how prior knowledge of monotonicity greatly improves the efficiency of the estimator.

Next we examine the application of a histogram estimator, a penalized likelihood estimator, and a penalized likelihood estimator constrained to be nondecreasing to estimate the intensity of metastatic display (regional to distant) for a data set from 152 patients with melanoma of the trunk. Given that all the patients considered ultimately displayed metastases, it is surprising that the estimated intensity function is nonincreasing.

In Figure 7.12 we observe a similar phenomenon for the intensity of metastatic display (local to distant) of 192 patients with primary trunk melanoma.

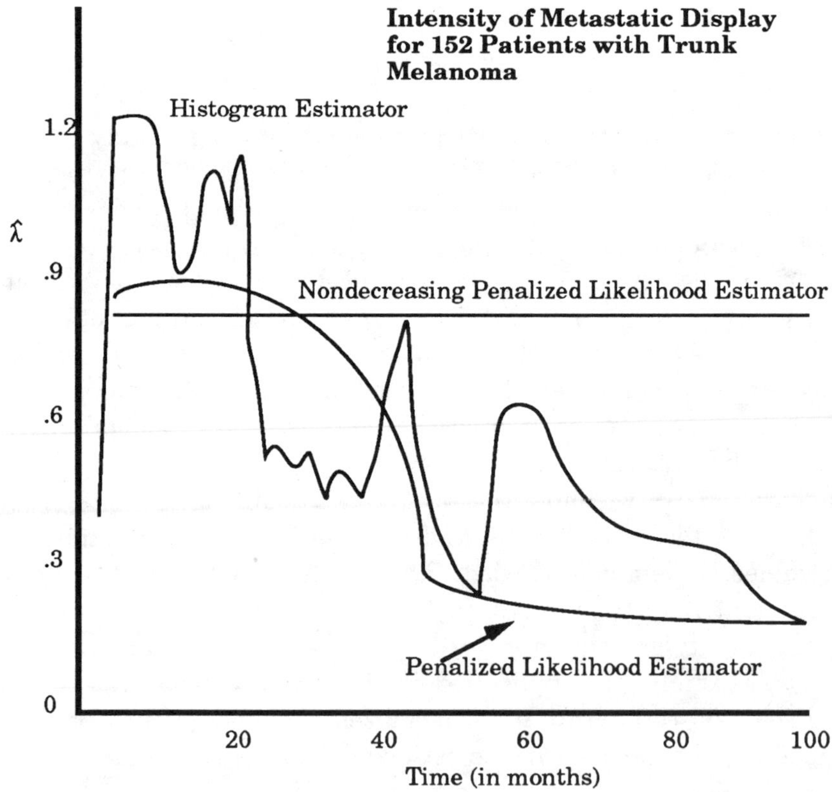

Figure 7.11.

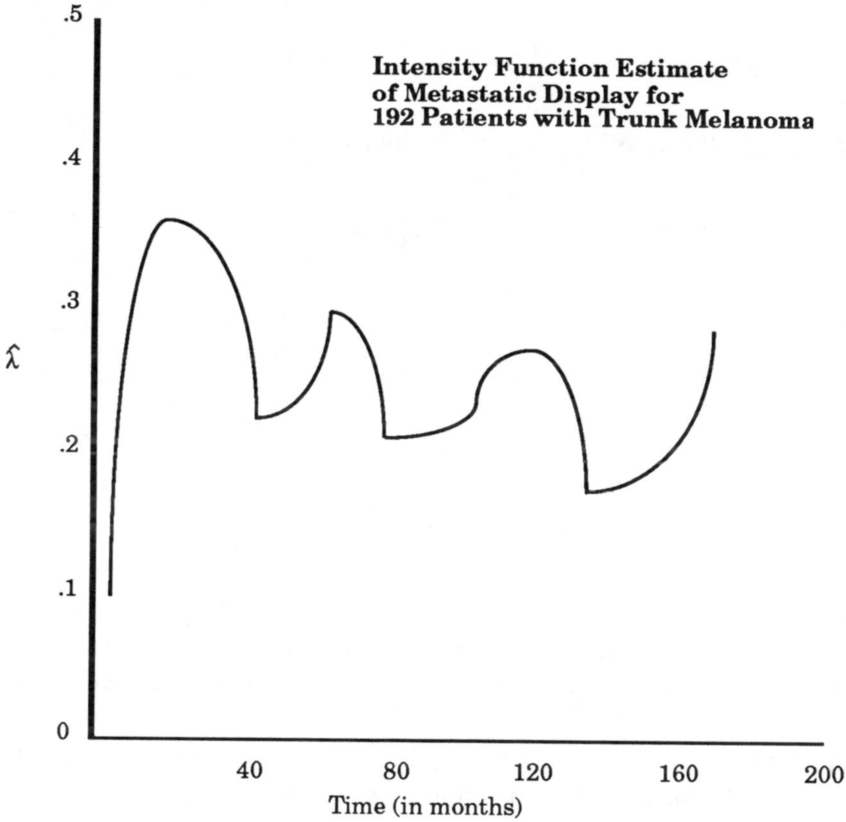

Figure 7.12.

Penalized Proportional Hazards Estimator. In order to incorporate covariate information in the estimation of the intensity function for each individual patient, we may use an estimator based on a log linear model of Cox [2]. Let us suppose that the intensity function for the $i'th$ patient is given by

$$\lambda_i(t) = \lambda(t)\exp(z_i^T\beta), \tag{55}$$

where $\lambda(t)$ is the same for all patients, z is a k-tuple of risk factors, β is a k-tuple of regression coefficients and

$$z_i^T\beta = \sum_{l=1}^{k}\beta_l z_{il}. \tag{56}$$

Let us then seek to solve the following problem:

$$\begin{aligned}\max_{(\lambda,\beta)\in S}Q(\lambda,\beta,\alpha,\gamma) &= \sum_{i=1}^{n}\sum_{j=1}^{m_i}\log(\lambda(t_{ij}))\\ &+\sum_{i=1}^{n} m_i(z_i^T\beta)\\ &-\sum_{i=1}^{n}\exp(z_i^T\beta)\int_0^{T_i}\lambda(\tau)d\tau - \|(\lambda,\beta)\|^2, \end{aligned} \tag{57}$$

where $z_i \in \Re^+$, and

$$\|(\lambda,\beta)\|^2 = \sum_{j=0}^{s}\alpha_j\int_0^T(\lambda^{(j)}(\tau))^2 + \gamma\sum_{i=1}^{k}\beta_i^2, \tag{58}$$

with $\alpha_0, \alpha_1, \ldots, \alpha_s, \gamma > 0$, and $S = [H^s(0,T)\bigcap\{\lambda|\lambda \geq 0\}] \times \Re_k^+$. It should be noted in passing that the solution to (57) is *not* separable into two problems, one involving only λ and the other involving only β.

To establish the existence and uniqueness of a solution to (57), we note that $H^s(0,T) \times \Re$ is a Hilbert space using the inner product implied by (58). S is a closed convex subset of $H^s(0,T) \times \Re_k$. The continuity of Q is clear. The second Gâteaux variation at (λ, β) in the direction (η_1, η_2) is given by

$$\begin{aligned}Q''(\lambda,\beta)((\eta_1,\eta_2),(\eta_1,\eta_2)) &= -\sum_{i=1}^{n}\sum_{j=1}^{m_i}\frac{\eta_1^2(t_{ij})}{\lambda^2(t_{ij})}\\ &-\sum_{i=1}^{n}(z_i^T\eta_2)^2\exp(z_i^T\beta)\int_0^{T_i}\lambda(\tau)d\tau\\ &-2\sum_{i=1}^{n}(z_i^T\eta_2)\exp(z_i^T\beta)\int_0^{T_i}\eta_1(\tau)d\tau\\ &-2\|\eta_1\|^2 - 2\|\eta_2\|^2\\ &< -\|\eta_1,\eta_2\|^2. \end{aligned} \tag{59}$$

Thus, we have established that Q is uniformly negative definite in S, and hence, finally, established the existence and uniquess of the solution to (57). It was in order to simplify the proof for the uniform negative definiteness of Q that we made the assumption that $z_i \in \Re^+$. No instability of the algorithm has been noted when we remove the constraint that $z_i \in \Re^+$.

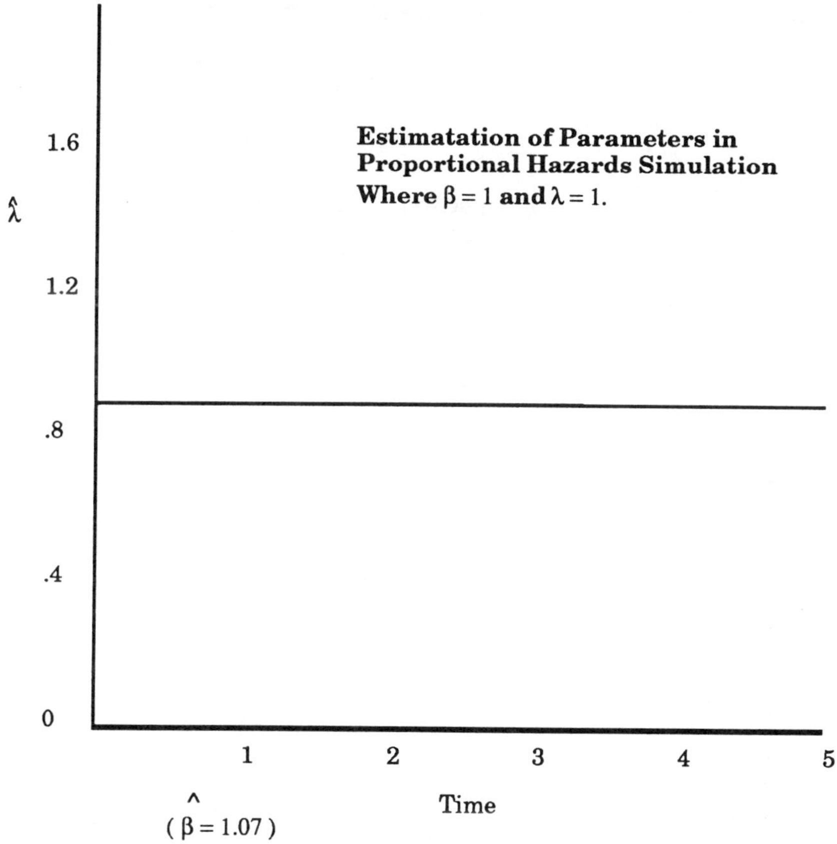

Figure 7.13.

Clearly, we have existence and uniqueness when we restrict (57) to the case where λ is nondecreasing, since $H^s(0,T)\bigcap\{\lambda|\lambda \geq 0, \lambda' \geq 0\} \times \Re^k_k$ is closed and convex. Again, we cut the time interval into equal subintervals with the estimated $\hat{\lambda}$ equal to a constant in each. In Figure 7.13, we show the solution to (57) for one hundred simulated patients where $\lambda(t) = 1$ for $t \in (0,5)$, $\beta = 1$, and z is normally distributed with mean and variance both equal to 1. Censoring occurs after the first metastasis or time $t = 5$, whichever comes first. We have used $\gamma = \alpha_0 = \alpha_1 = 0$ and $\alpha_2 = 1$.

In Figure 7.14, we use the proportional hazards algorithm (57) on a pooled data set of the melanoma data where z is coded as 0 when there is only local disease at presentation and 1 when there is regional disease. Again, we note that the very natural assumption of λ increasing in time does not seem to be substantiated by the data. We will use this exploratory result to build a model for the appearance of secondary tumors in Chapter 8.

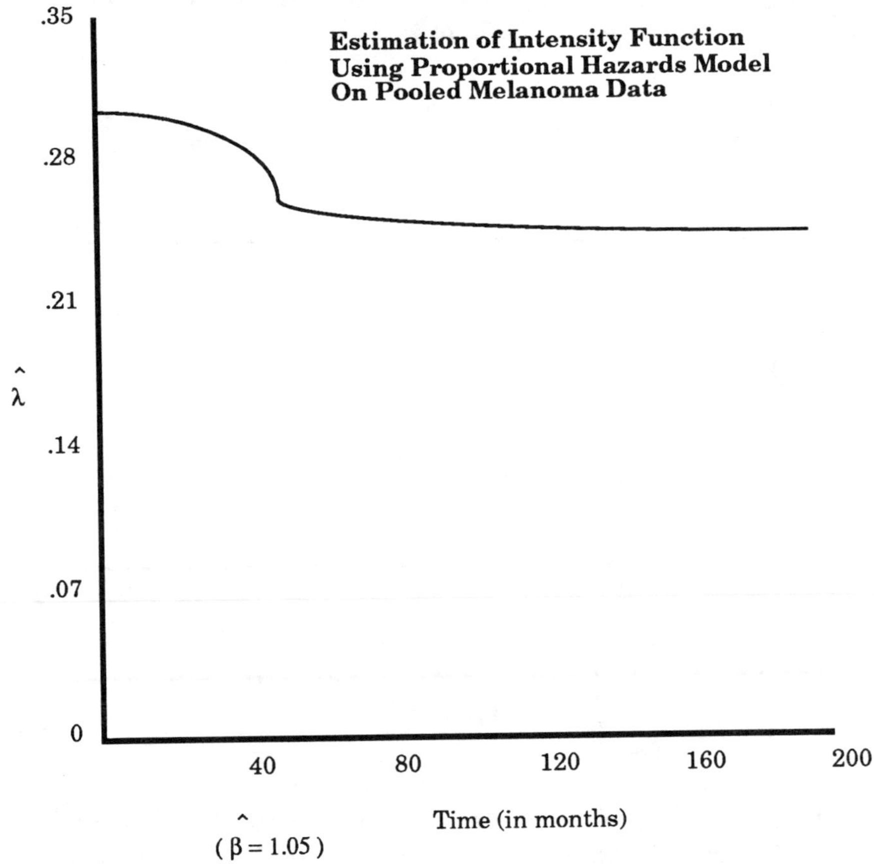

Figure 7.14.

References

[1] Bartoszyński, Robert, Brown, B.W., McBride, C. M. and Thompson, J.R. (1981). "Some nonparametric techniques for estimating the intensity function of a cancer related nonstationary Poisson process." *Annals of Statistics* 9: 1950-1060.

[2] Cox, D.R. (1972). "Regression models and life tables." *Journal of the Royal Statistical Society* 34: 187-200.

[3] Lions, J.L. and Magenes, E. (1972).*Non-homogeneous Boundary Value Problems and Applications.* New York: Springer-Verlag.

[4] Tukey, J.W. (1977). *Exploratory Data Analysis.* Reading: Addison-Wesley.

CHAPTER

8

Model Building: Speculative Data Analysis

8.1. Modeling the Progression of Cancer

Sometimes one may be content to stop with an exploratory analysis of data, such as a nonparametric density estimator. For example, if we seek an identification procedure for distinguishing between two different crop types, it may be that we will be willing to use a nonparametric density estimation based procedure without deep considerations as to why the density signatures are as they are. This may be particularly the case if we are carrying out a task which is only to be performed a few times.

In general, however, we should not be content to stop with an exploratory analysis. A nonparametric density estimator, a nonparametric regression, a nonparametric intensity function estimator, these will usually be first steps toward trying to understand better the mechanism driving a process.

For example, in Section 7.2, we noted that the hazard of an appearance of a secondary melanoma seemed, surprisingly, not to increase with time. Although the data base considered in Section 7.2 was based on melanoma, we have observed the nonincreasing hazard function phenomenon for several tumor types, including breast cancer. If we assume that secondary tumors are breakaway colonies from the primary tumor, and the breakaway tendency increases with size of the primary, then our roughly constant intensity function would appear to be strange indeed. If we wish to understand the mechanism of cancer progression, we need to conjecture a model and then test it against a data base.

One conjecture, consistent with a roughly constant intensity of display of secondary tumors, is that a patient with a tumor of a particular type is not displaying breakaway colonies only, but also new primary tumors due to suppression of a patient's immune system to attack tumors of a particular type. We can formulate axioms at the micro level which will incorporate the mechanism of new primaries. Such an axiomitization has been formulated by Bartoszyński, Brown and Thompson [3]. The first five axioms are consistent with the classical view as to metastatic progression. Hypothesis 6 is the mechanism we introduce to explain the nonincreasing intensity function of secondary tumor display.

Hypothesis 1. For any patient, each tumor originates from a single cell and grows at exponential rate α.

Hypothesis 2. The probability that the primary tumor will be detected and removed in $[t, t+\Delta t)$ is given by $bY_0(t)\Delta t + o(\Delta t)$, and until the removal of the primary, the probability of a metastasis in $[t, t+\Delta t)$ is $aY_0(t)\Delta t + o(\Delta t)$, where $Y_0(t)$ is the size of the primary tumor at time t.

Hypothesis 3. For patients with no discovery of secondary tumors in the time of observation, S, put $m_1(t) = Y_1(t) + Y_2(t) + \ldots$, where $Y_i(t)$ is the size of the $i'th$ originating tumor. After removal of the primary, the probability of a metastasis in $[t, t+\Delta t)$ equals $am_1(t) + o(\Delta t)$, and the probability of detection of a new tumor in $[t, t+\Delta t)$, is $bm_1(t) + o(\Delta t)$.

Hypothesis 4. For patients who do display a secondary tumor, after removal of the primary and before removal of Y_1, the probability of detection of a tumor in $[t, t+\Delta t)$ equals $bY_1(t) + o(\Delta t)$ while the probability of a metastasis being formed is $aY_1(t) + o(\Delta t)$.

Hypothesis 5. After removal of Y_1, the probability of a metastasis in $[t, t+\Delta t)$ is $am_2(t)\Delta t + o(\Delta t)$, while the probability of detection of a tumor is $bm_2(t)\Delta t + o(\Delta t)$, where $m_2(t) = Y_2(t) + \ldots$.

Hypothesis 6. The probability of a systemic occurrence of a tumor in $[t, t+\Delta t)$ equals $\lambda\Delta t + o(\Delta t)$, independent of the prior history of the patient.

Essentially, we shall attempt to develop the likelihood function for this model so that we can find the values of a, b, α, and λ which maximize the likelihood of the data set observed. It turns out that this is a formidable task indeed. The SIMEST algorithm which we shall develop later gives a quick alternative to finding the likelihood function. However, to give the reader some feel as to the complexity associated with model aggregation from seemingly innocent axioms, we shall give some of the details of getting the likelihood function. First of all, it turns out that in order to have any hope of obtaining a reasonable approximation to the likelihood function, we will have to make some further simplifying assumptions. We shall refer to the period prior to detection of the primary as Phase 0. Phase 1 is the period from detection of the primary to S', the first time of detection of a secondary tumor. For those patients without a secondary tumor, Phase 1 is the time of observation, S. Phase 2 is the time, if any, between S' and S. Now for the two simplifying axioms. T_0 is defined to be the (unobservable) time between the origination of the primary and the time when it is detected and removed (at time $t = 0$). T_1 and T_2 are the times until detection and removal of the first and second of the subsequent tumors (times to be counted from $t = 0$). We shall let X be the total mass of all tumors other than the primary at $t = 0$.

Hypothesis 7. For patients who do not display a secondary tumor, growth of the primary tumor, and of all tumors in Phase 1, is deterministically exponential with the growth of all other tumors treated as a pure birth process.

Hypothesis 8. For patients who display a secondary tumor, the growth of the following tumors is treated as deterministic: in Phase 0, tumors $Y_0(t)$ and $Y_1(t)$; in Phase 1, tumor $Y_1(t)$ and all tumors which originated in Phase 0; in Phase 2, all tumors. The growth of remaining tumors in Phases 0 and 1 is treated as a pure birth process.

We now define

$$H(s;t,z) = \exp\{\frac{az}{\alpha}e^{\alpha t}(e^s - 1)\log[1 + (e^{-\alpha t} - 1)e^{-s}] + \frac{\lambda}{\alpha}s - \frac{\lambda}{\alpha}\log[1 + e^{\alpha t}(e^s - 1)]\}, \tag{1}$$

and

$$p(t;z) = bze^{\alpha t}\exp\left[-\frac{bz}{\alpha}(e^{\alpha t} - 1)\right]. \tag{2}$$

Further, we shall define

$$w(y) = \lambda\left[\int_0^y e^{-\nu(u)}du - y\right], \tag{3}$$

where $\nu(u)$ is determined from

$$u = \int_0^{\nu} (a + b + \alpha s - ae^{-s})^{-1} ds. \tag{4}$$

Then, we can establish the following propositions, and, from these, the likelihood function:

$$p(T_0 > \tau) = \exp\left[-b \int_0^{\tau} e^{\alpha t} dt\right] = \exp\left[-\frac{b}{\alpha}(e^{\alpha\tau} - 1)\right]. \tag{5}$$

For patients who do not display a secondary tumor, we have

$$P(T_1 > S|X = x) = \exp\left[-x\nu(S) + w(S)\right]. \tag{6}$$

For patients who develop metastases, we have

$$\begin{aligned} P(T_1 > S) &= P(\text{ no secondary tumor in } (0, S)) \\ &= \int_0^{\infty} e^{w(s)} p(t; 1) H(\nu(s); t, 1) dt. \end{aligned} \tag{7}$$

Similarly, for patients who do display a secondary tumor, we have

$$\begin{aligned} &P(T_1 = S', T_2 > S) = \\ &\int_0^{\infty} \int_0^t e^{w(S-S')} p(t; 1) p(S'; e^{\alpha u}) (\lambda + a e^{\alpha(t-u)}) \\ &\exp\left[-\lambda(t-u) - \frac{a}{\alpha}(e^{\alpha(t-u)} - 1)\right] H(\nu(S - S'); S', e^{\alpha u}) \\ &\times H(\nu(S - S') e^{\alpha S'}; u, e^{\alpha(t-u)}) du dt \\ &+ \int_0^{\infty} \int_0^{S'} e^{w(S-S')} p(t; 1) \exp\left[-\lambda t - \frac{a}{\alpha}(e^{\alpha t} - 1)\right] \lambda e^{-\lambda u} \\ &\times p(S' - u; 1) H(\nu(S - S'); S' - u, 1) du dt. \end{aligned} \tag{8}$$

The propositions necessary for the likelihood function are, to say the least, labor intensive to obtain. Symbol manipulation programs such as Reduce, Mathematica, Maple, etc., simply do not have the capability of doing the work. Accordingly, it must be done by hand. Approximately one and a half man years were required to obtain the likelihood in this fashion.

The estimates for the parameter values from a data set consisting of 116 women who presented with primary breast cancer at the Curie-Sklodowska Cancer Institute in Warsaw (time units in months, volume units in cells) were $a = .17 \times 10^{-9}$, $b = .23 \times 10^{-8}$, $\alpha = .31$, and $\lambda = .0030$. Using these parameter values, we found excellent agreement between the proportion free of metastasis versus time obtained from the data and that obtained from the model, using the parameter values given above. When we tried to fit the model to the data with the constraint that $\lambda = 0$ (i.e., disregarding the systemic process as is generally done in oncology), the attempt failed. From the model, with the estimated parameters, one is immediately able to obtain estimates of other items of interest. For example, tumor doubling time is 2.2 months. The median time from primary origination to detection is 59.2 months and at this time the tumor consists of 9.3×10^7 cells. The probability of metastasis prior to detection of the primary is .069, and so on.

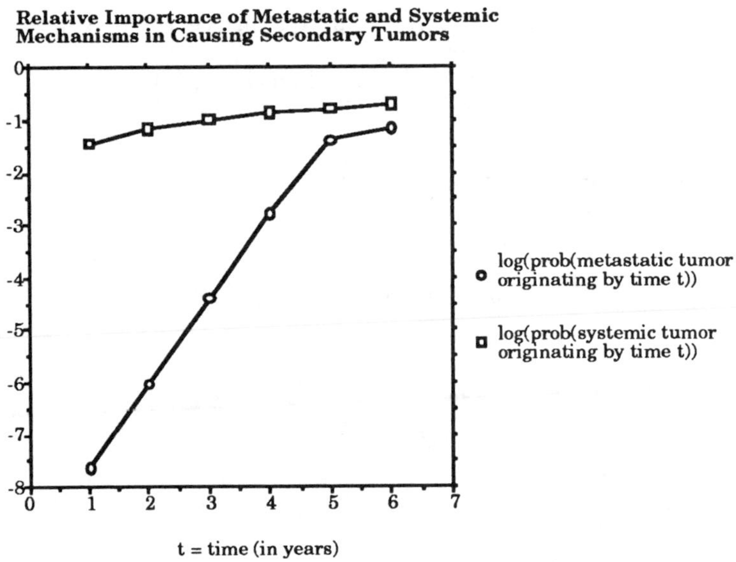

Figure 8.1.

And now for the bottom line, insofar as the relative importance of the systemic and metastatic mechanisms, in causing secondary tumors associated with breast cancer, it would appear from Figure 8.1 that the systemic is the more important. This result is surprising, but is consistent with what we have seen in our exploratory analysis of another tumor system (melanoma). It is interesting that it is by no means true that for all tumor systems, the systemic term has such dominance. For primary lung cancer, for example, the metastatic term appears to be far more important.

It is not clear how to postulate, in any meaningful fashion, a procedure for testing the null hypothesis of the existence of a systemic mechanism in the progression of cancer. We have already noted that when we suppress the systemic hypothesis, then we cannot obtain even a poor maximum likelihood fit to the data. However, someone might argue that a different set of nonsystemic axioms should have been proposed. Obviously, we cannot state that it is simply impossible to manage a good fit without the systemic hypothesis. However, it is true that the nonsystemic axioms we have proposed are a fair statement of traditional suppositions as to the growth and spread of cancer.

It is to be noted that in our development, we had to use data that was really oriented toward the life of the patient, rather than toward the life of a tumor system. This is due to the fact that human *in vivo* cancer data is seldom collected with an idea toward modeling tumor systems. For a number of reasons, including the difficulty mentioned in obtaining the likelihood function, deep stochastic modeling has not traditionally been employed by many investigators in oncology. Modeling frequently precedes the collection of the kinds of data of greatest use in the estimation of the parameters of the model. Anyone who has gone through a modeling exercise such as that covered in this section is very likely to treat such an exercise as a once in a lifetime experience. It simply is too frustrating to have to go through all the flailing around to come up with a quadrature approximation to the likelihood function. As soon as a supposed likelihood function has been found, and a corresponding parameter estimation algorithm constructed, the investigator begins a rather lengthy "debugging" experience. The algorithm's failure to work might be due to any number of reasons: e.g., an incorrect approximation to the likelihood function, a poor quadrature routine, a mistake in the code of the algorithm, inappropriateness of the model, etc. Typically, the debugging process is incredibly time consuming and painful.

If one is to have any hope for coming up with a successful model based investigation, then an alternative to the likelihood procedure for aggregation must be found.

8.2. SIMEST: A Simulation Based Algorithm for Parameter Estimation

In order to decide how best to construct an algorithm for parameter estimation, which does not have the difficulties associated with the classical "closed form" approach, we should try to see just what causes the difficulty with the classical method of aggregating from the microaxioms to the macro level, where the data lives. A glance at Figure 8.2 reveals the problem with the closed form approach.

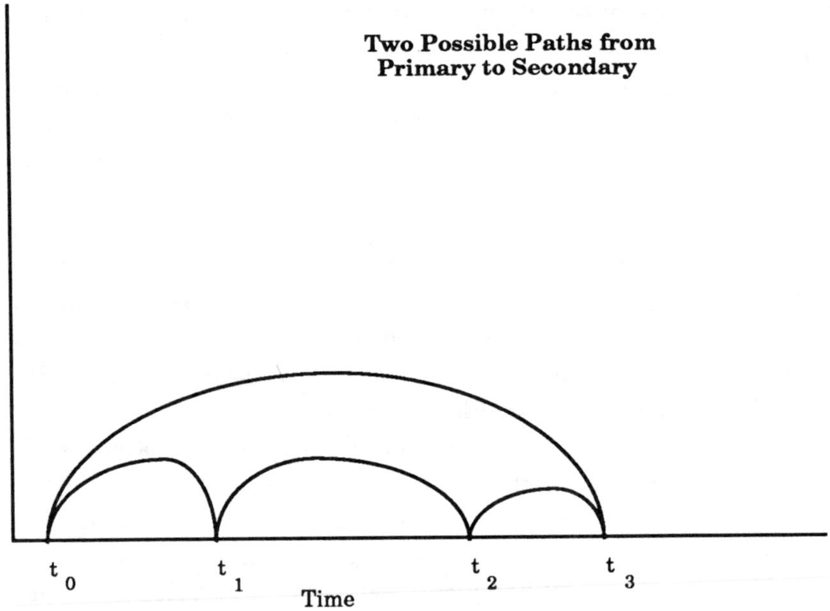

Figure 8.2.

The axioms in Section 8.1 are easy enough to implement in the forward direction. Indeed, they follow the natural forward formulation used since Poisson's work of 1837 [6]. But when we go through the task of finding the likelihood, we are essentially seeking all possible paths by which the observables could have been generated. The secondary tumor, originating at time t_3 could have been thrown off from the primary at time t_3 or it could have been thrown off from a tumor which itself was thrown off from another tumor at time t_2 which itself was thrown off from a tumor at time t_1 from the primary which originated at time t_0. The number of possibilities is, of course, infinite.

In other words, the problem with the classical likelihood approach in the present context is that it is a backwards look from a data base generated in the forward direction. To scientists before the present generation of fast, cheap computers, the backwards approach was, essentially, unavoidable unless one avoided such problems (a popular way out of the dilemma). However, we need not be so restricted.

Once we realize the difficulty when one uses a backwards approach with a forwardly axiomitized system, the way out of our difficulty is rather clear [1], [9]. We need to analyze the data using a forward formulation. The intuitively most obvious way to carry this out is to pick a guess for the underlying vector of parameters, put this guess in the micro-axiomitized model and simulate many times of appearance of secondary tumors. Then, we can compare the set of simulated quasi-data with that of the actual data. The greater the concordance, the better we will believe we have done in our guess for the underlying parameters. If we can quantitize this measure of concordance, then we will have a means for guiding us in our next guess. One such way to carry this out would be to order the secondary occurrences in the data set from smallest to largest and divide them into k bins, each with the same proportion of the data. Then, we could note the proportions of quasi-data points in each of the bins. If the proportions observed for the quasi-data, corresponding to parameter value Θ, were denoted by $\{\pi_j(\Theta)\}_{j=1}^k$, then a Pearson goodness of fit statistic would be given by

$$\chi^2(\Theta) = \sum_{j=1}^{k} \frac{(\pi_j(\Theta) - \frac{1}{k})^2}{\pi_j(\Theta)}. \tag{9}$$

The minimization of $\chi^2(\Theta)$ provides us with a means of estimating Θ.

Typically, the sample size, n, of the data will be much less than N, the size of the simulated quasi-data. With mild regularity conditions, assuming there is only one local maximum of the likelihood function (which function we of course do not know), Θ_0, as $n \to \infty$, then as $N \to \infty$, as n becomes large and k increases in such a way that $lim_{n\to\infty} k = \infty$ and $lim_{n\to\infty} k/n = 0$, the minimum χ^2 estimator for Θ_0 will have an expected mean square error which approaches the expected mean square error of the maximum likelihood estimator. This is, obviously, quite a bonus. Essentially, we will be able to forfeit the possibility of knowing the likelihood function, and still obtain an estimator with asymptotic efficiency equal to that of the maximum likelihood estimator. The price to be paid is simply a computer swift enough and cheap enough to carry out a very great number, N, of simulations, say 10,000. This ability to use the computer to get us out of the "backwards trap" is a potent but, as yet seldom used, bonus of the computer age.

First of all, we observe how the forward approach enables us to eliminate those hypotheses which were, essentially a practical necessity if a likelihood function was to be obtained. Our new axioms are simply:

Hypothesis 1. For any patient, each tumor originates from a single cell and grows at exponential rate α.

Hypothesis 2. The probability that the primary tumor will be detected and removed in $[t, t+\Delta t)$ is given by $bY_0(t)\Delta t + o(\Delta t)$. The probability that a tumor of size $Y(t)$ will be detected in $[t, t+\Delta t)$ is given by $bY(t)\Delta t + o(\Delta t)$.

Hypothesis 3. The probability of a metastasis in $[t, t+\Delta)$ is $a\Delta t \times$ (total tumor mass present).

Hypothesis 4. The probability of a systemic occurrence of a tumor in $[t, t+\Delta t)$ equals $\lambda\Delta t + o(\Delta t)$, independent of the prior history of the patient.

In order to simulate, for a given value of (α, a, b, λ), a quasi-data set of secondary tumors, we must first define:

t_D = time of detection of primary tumor;
t_M = time of origin of first metastasis;
t_S = time of origin of first systemic tumor;
t_R = time of origin of first recurrent tumor;
t_d = time from t_R to detection of first recurrent tumor;
t_{DR} = time from t_D to detection of first recurrent tumor.

Now, generating a random number u from the uniform distribution on the unit interval, as we show in Appendix IV, if $F(.)$ is the appropriate cumulative distribution function for a time, t, we set $t = F^{-1}(u)$. Then, assuming the tumor volume at time t is

$$v(t) = ce^{\alpha t}, \text{ where } c \text{ is the volume of one cell,} \tag{10}$$

we have

$$F_M(t) = 1 - \exp\left[-\frac{ac}{\alpha}e^{\alpha t_M}\right]. \tag{11}$$

Similarly, we have

$$\begin{aligned} F_D(t_D) &= 1 - \exp\left[-\int_0^{t_D} bce^{\alpha\tau}d\tau\right] \\ &= 1 - \exp[-\frac{bc}{\alpha}e^{\alpha t_D}], \end{aligned} \tag{12}$$

$$F_S = 1 - e^{-\lambda t_S}, \tag{13}$$

and

$$F_d(t_d) = 1 - \exp\left[-\frac{bc}{\alpha}e^{\alpha t_d}\right]. \tag{14}$$

Using the actual times of discovery of secondary tumors $t_1 \leq t_2 \leq ,\ldots, \leq t_n$ we generate k bins. In actual tumor situations, because of recording protocols, we may not be able to put the same number of secondary tumors in each bin. Let us suppose the observed proportions are given by $(p_1, p_2, \ldots, p_k)$. We shall generate N recurrences $s_1 < s_2 < \ldots < s_N$. The observed proportions in each of the bins will be denoted $\pi_1, \pi_2, \ldots, \pi_k$. The goodness of fit corresponding to (α, λ, a, b) will be given by

$$\chi^2(\alpha, \lambda, a, b) = \sum_{j=1}^{k} \frac{(\pi_j(\alpha, \lambda, a, b) - p_j)^2}{\pi_j(\alpha, \lambda, a, b)}. \tag{15}$$

Then, the following algorithm generates the times of detection of quasi-secondary tumors for the particular parameter value (α, λ, a, b).

Secondary Tumor Simulation (α, λ, a, b)

Enter (α, λ, a, b)
Generate t_D
$j = 0$
$i = 0$
Repeat until $t_M(j) > t_D$
$j = j + 1$
Generate $t_M(j)$
Generate $t_{dM}(j)$
$t_{dM}(j) \leftarrow t_{dM}(j) + t_M(j)$
If $t_{dM}(j) < t_D$, then $t_{dM}(j) \leftarrow \infty$
Repeat until $t_S > 10t_D$
$i = i + 1$
Generate $t_{dS}(i)$
$t_{dS}(i) \leftarrow t_{dS}(i) + t_S(i)$
$s \leftarrow \min [t_{dM}(j), t_{dS}(i)]$
Return s
End Repeat

The above algorithm does still have some simplifying assumptions. For example, we assume that metastases of metastases will not likely be detected before the metastases themselves. We assume that the primary will be detected before a metastasis, etc. Note, however, that the algorithm is much less restrictive in the simplifying assumptions which led to the terms of the likelihood in Section 8.1. Even more importantly, the Secondary Tumor Simulation algorithm can be discerned in a few minutes, whereas a likelihood argument is frequently the work of months.

Another advantage of the forward simulation approach is its ease of modification. Those who are familiar with "backwards" approaches based on the likelihood or the moment generating function are only too familiar with the experience of a slight modification causing the investigator to go back to the start and begin anew. This is again a consequence of the tangles required to be examined if a backwards approach is used. However, a modification of the axioms generally causes slight inconvenience to the forward simulator.

For example, let us add

Hypothesis 5. A fraction γ of the patients ceases to be at systemic risk at the time of removal of the primary tumor if no secondary tumors exist at that time. A fraction $1-\gamma$ of the patients remain at systemic risk throughout their lives.

To add the hypothesis is the matter of considerable work if we insist on using the classical aggregation approach of maximum likelihood. However, in the forward simulation method we simply add the following lines to the Secondary Tumor Simulation code.

Generate u from $U(0,1)$
If $u > \gamma$, then proceed as in the Secondary Tumor Simulation code
If $u < \gamma$, then proceed as in the Secondary Tumor Simulation code except replace the step "Repeat until $t_S > 10t_D$" with the step "Repeat until $t_S(i) > t_D$"

It is interesting to note that the implementation of SIMEST is generally faster on the computer than working through the estimation with the "closed form" likelihood. In the four parameter oncological example we have considered here, the running time of SIMEST was 10% of the likelihood approach. As a very practical matter, then, the simulation based approach would appear to majorize that of the "closed form likelihood" method in virtually all particulars. The running time for SIMEST can begin to become a problem as the dimensionality of the response variable increases past one. Up to this point, we have been working with the situation where the data consists of "failure times." In the systemic versus metastatic oncogenesis example, we managed to estimate four parameters based on this kind of one dimensional data. As a practical matter, for tumor data, the estimation of five or six parameters for failure time data is the most one can hope for. Indeed, in the oncogenesis example, we begin to observe the beginnings of singularity for four parameters, due to a near trade-off between the parameters a and b. Clearly, it is to our advantage to be able to increase the dimensionality of our observables. For example, with cancer data, it would be to our advantage to utilize not only the time from primary diagnosis and removal to secondary discovery and removal, but also the tumor volumes of the primary and the secondary. Such information enables one to postulate more individual growth rates for each patient. Thus, it is now appropriate to address the question of dealing with multivariate response data.

Gaussian Template Criterion. In many cases, it will be possible to employ a procedure using a criterion function similar to that of the Gaussian Template Test in Section 6.5. First, we transform the data $\{X_i\}_{i=1}^n$ by a linear transformation such that for the transformed data set $\{U_i\}_{i=1}^n$ the mean vector becomes zero and the covariance matrix becomes I.

$$U = AX + b. \tag{16}$$

Then, for the current best guess for Θ, we simulate a quasidata set of size N. Next, we apply the same transformation to the quasidata set $\{Y_j(\Theta)\}_{j=1}^N$. We compute the sample mean $\overline{Y}$ and covariance matrix Σ of the transformed data set. Assuming that both the actual data set and the simulated data set come from the same density, the likelihood ratio $\Lambda(\Theta)$ should increase as Θ gets closer to the value of Θ, say Θ_0, which gave rise to the actual data, where,

$$\Lambda(\Theta) = \frac{\prod_{i=1}^N \exp[-\frac{1}{2}(u_{1i}^2 + \ldots + u_{pi}^2)]}{\prod_{i=1}^N \sqrt{|\Sigma^{-1}|} \exp[-\frac{1}{2}\sum_{j,l=1}^p \sigma^{jl}(u_{ji} - \overline{Y_j})(u_{li} - \overline{Y_l})]}, \tag{17}$$

and σ^{jl} is the $j, l'th$ element of the inverse of Σ. As soon as we have a criterion fuction, we are able to develop an algorithm for estimating Θ_0.

Nearest Neighbor Criterion. A more robust, but computationally much more time-consuming procedure can be developed. Let the distance from the $i'th$ data point to its $m'th$ nearest neighbor be denoted by $d(i, m)$ and the distance of the $N(m/n)'th$ nearest neighbor from the $i'th$ data point be denoted by $D(i, m)$. Then, the naive nearest neighbor density estimator gives us the following goodness of fit statistic:

$$J_2(\Theta) = \sum_{i=1}^n \left(\frac{1}{d^k(i,m)} - \frac{1}{D^k(i,m)}\right)^2 d^{2k}(i,m). \tag{18}$$

We recall that the dimension of the data and that of the underlying parameter are seldom the same. It should also be observed that, by virtue both of the intrinsic noisiness of the data, and the fact that we have introduced noise when we generated the quasidata set, the standard algorithms of deterministic optimization theory are not likely to perform very satisfactorily. In Appendix III we describe two algorithms useful in simulation based estimation.

In the discussion of metastasis and systemic occurrence of secondary tumors, we have used a model supported by data to try to gain some insight into a part of the complexities of the progression of cancer in a patient. Perhaps this sort of approach should be termed "speculative data analysis." In the current example, we have been guided by a nonparametric intensity function estimate, which was surprisingly nonincreasing, to conjecture a model, which enabled us to test systemic origin against metastatic origin on something like a level playing field. The fit without the systemic term was so bad that anything like a comparison of goodness of fit statistics was unnecessary. It should be observed here that the very idea of statisticians making a conjecture as to the mechanism of the cancer process is (unfortunately) rather rare. Statisticians in many fields, including biometry, are generally content to allow physicians, say, to make the conjectures, which the statisticians then test. This approach has been dubbed by Tukey "confirmatory data analysis." The dichotomy between "exploratory data analysis" and "confirmatory data analysis" should not be sharp. The traditional scientific method of conjecture, test, reformulation of conjecture, etc., is probably the more suitable manner of attack.

8.3. Adjuvant Chemotherapy: An Oncological Example with Skimpy Data.

In the United States, for most solid tumors, the primary method of treatment of a tumor is surgical removal. But if the tumor has metastasized, such a removal is not sufficient for a cure, since the metastatic colonies provide a self-sustaining attack on the human body. For this reason, it has become a common protocol for some solid tumors to follow the surgery with a regime of chemotherapy in the hopes that the chemotherapeutic regime will destroy any metastatic outcroppings. An example of such a tumor system is breast cancer. If, during analysis of the lymph nodes of a patient who has undergone removal of a breast cancer, cancer cells are found present, then it is customary to follow the surgery with "adjuvant (assisting) chemotherapy." Survival data gives strong indication that such a procedure increases the life expectancy of patients with this condition. However, in far too many cases, patients who have had the adjuvant regime still later display metastases not destroyed by the chemotherapy. The reason for this phenomenon is thought to be the random mutation of some cells to resistance to the chemotherapeutic agent. This resistance might have developed within a cell colony within the primary tumor from which a metastatic cell dislodged. Or it might have develped within a cell colony of a metastatic tumor not originally resistant. If the probabilities of both metastases and the development of resistant cells by mutation are proportional to the number of tumor cells present, then the importance of early detection and removal of the primary is intuitively obvious. Unfortunately, since statisticians have not been very active in attempting to model the processes of resistance and metastasis, data which would enable us to estimate the tendencies to develop resistant strains and metastases is not generally collected. One of the problems with the clear cut exploratory-confirmatory procedures is the fact that, absent a model, the right kind of data is seldom collected. Some would argue that in such data skimpy situations the statistical investigator should remain mute until enough of the right sort of data has been collected. Unfortunately, it is frequently the case that unless we are willing to carry out some modeling early on, the right sort of data will never be collected.

Thus, we shall examine the adjuvant chemotherapy example realizing that we do not have enough of the right sort of data to make even the weak model-data based conjectures that we did in the systemic-metastatic example. Let us suppose, first of all, that the processes of resistance and metastasis are independent of each other. Suppose further that

$$\text{(19)} \qquad \text{Prob(resistance develops in } [t, t+\Delta t)\text{)} = \beta n(t)\Delta t$$

where β is the per cell rate of development to resistance and $n(t)$ is the number of tumor cells at time t. Now the rate of tumor growth is generally assumed (with good experimental justification) to be exponential with growth rate α. We can express the time axis in arbitrary units, and hence we lose no generality if we simply take $\alpha = 1$.

Similarly, we shall write the metastatic mechanism as

$$\text{(20)} \qquad \text{Prob(metastasis develops in } [t, t+\Delta t)\text{)} = \mu e^t \Delta t.$$

Next, as in [10], let us consider the following two events:

$A(N)$ = the event that by the time the total tumor mass equals N cells a nonresistant metastasis develops in which a resistant population subsequently develops (before a total tumor mass of N cells).

$B(N)$ = the event that by the time the total tumor mass equals N cells a resistant clone develops from which a metastasis subsequently develops (before a total tumor mass of N cells).

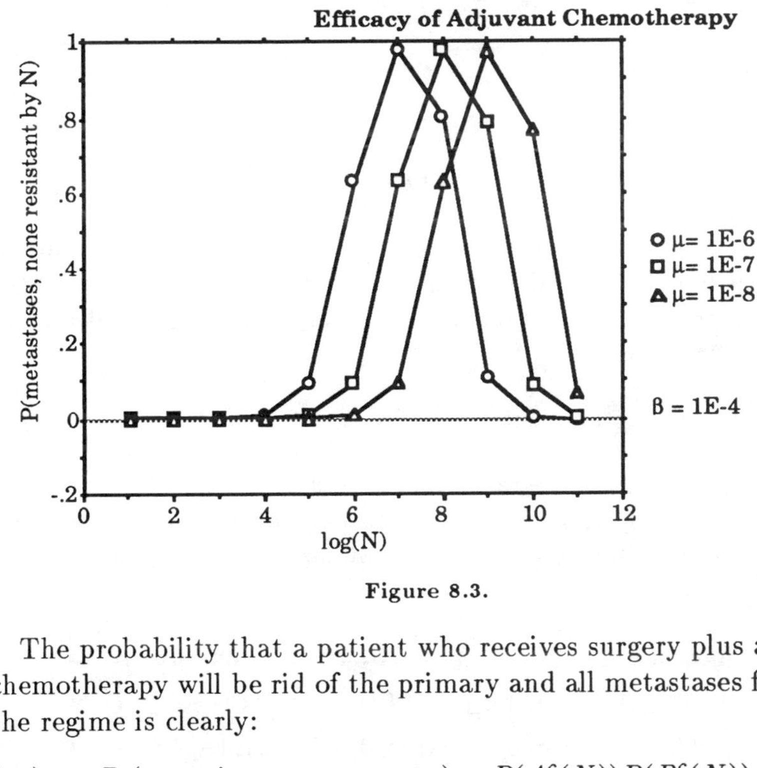

Figure 8.3.

The probability that a patient who receives surgery plus adjuvant chemotherapy will be rid of the primary and all metastases following the regime is clearly:

(21) $\quad P$ (no resistant metastases) $= P(A^c(N))P(B^c(N))$.

We shall demonstrate how to obtain $P(A^c(N))$. The derivation of $P(B^c(N))$ proceeds in exactly the same way. Now,

(22) $\quad P$ (metastasis occurs in $[t, t+\Delta t)$ followed by a resistant subclone before T= $\log(N)$)

$$= \mu e^t \left[1 - \exp(-\int_0^{T-t} \beta e^\tau d\tau)\right] \Delta t = \mu e^t \left[1 - e^\beta e^{-\beta N/n}\right] \Delta t.$$

So

$$P(A^C(N)) = \exp\left[-\int_0^T \mu e^t(1 - e^{\beta} e^{-\beta N/n})dt\right]$$

$$(23) \qquad = \exp\left[\mu - \mu e^{\beta(1-N)} + \mu\beta e^{\beta} N\{Ei(-\beta) - Ei(-\beta N)\}\right].$$

Here, the exponential integral $Ei(x)$ is defined for negative x by

$$(24) \qquad Ei(x) = -\int_{-x}^{\infty} \frac{e^{-t}}{t} dt.$$

Similary, we have

$$(25) \qquad P(B^C(N)) = \exp[\beta - \beta e^{\mu} e^{-\mu N} + \mu\beta e^{\mu} N\{Ei(-\mu) - Ei(-\mu N)\}].$$

For N very large and μ and β very small (and, typically the number of tumor cells present at detection of the primary is very large, and the per cell tendency to metastasize and/or to develop resistance is very small), we have

$$(26) \qquad P(\text{no resistant metastases by } N\ ) = (\mu\beta\gamma^2)^{\mu\beta N},$$

where $\gamma = 1.781072$, i.e., e raised to Euler's constant.

To determine the probability that a patient will be cured by surgery plus adjuvant chemotherapy, but not by surgery alone, we need to compute

$$(27) \qquad P(\text{metastases, none resistant by total tumor size } N\ ) = P(\text{no resistant metastases by } N) - P(\text{no metastases by N}) = (\mu\beta\gamma^2)^{\mu\beta N} - \exp[-\mu(N-1)].$$

It is interesting to note that although the probability of no resistant metastases is symmetrical in β and μ, that is not the case for the probability of an improved chance of cure by adjuvant therapy. In Figure 8.3 we note, for some particular values of μ and β, the advantages available to a patient due to adding adjuvant chemotherapy to a surgical regime as a function of base ten logarithms of tumor size (in cells). For a very small tumor, there appears little benefit to the adjuvant regime, since the tumor has probably not metastasized at all. For large tumors, the benefit is also small, since the tumor has probably thrown off metastases which have become resistant to chemotherapy. Estimates for μ have been obtained [3]. We have very skimpy information concerning values of β. Values from .0001 to .000001 are typical. In general, it appears that the force of metastasis is less than that for resistance, perhaps one hundredth as high. But one must frankly admit that we must deal, at this point, with estimates of β and μ which are not very reliable.

Now although one can, with some labor, develop (26), it was not developed until 1989 [10]. As we have noted, the axiomitization of the metastasis-resistance phenomenon is not very difficult to perceive. But it is generally quite difficult to proceed from the axioms to the "closed form." It is not a particularly good idea to leave scientific advance to cleverness in the manipulation of formulae. If we had chosen the simulation based route to obtain Figure 8.3, our task would have been quite easy.

For example, we immediately have

$$P(\text{no metastases by } N\ ) = e^{-\mu(N-1)}. \tag{28}$$

Then, using the number of tumor cells as our "clock," we may write the distribution function

$$F(n) = 1 - e^{-\mu(n-1)}. \tag{29}$$

Thus, we can simulate a series of "times" at which metastases occur for a patient who has his primary tumor discovered and removed at "time" N, using

$$n_{j+1} = n_j - \frac{1}{\mu}\log(1-u), \tag{30}$$

where u is a uniform variate on the unit interval.

For each simulated metastasis, we can then compute (in closed form) the probability that no resistant clone develops before the primary is removed and adjuvant chemotherapy is instituted at "time" N:

$$P(\text{no resistance before primary size } N) = e^{-\beta((N/n)-1)}. \tag{31}$$

This gives us immediately a means of obtaining approximations for (26) and hence Figure 8.3 as we show using the flowchart in Figure 8.4.

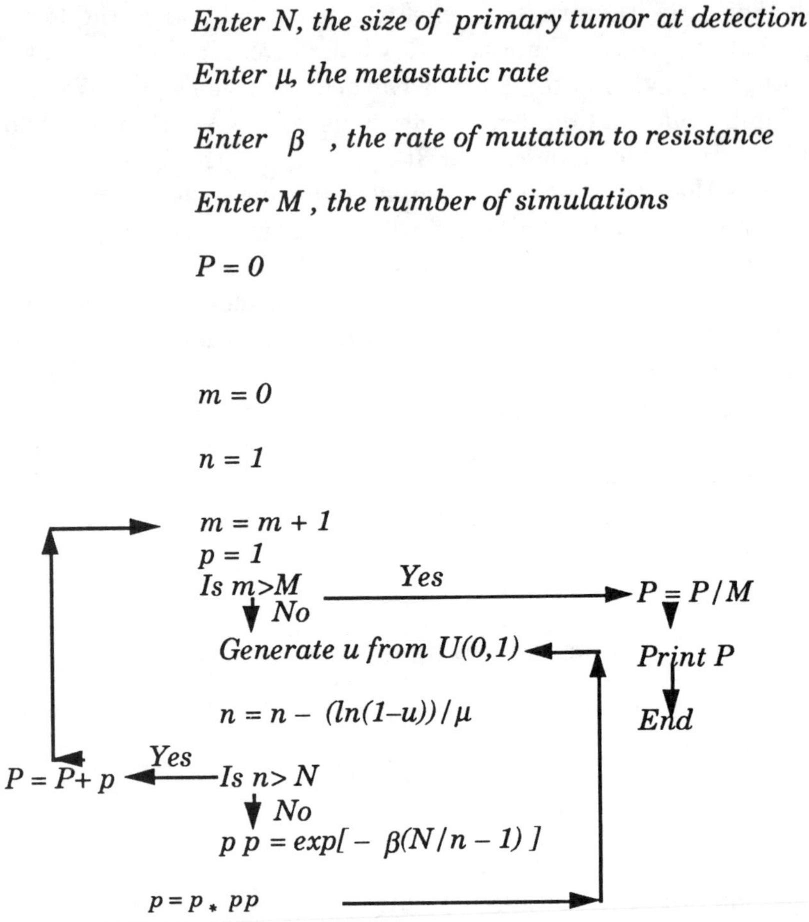

Figure 8.4.

Naturally, there are advantages here to having the closed form solution (26), since it gives us a holistic perception of the underlying process. However, the simulation route is direct and relatively painless. The closed form solution took some weeks of work.

In most nontrivial cases, such as the systemic-metastatic sources of secondary tumors in Section 8.2, the "closed form" solution is itself so complicated that it is good for little other than as a device for pointwise numerical evaluation. The simulation route should generally be the method of approach for nontrivial time-based modeling. As we have argued, the advantage of the simulation based approach is that it enables us to work in the forward direction of the axioms. The major motivation for von Neumann to build the digital computer was the ability to perform simulations. Furthermore, simulation based algorithms generally have the best compatibility with parallel devices. Unfortunately, at the present time, the use of the modern digital computer for simulation based modeling and computation is an insignificant fraction of total computer usage.

8.4. Modeling the AIDS Epidemic.

We have observed in Section 1.2 the important exploratory work of John Graunt in developing the first distribution function. The data used was collected in connection with the bubonic plague. It is interesting to note that Graunt's study was carried out 127 years after the keeping of records was commissioned by Thomas Cromwell. The decision to collect data without any consideration of what is to be done with it is an unfortunate and continuing tendency of bureaucrats, particularly where epidemics are concerned. Graunt's work was not commissioned by the Crown and although it had valuable consequences for statistical science, the analysis was sufficiently late in coming that it is clear that the data collection achieved nothing either in control of the plague in Tudor London or in dealing with its consequences.

Prior to the AIDS epidemic, the last contagious disease in the United States which captured the attention of the public was that of polio (infantile paralysis). We still do not have a very good feel as to the conditions which caused this epidemic to be so prevalent in the United States for a period covering roughly 1915 to 1955. There was a particularly virulent period from the end of World War II until the introduction of the Salk vaccine in 1955. Individuals who had the disease as children at that time and recovered at that time frequently exhibit a tendency in middle age to develop symptoms similar to those associated with multiple sclerosis.

It is interesting to note the general inertia of public officials in dealing with the polio epidemic in the period 1946 to 1955. It was intuitively perceived by many citizens that the disease was associated with the crowding together of children during the hot weather months. Consequently, it would appear that certain natural steps would have been taken. These would include the closing of summer schools, swimming pools, and cinemas. Such steps were seldom taken. This was the golden age of the kiddie matinee, and municipalities vied to demonstrate how many swimming pools they could build. Parents who kept their children away from matinees and swimming parties did so without any encouragement from the civil authorities whose business it was to control epidemics. The merchants who owned facilities for the amusement of children formed a powerful lobby against closing anything. As is generally the case, there was no constituency for taking prudent practical steps to break the transmission chain. Roughly 25,000 Americans each year paid the price for the failure of public officials to take prudent action.

The control of epidemics by some sort of separation of infectives from susceptibles is a part of many tribal religions. The quarantine protocols of Moses for dealing with leprosy were generally sophisticated and effective. Moses never had to contend with merchants and other pressure groups (well, hardly ever, and when he did have to contend, he was not terribly accommodating).

In the matter of the present AIDS epidemic in the United States, a great deal of money is being spent on AIDS. Roughly 50 times is being spent in research on a per fatality basis on AIDS as on cancer. However, practically nothing in the way of steps for stopping the transmission of the disease is being done (beyond education in the use of condoms). Indeed, powerful voices in the Congress speak against any sort of government intervention. On April 13, 1982, Congressman Henry Waxman [7] stated in a meeting of his Subcommittee on Health and the Environment, "I intend to fight any effort by anyone at any level to make public health policy regarding Kaposi's sarcoma or any other disease on the basis of his or her personal prejudices regarding other people's sexual preferences or life styles." We do not even have a very good idea as to what fraction of the target population in this country is HIV positive.

Let us see what we might be able to discover about the epidemic by the use of some exploratory data analysis and some model building. First of all, it should be noted that the United States has, since the earliest period of the epidemic, furnished the vast majority of AIDS cases in the developed world. A popular belief was that this was due to a time lag between outbreak of the epidemic in the United States and other countries. Assurances that the incidence of AIDS in the United States relative to that in the rest of the first world would plummet now begin to ring a bit hollow. In Figure 8.5 we show WHO relative incidence figures for 1988 and 1990. We note that the curves are not plummeting down to the "1" line. It appears that the epidemic in the United States has a degree of virulence very substantially greater than that in the rest of the developed world. It is interesting to note that in the United Kingdom, which has an AIDS rate less than one twelfth of that in the United States, the average gay male has eight different partners per year.

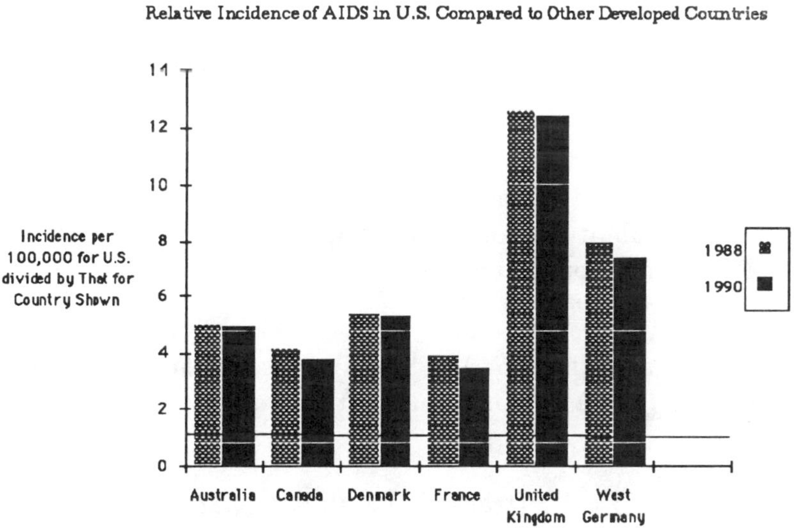

Figure 8.5.

To develop a model for AIDS could, quite easily, involve hundreds of variables. The difficulty with such an approach is that, whereas a complex model might satisfy our understandable desire for completeness, it would almost certainly be nearly useless. We have neither the data nor the understanding to justify such a model for AIDS. Then, one could argue, as some have, that we should disdain modeling altogether until such time as we have sufficient information. But to give up modeling until "all the facts are in" would surely push us pass the time where our analyses would be largely irrelevant. There is every reason to hope that at some future time we will have a vaccine and/or a treatment for AIDS. This was the case, for example, with polio. Do we now have the definitive models for polio? The time for developing models of an epidemic is during the period when they might be of some use.

We shall, accordingly, present a simple and, admittedly, incomplete model for the epidemic which may nonetheless be useful. Such a model was first developed in 1983 [8] and the basic conclusions of that model have not been negated by subsequent information. We disregard the influence of i.v. drug use in the transmission of the disease and restrict ourselves exclusively to a model for male homosexual transmission. Accordingly, we shall begin with a classical contact formulation:

$$P(\text{transmission from infective in } [t, t+\Delta t)) = k\alpha\Delta t \frac{X}{X+Y}, \tag{32}$$

where

k = number of contacts per month;
α = probability of contact causing AIDS;
X = number of susceptibles;
Y = number of infectives.

We shall then seek the expected total increase in the infective population during $[t, t+\Delta t)$ by multiplying the above by the total number of infectives.

$$\Delta E(Y) = YP(\text{transmission in } [t, t+\Delta t) \approx \Delta Y. \tag{33}$$

Letting $\Delta t \to 0$, we have

$$\frac{dY}{dt} = \frac{k\alpha XY}{X+Y}, \tag{34}$$

$$\frac{dX}{dt} = -\frac{k\alpha XY}{X+Y}. \tag{35}$$

There are other factors which must be added to the model such as immigration into the susceptible population, λ, and emigration, μ, from both the susceptible and infective populations, as well as a factor, γ, to allow for marginal increase in the emigration from the infective population due to sickness and death. Thus, we have the improved differential equation model

$$\frac{dY}{dt} = \frac{k\alpha XY}{X+Y} - (\gamma + \mu)Y, \tag{36}$$

$$\frac{dX}{dT} = -\frac{k\alpha XY}{X+Y} + \lambda - \mu X. \tag{37}$$

If one uses the early stages of the disease $X/(X+Y) \approx 1$, then we have

$$Y \approx Y(0)\exp[(k\alpha - \mu - \gamma)t]. \tag{38}$$

Using such an approximation, it was determined [8] that in 1983 $k\alpha \approx .25$. But since the time units are in months, given the best information available as to k, it was clear that the per contact probability of an infective transmitting the disease, α was low, probably no more than .01. This is much lower than the per contact probability of transmitting other major venereal diseases, and gives immediate indication of why the epidemic manifested itself predominantly in the homosexual population rather than in the heterosexual population. For such a low per contact probability, it is likely that the volume of the virus assault on a susceptible is important. Hence, active to passive transmission appears much the more likely route. In Figures 8.6 and 8.7, we demonstrate the transmission patterns for the heterosexual and homosexual modes.

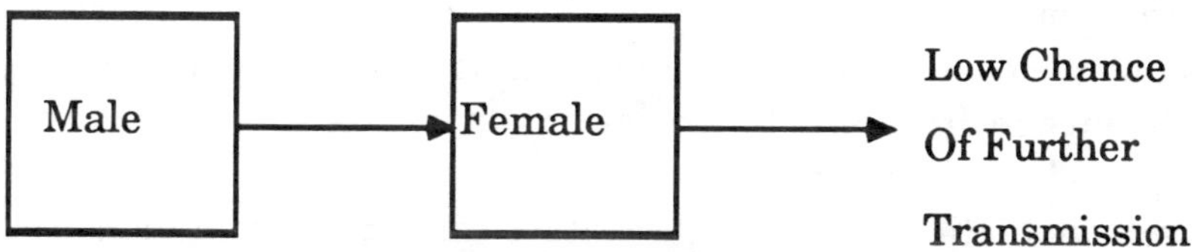

Heterosexual Transmission Pattern

Figure 8.6.

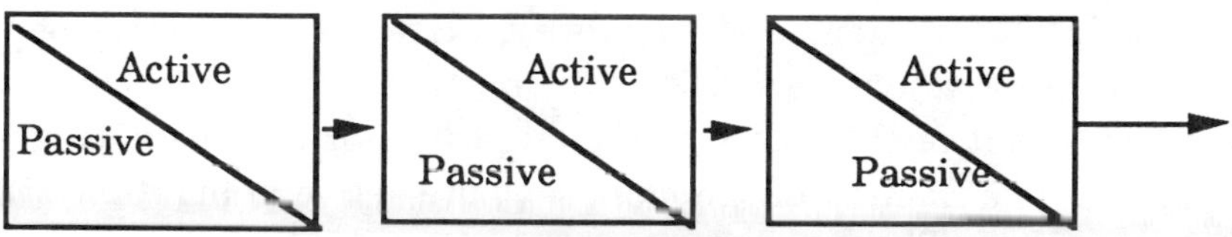

Homosexual Transmission Pattern

Figure 8.7.

Naturally, we ought not use the exponential growth model indefinitely, since, as the number of susceptibles decreases, so also will the growth rate of the epidemic. The doubling time of the cases in the United States which was around five months in the early 1980's is now over a year. The model developed here applies to the epidemic in the United Kingdom, say, as well as to the United States. Perhaps the k values, i.e., per monthly number of contacts, is higher here than there. One of the maxims of Vilfredo Pareto was that the catastrophically many failures in a system tend to be due to a small number of causes, rather than a general malaise across the system. We know that the "bathhouse culture" is far more prevalent in the United States than in other developed countries. Those individuals who frequent the gay bathhouses tend to experience multiple anonymous contacts with multiple partners in a single session. Public officials in the United States have shown great reluctance to closing down these establishments as a public health policy (in spite of the fact that any talk of opening heterosexual establishments oriented toward similar promiscuity would probably be unacceptable in most cities). As this book goes to press, the gay bathhouses are operating with essentially no public interference throughout the United States. Such is not the case generally in the rest of the developed world. One argument against closing the bathhouses is that since most gays do not frequent them, any closure would simply purchase a marginal lowering of the overall average contact rate at an unacceptable price of interference in freedom of association. In our heterogeneous contact rate model below, we will demonstrate that a small fraction of the target population with a high contact rate can drive the disease across the epidemic threshold, even if the overall average contact rate of the target population is unchanged.

Our heterogeneous model will use the subscript "1" to denote the majority, less sexually active portion of the target population, and the subscript "2" to denote the minority, more sexually active portion. The more active population will be taken to have a contact rate τ times that of the rate k of the majority portion of the target population. The fraction of the more sexually active population will be taken to be p. Then we have:

$$
\begin{aligned}
(39) \quad \frac{dY_1}{dt} &= \frac{k\alpha X_1(Y_1+\tau Y_2)}{X_1+Y_1+\tau(Y_2+X_2)} - (\gamma+\mu)Y_1; \\
\frac{dY_2}{dt} &= \frac{k\alpha\tau X_2(Y_1+\tau Y_2)}{X_1+Y_1+\tau(Y_2+X_2)} - (\gamma+\mu)Y_1; \\
\frac{dX_1}{dt} &= -\frac{k\alpha X_1(Y_1+\tau Y_2)}{X_1+Y_1+\tau(Y_2+X_2)} + (1-p)\lambda - \mu X_1; \\
\frac{dX_2}{dt} &= -\frac{k\alpha\tau X_2(Y_1+\tau Y_2)}{X_1+Y_1+\tau(Y_2+X_2)} + p\lambda - \mu X_2.
\end{aligned}
$$

We note that even with a simplified model such as that considered in (39), we appear to be hopelessly overparameterized. There is little chance that we shall have reliable estimates of all of: $k, \alpha, \gamma, \mu, \lambda, p, \tau$. One of the tricks frequently available to the modeler is to express the problem in such a form that most of the parameters will cancel. For the present case, we will attempt to determine the $k\alpha$ value necessary to sustain the epidemic for the heterogeneous case when the number of infectives is very small.

If $Y_1 = Y_2 = 0$, then the equilibrium values for X_1 and X_2 are $(1-p)(\lambda/\mu)$ and $p(\lambda/\mu)$, respectively. Expanding the right-hand sides of (39) in a Maclaurin series, we have (using lower case symbols for the perturbations from 0),

$$
\begin{aligned}
(40) \quad \frac{dy_1}{dt} &= \left[\frac{k\alpha(1-p)}{1-p+\tau p} - (\gamma+\mu)\right] y_1 + \frac{k\alpha(1-p)}{1-p+\tau p} y_2 \\
\frac{dy_2}{dt} &= \frac{k\alpha\tau p}{1-p+\tau p} y_1 + \left[\frac{k\alpha\tau^2 p}{1-p+\tau p} - (\gamma+\mu)\right] y_2.
\end{aligned}
$$

The solutions to a system of the form

$$
\begin{aligned}
(41) \quad \frac{dy_1}{dt} &= ay_1 + by_2, \\
\frac{dy_2}{dt} &= cy_1 + dy_2,
\end{aligned}
$$

are given by

$$
\begin{aligned}
(42) \quad y_1(t) &= c_1 e^{r_1 t} + c_2 e^{r_2 t}, \\
y_2(t) &= c_1 \frac{r_1-d}{c} e^{r_1 t} + c_2 \frac{r_2-d}{c} e^{r_2 t},
\end{aligned}
$$

where

$$r_1 = \frac{a+b+\sqrt{(a+d)^2-4(ad-bc)}}{2}, \tag{43}$$

$$r_2 = \frac{a+b-\sqrt{(a+d)^2-4(ad-bc)}}{2}. \tag{44}$$

For the epidemic not to be sustained, we must have r_1 and r_2 to be negative. This will be achieved if

$$k\alpha < (\gamma+\mu)\frac{\tau p+1-p}{\tau^2 p+1-p}. \tag{45}$$

Now then, we are in a position to solve the following problem. For the heterogeneous contact case with a subpopulation with parameters $k, \alpha, \gamma, \mu, \lambda$ with fraction p having sexual contact rate τ times that of the subpopulation of fraction $1-p$, find the overall average contact rate necessary to cross the epidemic threshold, i.e., to sustain the epidemic. Then find what the contact rate would have to be for the epidemic to be sustained in the homogeneous population situation. After some calculation, we find that the ratio of the homogeneous contact rate divided by that of the heterogeneous contact rate is given by

$$Q = \frac{1-p+\tau^2 p}{(1-p+\tau p)^2}. \tag{46}$$

In Figure 8.8, we note a plot of this "bathhouse enhancement factor." Note that the addition of the bathhouses to the transmission picture had roughly the same effect as if all the members of the target population had doubled their contact rate. And remember that the picture has been corrected to discount any increase in the overall contact rate which occurred as a result of the addition of the bathhouses. In other words, the enhancement factor is totally due to heterogeneity.

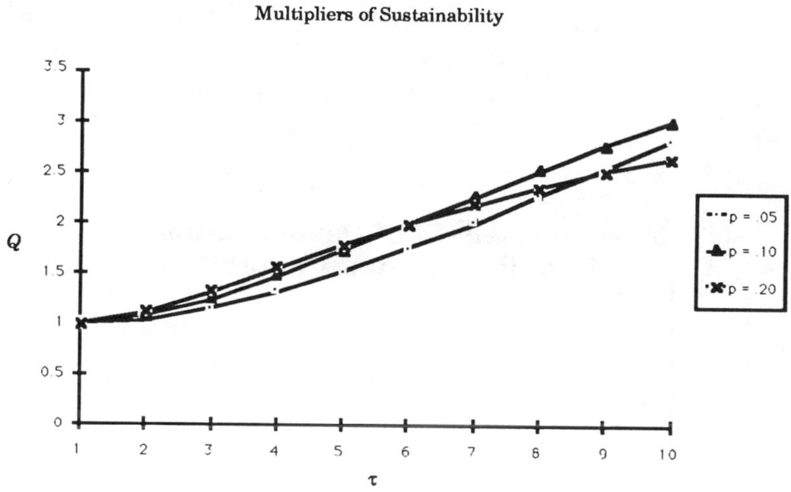

Figure 8.8.

We note that we have been able here to reduce the parameters necessary for consideration from seven to two. This is fairly typical for model based approaches: the dimensionality of the parameter space may be reducible in answering specific questions. It is shown elsewhere [11] how even more complex models of the AIDS epidemic display precisely the enhancement factor shown in Figure 8.8.

Naturally at the time this book goes to press, the AIDS epidemic in the United States no longer has only a small proportion of the gay population infective. Accordingly, one might well ask the question as to whether bathhouse closings at this late date would have any benefit. To deal with this question, we unfortunately lose our ability to disregard five of the seven parameters, and must content ourselves with picking reasonable values for those parameters. A detailed analysis is given in [12]. Here we shall content ourselves with looking at the case where the contact rate before the possible bathhouse closings is given by

$$(k\alpha)_{overall} = (1 - p + \tau p)(\gamma + \mu). \tag{47}$$

Furthermore, we shall take $\mu = 1/(180 \text{ months})$ and λ = 16,666 per month. (We are assuming a target population, absent the epidemic, of roughly 3,000,000.) For a given fraction of infectives in the target population of π, we ask what is the ratio of contact rates causing elimination of the epidemic for the closings case divided by that without closings. Such a picture is given in Figure 8.9. It would appear that as long as the proportion of infectives is no greater than 40% of the target population, there would be a benefit from bathhouse closings.

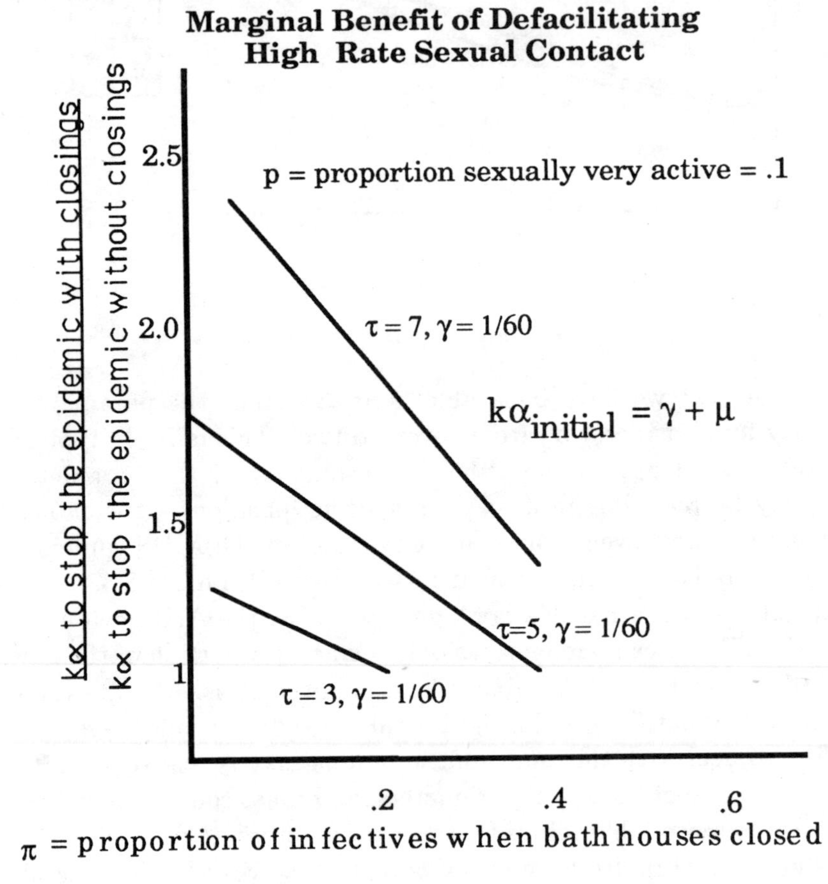

Figure 8.9.

Next, we look at the possible effects on the AIDS epidemic of administering a drug, such as AZT, to the entire infective population. We shall assume that the drug increases life expectancy by two years. In Figure 8.10, we demonstrate the change in the percent infective if the drug also increases the period of infectivity by two years for various thresholds of proportion infective at the time that the drug is administered. The curves asymptote to $1.4 = 84/60$, just as they should. The greater pool of infectives in the target population can, under some circumstances, create a kind of "Typhoid Mary" effect, where long lived infectives wander around spreading the disease. It may be desirable to obtain the cooperation of infectives receiving AZT, that they cease being sexually active with susceptibles as a condition of their receiving the drug.

Our discussions in this section have stepped well outside the boundaries of exploratory or confirmatory data analysis. For better or worse, an attack on most important real world problems requires some such sort of model building (i.e., speculative data analysis). Without a willingness to take the speculative step, statisticians and other quantitative workers are, in effect, relegating the solution of such problems to those willing to solve those problems based on considerations other than science and logic.

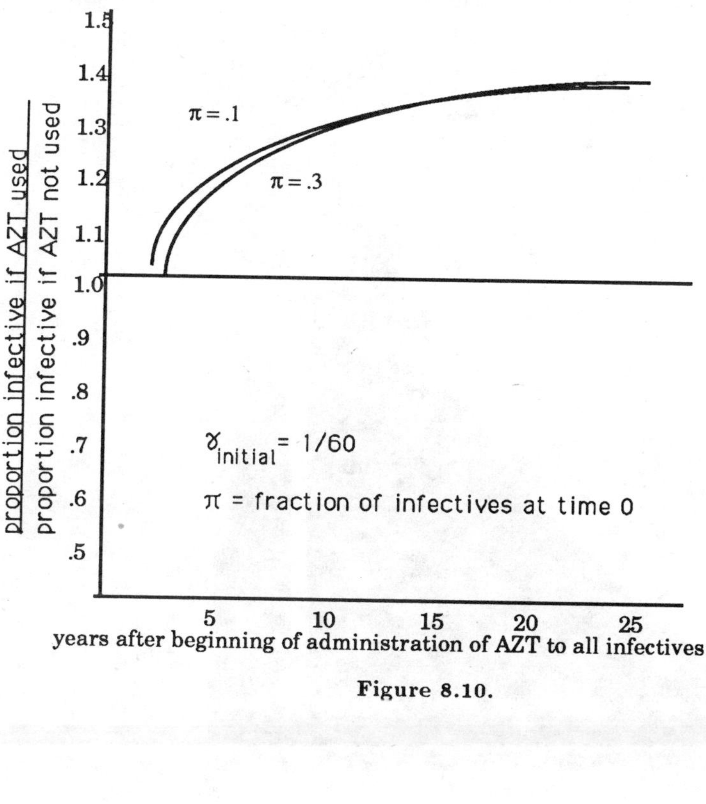

Figure 8.10.

8.5. Does Noise Prevent Chaos?

The topic of chaos theory casts doubt on the very essence of the scientific method and raises important questions about reality itself. We shall begin our discussion with Lorenz's discussion [5] of the logistic model. The logistic model, developed by Verhulst in 1844 to describe the growth of a population subject to a fixed food supply, has the basic form

$$\frac{dX}{dt} = X(\alpha - X) \tag{48}$$

where α essentially represents the limit, in population units, of the food supply. Lorenz has examined a discrete version of the logistic model:

$$X_n = X_{n-1}(a - X_{n-1}). \tag{49}$$

Using $X_0 = a/2$, he considers the time average of the modeled population size:

$$\bar{X} = \lim_{N\to\infty} \frac{1}{N} \sum_{n=1}^{N} X_n. \tag{50}$$

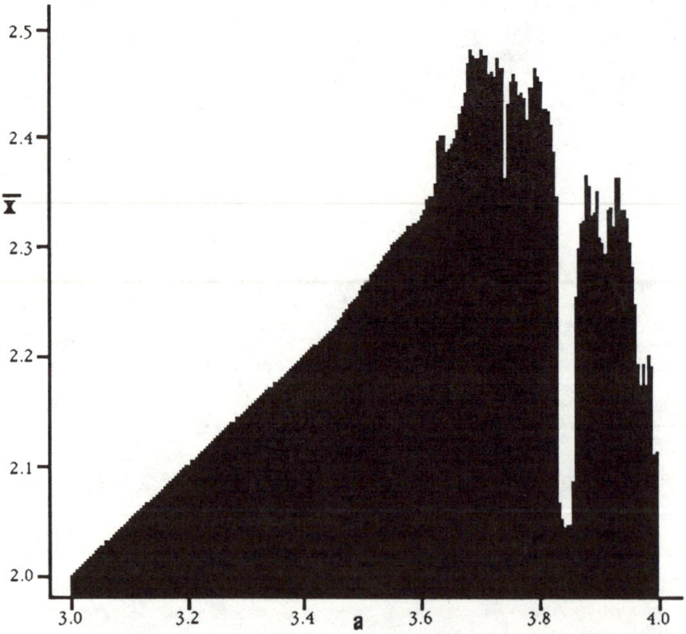

Figure 8.11.

For a values below $1+\sqrt{6}$, the graph of $\bar{X}$ behaves quite predictably. Above this value, great instability appears as we show in Figure 8.11.

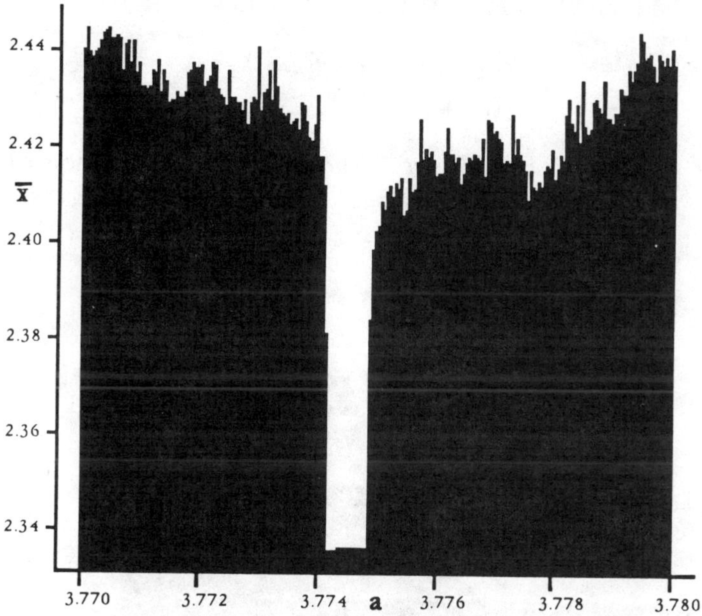

Figure 8.12.

We note in Figure 8.12 how this *fractal* structure is maintained at any level of microscopic examination we might choose. Let us look at Figure 8.12 in the light of real world ecosystems. Do we know of systems where we increase the food supply slightly and the supported population crashes, we increase it again and it soars, etc.? What should be our point of view concerning a model which produces results greatly at variance with the real world? In the case of chaos theory (and catastrophe theory and fuzzy set theory) it is frequently the practice of enthusiasts to question not the model but the reality.

Returning to the kinds of systems the logistic model was supposed to describe, we could axiomitize by a birth and death process as follows:

$$
\begin{aligned}
(51) \qquad P(\text{birth in } [t, t+\Delta t)\ ) &= \beta(\gamma - X)X\Delta t \\
P(\text{death in } [t, t+\Delta t)\ ) &= \eta X \Delta t.
\end{aligned}
$$

No doubt Verhulst would have agreed that what he had in mind to do was to aggregate from such microaxioms but had not the computational ability to do so. (48) was the best he could do. We have the computing power to deal directly with the birth and death model. However, we can make essentially the same point by adding a noise component to the logistic model. We do so as follows:

$$(52)\qquad \begin{aligned} X_n &= X_{n-1}(a - X_{n-1}) + \mu_{n-1}X_{n-1} \\ &= X_{n-1}([a + \mu_{n-1}] - X_{n-1}). \end{aligned}$$

where μ_{n-1} is a random variable from the uniform distribution on $(-\tau, \tau)$.

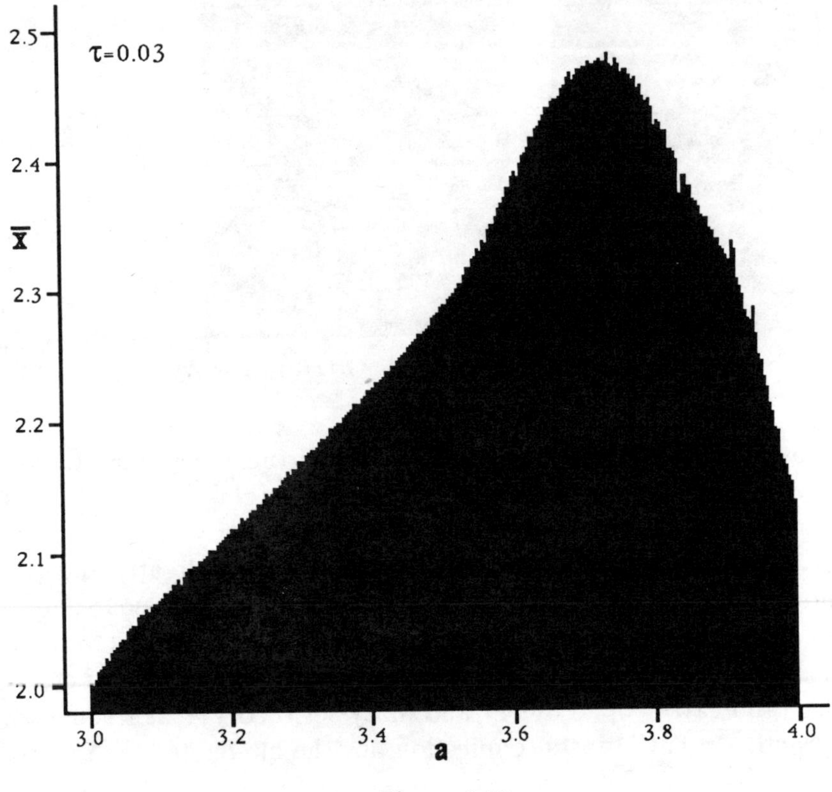

Figure 8.13.

As a convenience, we add a bounceback effect for values of the population less than zero. Namely, if the model drops the population to $-\epsilon$, then we record it as $+\epsilon$. In Figure 8.13, we note that the stochastic model produces no chaos (the slight fuzziness is due to the fact that we have averaged only 5,000 successive X_n values). Nor is there fractal structure at the microscopic level as we show in Figure 8.14 (using 70,000 successive X_n values).

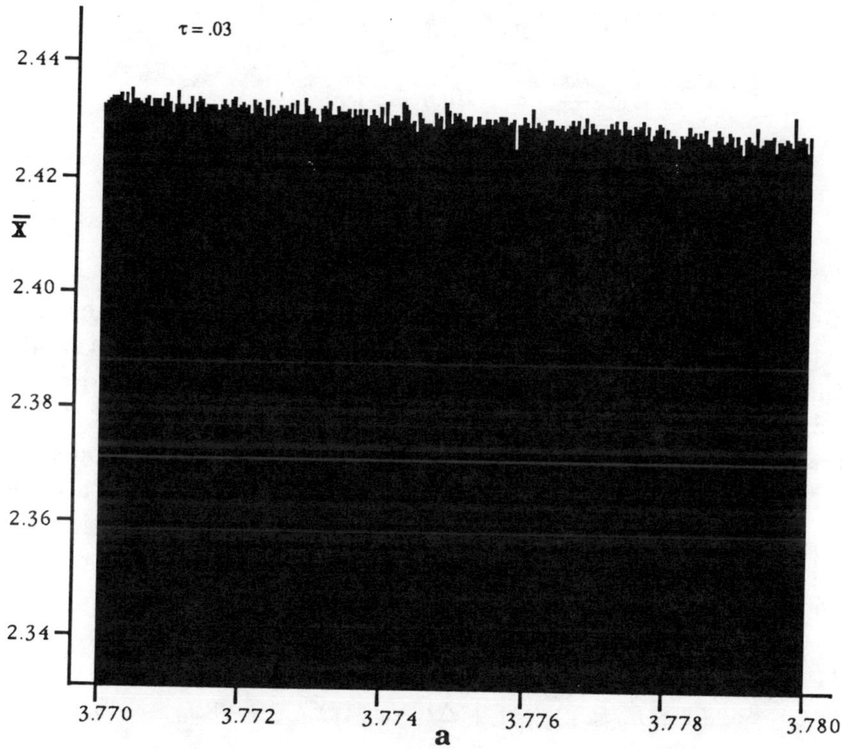

Figure 8.14.

Clearly, the noisy version of (48) is closer to the real world than the purely deterministic model. The food supply changes; the reproductive rate changes; the population is subject to constant change. However, the change itself induces stability into the system. The extreme sensitivity to a in Figure 8.11 is referred to as the "butterfly effect." The notion is that if (48) described the weather of the United States, then one butterfly flying across a backyard could dramatically change the climate of the nation. Such an effect, patently absurd, is a failure of the model (48), not of the real world.

Next, we consider a three variable convection model, which Lorenz [4] discovered was extremely sensitive to the intial value of (x, y, z).

$$\begin{aligned}\frac{dx}{dt} &= 10(y-x) \\ \frac{dy}{dt} &= -xz+28x-y \\ \frac{dz}{dt} &= xy-\frac{8}{3}z.\end{aligned} \tag{53}$$

Experience teaches us that, with few exceptions, the models we use to describe reality are only rough approximations. Whenever a model is thought to describe a process completely, we tend to discover, in retrospect, that factors were missing, that perturbations and shocks entered the picture which had not been included in the model. A common means of trying to account for such phenomena is to assume that random shocks of varying amplitudes are constantly being applied to the system. Let us, accordingly, consider a discretized noisy version of (53) :

$$\begin{aligned}(54) x_n &= (1+\mu_{x,n-1})x_{n-1}+\Delta t 10(y_{n-1}-x_{n-1}) \\ y_n &= (1+\mu_{y,n-1})y_{n-1}+\Delta t(-x_{n-1}z_{n-1}+28x_{n-1}-y_{n-1}) \\ z_n &= (1+\mu_{z,n-1})z_{n-1}+\Delta t\left(x_{n-1}y_{n-1}-\frac{8}{3}z_{n-1}\right),\end{aligned}$$

where the $\mu's$ are independently drawn from a uniform distribution on the interval from $(-\tau,\tau)$.

In Figure 8.15, we show a plot of the system for 2,000 steps using $\Delta t = .01$ and $\tau = 0.0$. We observe the nonrepeating spiral which characterizes this chaotic model. The point to be made here is that, depending on where one starts the process, the position in the remote future will be very different. Lorenz uses this model to explain the rather poor results one has in predicting the weather. Many have conjectured that such a model corresponds to a kind of uncertainty principle operating in fields ranging from meteorology to economics. Thus, it is opined that in such fields, although a deterministic model may accurately describe the system, there is no possibility of making meaningful long range forecasts, since the smallest change in the initial conditions dramatically changes the end result. The notion of such an uncertainty principle has brought comfort to such people as weather scientists and econometricians who are renowned for their inability to make useful forecasts. What better excuse for poor performance than a mathematical basis for its inevitability?

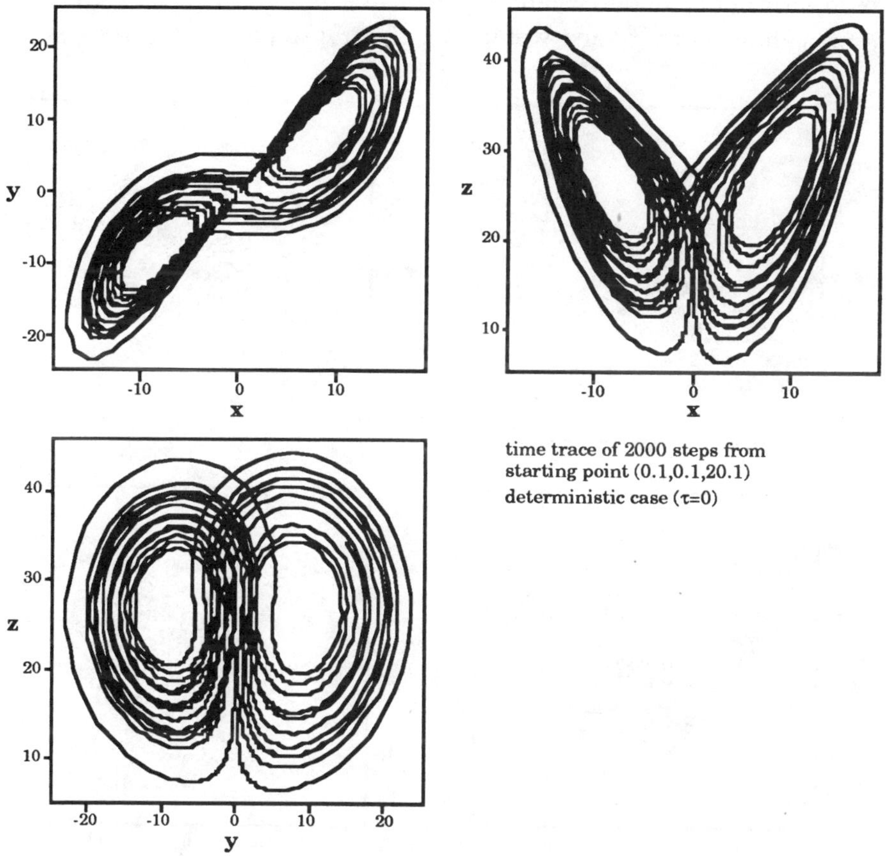

Figure 8.15.

The philosophical implications of (53) are truly significant. Using (53), we know precisely what the value of (x, y, z) will be at any future time, if we know precisely what the initial values of these variables are (an evident impossibility). We also know that the slightest change in these initial values will dramatically alter the result in the remote future. Furthermore, (53) essentially leads us to a dead end. If we believe this model, then it cannot be improved, for if at some time a better model for forecasting were available so that we really could know what (x, y, z) would be at a future time, then, since the chaos spirals are nonrepeating, we would be able to use our knowledge of the future to obtain a precise value of the present.

If one accepts the ubiquity of chaos in the real world (experience notwithstanding), then one is essentially driven back to presocratic modalities of thought, where experiments were not carried out, since it was thought that they would not give reproducible results.

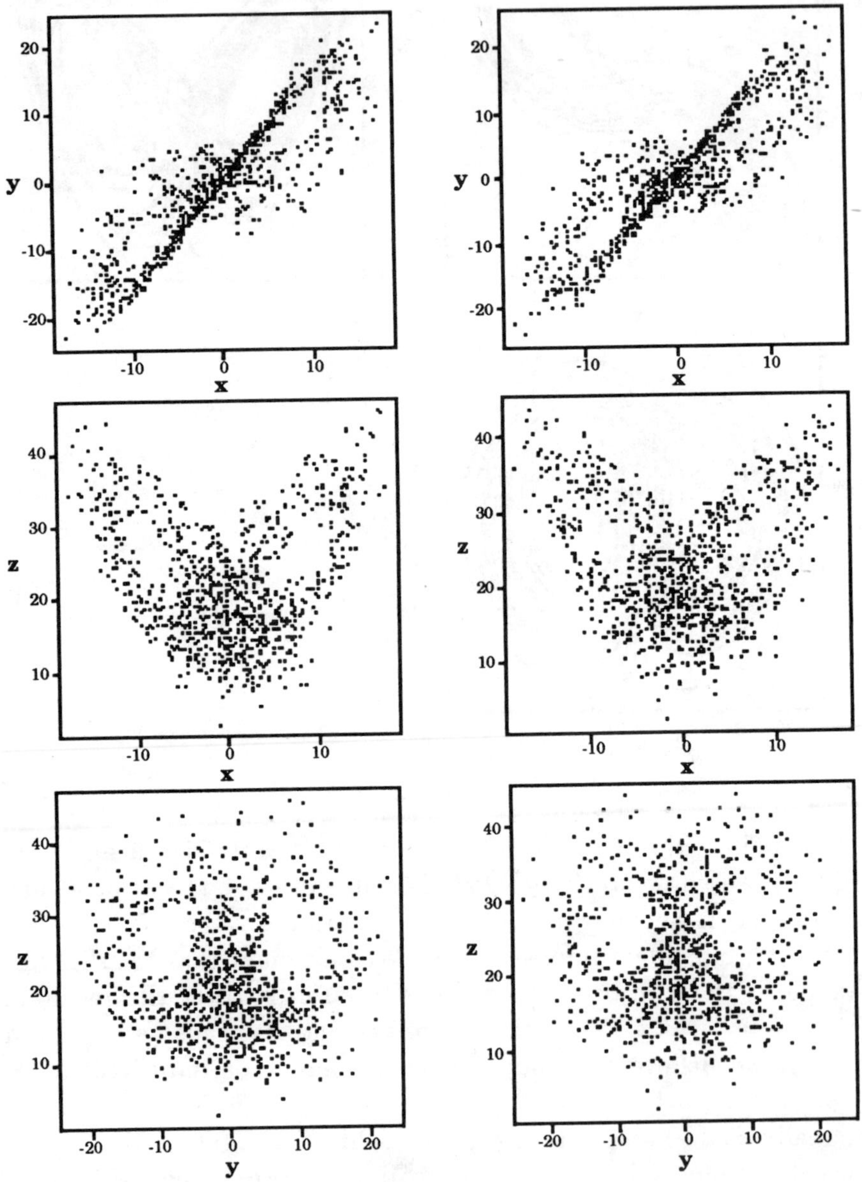

Figure 8.16.

We consider in Figure 8.16 the final point at the 2,500'th step of each of 1,000 random walks using the two inital points (0.1, 0.1, 20.1) and (−13.839, −6.66, 40.289), with $\tau = 0.0001$. These two starting points are selected since, in the deterministic case, the end results are dramatically different. We note that, just as we would expect in a model of the real world, the importance of the initial conditions diminishes with time, until, as we see in Figure 8.16, the distribution of end results is essentially independent of the initial conditions.

The subject of chaos theory, like fuzzy set theory and catastrophe theory before it, tends to be anti-rational and anti-stochastic, and like these earlier theories probably bears little relation to the real world. It is unlikely that most scientists will be drawn to a theory which promises such a chaotic world that, if it were the real one, one would not even be able to have a conversation on the merits of the notion. The maxim of deconstructionism: "If facts do not conform to theory, then so much the worse for the facts," does not work very well in the sciences. Nevertheless, chaos theory should cause us to rethink aggregation to a supposed mean trace deterministic model. In many cases we will not get in trouble if we go down this route. For example, for most of the conditions one would likely put in a logistic model, one would lie in the smooth part of the curve (i.e., before $a = 1 + \sqrt{6}$). On the other hand, as we saw in looking at the logistic model for $a > 1+\sqrt{6}$, if we consider a realistic model of a population, where the model must constantly be subject to noise, since that is the way the real world operates, then if we aggregate by simulating from the stochastic model as we did in Figure 8.13, we still obtain smooth, sensible results, even after the Verhulst aggregate (48) has failed. In Section 8.2, we demonstrated how simulation could be used as a way of avoiding the work necessary to obtain the "closed form." But, as we have seen using two stochastic versions of examples from the work of Lorenz, there are times when the "closed form" is itself unreliable, even though the simulation based aggregate, proceeding as it does from the microaxioms, is not. As rapid computing enables us to abandon more and more the closed form, we will undoubtedly find that simulation and stochastic modeling expand our ability to perceive the world as it truly is. A more complete analysis of simulation and chaos is given in [13].

References

[1] Atkinson, E.N., Bartoszyński, Robert, Brown, B.W., and Thompson, J.R. (1983). "Simulation techniques for parameter estimation in tumor related stochastic processes." *Proceedings of the 1983 Computer Simulation Conference.* New York: North Holland. 754-757.

[2] Bartoszyński, Robert, Brown, B.W., McBride, C.M., and Thompson, J.R. (1981). "Some nonparametric techniques for estimating the intensity function of a cancer related nonstationary Poisson process." *Annals of Statistics* 9: 1050-1060.

[3] Bartoszyński, Robert, Brown, B.W. and Thompson, J.R. (1982). "Metastatic and systemic factors in neoplastic progression." *Probability Models and Cancer.* Eds. L. LeCam and J. Neyman. New York: North Holland. 253-264.

[4] Lorenz, E.N. (1963). "Deterministic nonperiodic flow." *Journal of the Atmospheric Sciences* 20: 130-141.

[5] Lorenz, E.N. (1964). "The problem of deducing the climate from the governing equations. *Tellus* 16: 1-11.

[6] Poisson, S.D. (1837). *Recherches sur la probabilité des jugements en matière criminelle et en matière civile, précedées des réglés générales du calcul des probabilités.* Paris.

[7] Shilts, Randy. *And the Band Played On: Politics, People, and the AIDS Epidemic.* New York: St. Martin's. 144.

[8] Thompson, J.R. (1984). "Deterministic versus stochastic modeling in neoplasia." *Proceedings of the 1984 Summer Computer Simulation Conference.* 822-825.

[9] Thompson, J.R., Atkinson, E.N., and Brown, B.W. (1987). "SIMEST: an algorithm for simulation based estimation of parameters characterizing a stochastic process." *Cancer Modeling.* Eds. Thompson, J. and Brown, B. New York: Marcel Dekker. 387-415.

[10] Thompson, J.R. *Empirical Model Building.* New York: John Wiley and Sons. 35-43.

[11] Thompson, J.R. (1989). "AIDS: the mismanagement of an epidemic." *Computers and Mathematics with Applications.* 18: 965-972.

[12] Thompson, J.R. and Go, K.W. "AIDS: modeling the mature epidemic in the gay community." To appear in *Mathematical Population Dynamics.* Eds. Arino, O., Axelrod, D. and Kimmel, M., New York: Marcel Dekker.

[13] Thompson, J.R., Stivers, D.N., and Ensor, K.B. "SIMEST: a technique for model aggregation with considerations of chaos." To appear in *Mathematical Population Dynamics.* Eds. Arino, O., Axelrod, D. and Kimmel, M., New York: Marcel Dekker.

APPENDIX

I
An Introduction to Mathematical Optimization Theory

I.1. Hilbert Space

In this section we review some of the basic properties of Hilbert spaces. For our purposes all vector spaces are real vector spaces. We only prove results which cannot be immediately found in standard analysis texts, for example, Dunford and Schwartz [2], Goffman and Pedrick [3], Royden [11], and Taylor [13]. For related texts with a view toward applications the reader is also referred to Daniel [1], Goldstein [4], Kantorovich and Akilov [5], Luenberger [7], and Rall [10].

Definition 1. The pair $(H, \langle\cdot,\cdot\rangle)$ is called an *inner product space* if H is a vector space and $\langle\cdot,\cdot\rangle : H \times H \to R$ satisfies the properties:

(i) $\langle x, x\rangle \geq 0$ with equality if and only if $x = 0$.

(ii) $\langle x, y\rangle = \langle y, x\rangle \ \forall x, y \in H$.

(iii) $\langle \alpha x + \beta y, z\rangle = \alpha\langle x, z\rangle + \beta\langle y, z\rangle \ \forall \alpha, \beta \in R$ and $\forall x, y, z \in H$.

Definition 2. By the norm on the inner product space H we mean $\|\cdot\| : H \to R$ defined by

(1)
$$\|x\| = \langle x, x\rangle^{1/2}$$

For the sake of simplicity, when the inner product is understood, the inner product space $(H, \langle\cdot,\cdot\rangle)$ is simply denoted by H. Moreover, when it is necessary to differentiate between norms or inner products in different spaces, we will write $\|\cdot\|_H$ or $\langle\cdot,\cdot\rangle_H$.

Proposition 1. (properties of inner product and norm.)

(i) $\|x\| \geq 0$ with equality if and only if $x = 0$.

(ii) $\|\alpha x\| = |\alpha| \, \|x\|, \ \forall \alpha \in R$ and $\forall x \in H$.

(iii) $|\langle x, y\rangle| \le \|x\| \, \|y\| \; \forall x, y \in H$ (Cauchy–Schwarz).
(iv) $|\, \|x\| - \|y\| \,| \le \|x + y\| \le \|x\| + \|y\| \; \forall x, y \in H$ (triangle inequality).

Definition 3. When H and J are inner product spaces, then $f: H \to J$ is called an *operator* and, if $J = R$, the operator f is said to be a *functional.* Moreover, the operator f is said to be *linear* if

$$f(\alpha x + \beta y) = \alpha f(x) + \beta f(y), \qquad \forall \alpha, \beta \in R \qquad \text{and} \qquad \forall x, y \in H.$$

The norm on H is not a linear functional; however, the functional $f(\cdot) = \langle x, \cdot\rangle$ for fixed $x \in H$ is linear.

Let H be an inner product space.

Definition 4. A sequence $\{x^m\} \subset H$ is said to converge to $x^* \in H$ (denoted $x^m \to x$), if given $\epsilon > 0$ there exists an integer N, such that $\|x^m - x^*\| \le \epsilon$ whenever $m \ge N$.

Definition 5. A sequence $\{x^m\} \subset H$ is said to be a *Cauchy sequence* if given $\epsilon > 0$ there exists an integer N, such that $\|x^n - x^m\| \le \epsilon$ whenever $n, m \ge N$.

Definition 6. An inner product space H is said to be *complete* if every Cauchy sequence in H converges to a member of H. A complete inner product space is called a *Hilbert space.*

By the *dimension* of an inner product space we mean the vector space dimension, i.e., the cardinality of an algebraic (Hamel) basis. As we shall see, the existence theory for solutions of optimization problems is very much dependent on the Hilbert space (completeness) structure. The jth derivative of f is denoted by $f^{(j)}$ with $f^{(0)}$ denoting f. On occasion we use f' for $f^{(1)}$ and f'' for $f^{(2)}$.

Example 1. The vector space

$$R^q = \{x = (x_1, \ldots, x_q): x_i \in R\}$$

with inner product

$$\langle x, y\rangle = \sum_{i=1}^{q} x_i y_i \tag{2}$$

is a q-dimensional Hilbert space.

Example 2. The vector space

$$L^2(a, b) = \{f:(a, b) \to R : f \text{ is Lebesgue square integrable}\}$$

with inner product

$$\langle f, g \rangle = \int_a^b f(t)g(t)\,dt \tag{3}$$

is an infinite dimensional Hilbert space if we identify members of $L^2(a, b)$, which differ on a set of Lebesgue measure zero.

Example 3. The vector space

$$C(a, b) = \{\text{all continuous } f:(a, b) \to R\}$$

with inner product given by (3) is an infinite dimensional inner product space.

Example 4 (Sobolev spaces on the real line.) For $s = 1, 2, \ldots$ the vector space

$$H^s(-\infty, \infty) = \{f : f^{(j)} \in L^2(-\infty, \infty) \text{ for } j = 0, \ldots, s\}$$

with inner product

$$\langle f, g \rangle = \sum_{j=0}^{s} \langle f^{(j)}, g^{(j)} \rangle_{L^2(-\infty, \infty)}$$

is an infinite dimensional Hilbert space. We may think of $H^0(-\infty, \infty)$ as $L^2(-\infty, \infty)$.

Example 5 (Restricted Sobolev Spaces.) Let (a, b) be a finite interval. Then the vector space

$$H_0^S(a, b) = \{f : f^{(j)} \in L^2(a, b), j = 0, \ldots, s \quad \text{and} \quad f^{(j)}(a) = f^{(j)}(b) = 0, j = 0, \ldots, s-1\}$$

with inner product

$$\langle f, g \rangle = \langle f^{(s)}, g^{(s)} \rangle_{L^2(a, b)} \tag{5}$$

is an infinite dimensional Hilbert space.

Remark. Actually, the derivatives $f^{(j)}$ in the definition of the Sobolev spaces are taken in the sense of distributions. The main consequence of this for our purposes is that the sth derivative only exists almost everywhere (i.e., on the complement of a set of Lebesgue measure zero); however, the function and its first $s - 1$ derivatives will be absolutely continuous. For further details see Lions and Magenes [6] or Oden and Reddy [9].

Lemma 1. The inner product space given in Example 3 is not a Hilbert space.

Proof. Let $(a, b) = (0, 1)$ and define $\{f_n\} \subset C(0, 1)$ as follows

$$(6) \qquad f_n(t) = \begin{cases} 0 & 0 \le t \le \frac{1}{2} \\ n\left(t - \frac{1}{2}\right) & \frac{1}{2} \le t \le \frac{1}{2} + \frac{1}{n} \\ 1 & \frac{1}{2} + \frac{1}{n} \le t \le 1 \end{cases}$$

It is not difficult to see that if $m \ge n$, then $\|f_n - f_m\|^2 \le \frac{1}{n}$. Hence, $\{f_n\}$ is a Cauchy sequence in $C(0, 1)$. However, in $L^2(0, 1)$ the sequence $\{f_n\}$ converges to the discontinuous step function

$$(7) \qquad f^*(t) = \begin{cases} 0 & 0 \le t < \frac{1}{2} \\ 1 & \frac{1}{2} \le t < 1 \end{cases}$$

Since f^* does not differ from a continuous function on a set of Lebesgue measure zero, the sequence $\{f_n\}$ defined by (6) has no limit in $C(0, 1)$. ∎

Proposition 2. Any finite dimensional inner product space H is a Hilbert space and is congruent to R^q (q = dimension of H), i.e., there is a one-one correspondence between the elements of H and R^q which preserves the linear structure and the inner product.

Since the main existence theorem we are striving toward holds only in Hilbert space, we lose little generality by working in Hilbert space from now on. Toward this end, let H and J be Hilbert spaces.

Definition 7. An operator $f: H \to J$ is said to be *continuous at* $x \in H$ if $\{x^n\} \subset H$ and $x^n \to x$ implies $f(x^n) \to f(x)$ in J. Moreover, f is said to be *continuous in* $S \subset H$ if it is continuous at each point in S.

Definition 8. A linear operator $f: H \to J$ is said to be *bounded* if there exists a constant M, such that

$$(8) \qquad \|f(x)\|_J \le M\|x\|_H, \qquad \forall x \in H.$$

The vector space of all bounded operators from H into J is denoted by $[H, J]$. We denote $[H, R]$ also by H^* and call it the *dual* or *conjugate* of H. By the *operator norm of* $f \in [H, J]$ we mean

$$(9) \qquad \|f\|_{[H, J]} = \sup\{\|f(x)\|_J : \|x\|_H = 1\}.$$

Proposition 3. Consider the linear operator $f: H \to J$. Then the following are equivalent:

(i) f is bounded.

(ii) f is continuous in H.

(iii) f is continuous at one point in H.

The Cauchy–Schwarz inequality, (iii) of Proposition 1, shows that for fixed x contained in the Hilbert space H the linear functional $x^*(\cdot) = \langle x, \cdot \rangle$ is bounded and, consequently, $x^* \in H^*$. The following very important theorem shows that actually all members of H^* arise in this fashion.

Theorem 1 (Riesz Representation Theorem.) If $f \in H^*$, then there exists a unique $x_f \in H$, such that

$$f(x) = \langle x_f, x \rangle, \qquad \forall x \in H. \tag{10}$$

Moreover, the vector space H^* becomes a Hilbert space with the inner product

$$\langle f, g \rangle_{H^*} = \langle x_f, x_g \rangle_H \qquad \forall f, g \in H^*, \tag{11}$$

and the operator norm on H^* coincides with the norm induced by the inner product (11).

Proposition 4. All linear functionals on finite dimensional inner product spaces are continuous.

Proposition 5. There exist linear functionals on infinite dimensional inner product spaces which are not continuous.

Proof. Consider $C(-1, 1)$ given by Example 3. Let $\delta(f) = f(0)$ for $f \in C(-1, 1)$. Clearly, δ is a linear functional. Construct

$$f_n(t) = \begin{cases} 0 & -1 \le t \le -\dfrac{1}{n} \\ 1 + nt & -\dfrac{1}{n} \le t \le 0 \\ 1 - nt & 0 \le t \le \dfrac{1}{n} \\ 0 & \dfrac{1}{n} \le t \le 1 \end{cases} \tag{12}$$

We have that $f_n \in C(-1, 1)$ and that $f_n \to 0$ in $C(-1, 1)$. To see this observe that

$$\|f_n - 0\|_{C(-1,1)}^2 = \int_{-1}^{1} f_n(t)^2\, dt \le \int_{-1}^{1} f_n(t)\, dt = \frac{1}{n}. \tag{13}$$

However,

$$1 = \delta(f_n) \nrightarrow \delta(0) = 0. \quad \blacksquare \tag{14}$$

Remark. Although we did not demonstrate Proposition 5 for Hilbert space it does hold in this case; however, a constructive proof is not possible.

Definition 9. A sequence $\{x^n\}$ in a Hilbert space H is said to *converge weakly* to $x^* \in H$ if $f(x^n) \to f(x^*)\ \forall f \in H^*$.

Proposition 6. If $\{x^n\}$ converges to x^* in H, then it converges weakly to x^*. Moreover, if H is finite dimensional, then $\{x^n\}$ converges to x^* if and only if $\{x^n\}$ converges weakly to x^*.

Definition 10. An operator $f: H \to J$ is said to be *weakly continuous* if $\{x^n\} \subset H$ converges weakly to $x^* \in H$, then $\{f(x^n)\}$ converges weakly to $f(x^*)$ in J.

Clearly, weak continuity implies continuity, and by definition members of H^* are weakly continuous.

Definition 11. The set $S \subset H$ is *closed*, respectively, *weakly closed*, if $\{x^n\} \subset S$ converges, respectively, converges weakly, to $x^* \in H$ implies that $x^* \in S$.

It is clear that weakly closed subsets of H are closed.

Definition 12. The set $S \subset H$ is *compact*, respectively, *weakly compact*, if $\{x^n\} \subset S$ implies that $\{x^n\}$ contains a subsequence which is convergent, respectively, weakly convergent, to a member of S.

It follows that compact subsets of H are weakly compact. However, when working with infinite dimensional Hilbert spaces, compact subsets are rare and hard to come by. Consequently, any theory which requires compactness is essentially a finite dimensional theory. In particular, we have the following characterization of compactness in finite dimensions.

Proposition 7. Consider $S \subset R^q (q < \infty)$. Then S is compact if and only if S is closed and bounded (i.e., there exists M, such that $\|s\| \le M\ \forall s \in S$).

Proposition 7 does not hold for infinite dimensional Hilbert spaces; the missing ingredient is convexity.

Definition 13. A subset S of H is said to be *convex* if $s_1, s_2 \in S$ implies $ts_1 + (1 - t)s_2 \in S\ \forall t \in [0, 1]$.

Proposition 8. Consider convex $S \subset H$. Then S is weakly closed if S is also closed. Moreover, S is weakly compact if S is also closed and bounded.

Proposition 8 leads us to anticipate the strong role of convexity in infinite dimensional optimization problems.

I.2. *Reproducing Kernel Hilbert Spaces*

In statistics we are usually dealing with probability density functions which are by definition nonnegative. In order to enforce this constraint, we must work with the point evaluation functionals introduced in the previous section. Moreover, if we are to utilize compactness or weak compactness the collection of functions of interest (probability density functions) must be a closed set, which in turn essentially requires that point evaluation be a continuous operation. This is the subject of the present section. We would like to isolate function spaces which are sufficiently large or rich (i.e., infinite dimensional) and still have the property that point evaluation is a continuous operation.

Definition 14. A Hilbert space H of functions defined on a set T is said to be a *proper functional Hilbert space* if for every $t \in T$ the point evaluation functional at t is continuous, i.e., there exists M_t, such that

$$|f(t)| \leq M_t \|f\|, \qquad \forall f \in H. \tag{15}$$

Definition 15. A Hilbert space $H(T)$ of functions defined on a set T is said to be a *reproducing kernel Hilbert space* (RKHS) if there exists a *reproducing kernel functional* $K(\cdot,\cdot)$ defined on $T \times T$ with the properties

(16)

(i) $K(\cdot, t) \in H(T), \qquad \forall t \in T$

(ii) $f(t) = \langle f, K(\cdot, t)\rangle, \forall f \in H(T)$ and $\forall t \in T$.

Proposition 9. A Hilbert space of functions defined on a set T is a proper functional Hilbert space if and only if it is a reproducing kernel Hilbert space.

Proof. The Cauchy–Schwarz inequality (Proposition 1) shows that (ii)⇒(i); while the Riesz representation theorem (Theorem 1) shows that (i)⇒(ii). ∎

Proposition 9 is one of the reasons for the large emphasis on RKHS in statistical applications.

Proposition 10. The restricted Sobolev space $H_0^s(a, b)$ given in Example 5 is a RKHS.

Proof. For $f \in H_0^s(a, b)$ we may write, using Cauchy–Schwarz in $L^2(a, b)$,

$$|f(t)| = \left|\int_a^t f'(u)\, du\right| \leq \int_a^b |f'(u)|\, du \leq (b-a)\|f'\|_{L^2(a,b)}. \tag{17}$$

Now, by squaring (17), integrating in t over (a, b), and taking square roots we obtain

$$\|f\|_{L^2(a,b)} \leq (b-a)^{3/2}\|f'\|_{L^2(a,b)}. \tag{18}$$

Similar arguments show that for $j = 1, 2, \ldots, s-1$

$$\|f^{(j)}\|_{L^2(a,b)} \leq (b-a)^{3/2}\|f^{(j+1)}\|_{L^2(a,b)}. \tag{19}$$

Now, combining (17) and (19) gives

$$|f(t)| \leq (b-a)^{(3s-1)/2}\|f^{(s)}\|_{L^2(a,b)}.$$

The result now follows from (15) and Proposition 9. ∎

Proposition 11. The integration functional defined by $Q(f) = \int_a^b f(t)\,dt$ is continuous on $H_0^s(a, b)$.

Proof. A straightforward integration by parts and the Cauchy–Schwarz inequality in $L^2(a, b)$ gives for $f \in H_0^s(a, b)$

$$\left|\int_a^b f(t)\,dt\right| = \left|\int_a^b \frac{t^s}{s!} f^{(s)}(t)\,dt\right| \leq \left\|\frac{t^s}{s!}\right\|_{L^2(a,b)} \|f^{(s)}\|_{L^2(a,b)}. \tag{21}$$

The result now follows from Proposition 3. ∎

Proposition 12. The Sobolev space $H^s(-\infty, \infty)$ given in Example 4 is a RKHS.

Proof. The proof will require the introduction of the "fractional order Sobolev spaces." Toward this end for any $s \in (-\infty, \infty)$ let

$$H^s(-\infty, \infty) = \{v \in L : (1 + w^2)^{s/2}\hat{v}(w) \in L^2(-\infty, \infty)\}, \tag{22}$$

where L is the space of Schwartz's tempered distribution and $\hat{v}$ denotes the Fourier transform of v. The inner product on $H^s(-\infty, \infty)$ is given by

$$\langle u, v\rangle = \langle (1 + w^2)\hat{u}, \hat{v}\rangle_{L^2(-\infty,\infty)}. \tag{23}$$

It is shown in Lions and Magenes [6] that the inner product space $H^s(-\infty, \infty)$ given by (22) and (23) is a Hilbert space, and for $s = 1, 2, \ldots$ it is isomorphic to the Sobolev space $H^s(-\infty, \infty)$ given in Example 4. Hence, we need not worry about any ambiguity in notation. Furthermore, they show that the dual of $H^s(-\infty, \infty)$ is $H^{-s}(-\infty, \infty)$. Now, $H^s(-\infty, \infty)$ will be a RKHS if the Dirac δ distribution δ_t is in the dual. Recall that δ_t is defined to be the distribution with the property that

$$u(t) = \int_{-\infty}^{\infty} \delta_t(r)u(r)\,dr. \tag{24}$$

The integration in (24) is taken in the sense of distributions. It follows that we want

$$(1 + w^2)^{-s/2}\, \hat{\delta}_t \in L^2(-\infty, \infty). \tag{25}$$

Since the Fourier transform of δ_t is a constant, (25) is equivalent to

$$(1 + w^2)^{-s/2} \in L^2(-\infty, \infty). \tag{26}$$

Clearly, (26) holds if and only if $s > \frac{1}{2}$. ∎

I.3. Convex Functionals and Differential Characterizations

Throughout this section, H will denote a Hilbert space. A majority of the following material can be found in Ortega and Rheinboldt [8]. For a more detailed account of differentiation, see also Tapia [12].

Definition 16. Let S be a convex subset of H and consider $f:H \to R$. Then

(i) f is *convex* on S if

$$tf(x) + (1 - t)f(y) \geq f(tx + (1 - t)y), \qquad \forall t \in [0, 1], \qquad \forall x, y \in S. \tag{27}$$

(ii) f is *strictly convex* if strict inequality holds in (1) for $0 < t < 1$ and $x \neq y$.

(iii) f is *uniformly convex* on S if there exists $C > 0$, such that

$$tf(x) + (1 - t)f(y) - f(tx + (1 - t)y) \geq Ct(1 - t)\|x - y\|^2 \qquad \forall t \in [0, 1] \quad \text{and} \quad \forall x, y \in S. \tag{28}$$

It should be obvious that uniform convexity implies strict convexity and that strict convexity implies convexity. We will now develop reasonable criteria for deciding when a functional is uniformly convex, strictly convex or convex.

Definition 17. Consider $f:H \to R$. Given $x, \eta_1, \ldots, \eta_n \in H$ by the *nth Gâteaux variation of S at x in the directions $\eta_1, \ldots, \eta_n$* we mean

$$f^{(n)}(x)(\eta_1, \ldots, \eta_n) = \lim_{t \to 0} t^{-1}[f^{(n-1)}(x + t\eta_n)(\eta_1, \ldots, \eta_{n-1}) - f^{(n-1)}(x)(\eta_1, \ldots, \eta_{n-1})], \tag{29}$$

with $f^{(0)}(x) = f(x)$. Moreover, we say *f is m times Gâteaux differentiable* in $S \subset H$ if the first m Gâteaux variations exist at x and are linear in η_i $\forall i$ and for each fixed x in S. We also say that f is *continuously differentiable* in S if it is Gâteaux differentiable in S, $f^{(1)}(x) \in H^*$ $\forall x \in S$ and $f^{(1)}:S \subset H \to H^*$

is a continuous operator. In the case that $f^{(1)}(x) \in H^*$ the Riesz representer of $f^{(1)}(x)$ is denoted by $\nabla f(x)$ and is called the *gradient* of f at x (i.e., $f^{(1)}(x)(\eta) = \langle \nabla f(x), \eta \rangle \forall \eta \in H$).

Proposition 13. If $\Phi(t) = f(x + t\eta)$, then

$$\Phi^{(n)}(0) = f^{(n)}(x)(\eta, \ldots, \eta). \tag{30}$$

Proof. Let $n = 1$. Then

$$\Phi'(t) = \lim_{\Delta t \to 0} \left[\frac{\Phi(t + \Delta t) - \Phi(t)}{\Delta t} \right] \tag{31}$$

$$= \lim_{\Delta t \to 0} \left[\frac{f(x + t\eta + \Delta t\eta) - f(x + t\eta)}{\Delta t} \right] \tag{32}$$

$$= f'(x + t\eta)(\eta); \tag{33}$$

and (30) holds for $n = 1$. Assume that (30) holds for $n = 1, \ldots, k - 1$. Then

$$\Phi^{(k)}(t) = \lim_{\Delta t \to 0} \left[\frac{\Phi^{(k-1)}(t + \Delta t) - \Phi^{(k-1)}(t)}{\Delta t} \right] \tag{34}$$

$$= \lim_{\Delta t \to 0} \left[\frac{f^{(k-1)}(x + t\eta + \Delta t\eta)(\eta, \ldots, \eta) - f^{(k-1)}(x + t\eta)(\eta, \ldots, \eta)}{\Delta t} \right] \tag{35}$$

$$= f^{(k)}(x + t\eta)(\eta, \ldots, \eta); \tag{36}$$

hence (30) holds for k and by induction for all n. ∎

Example 6. Consider $f: H \to R$ defined by

$$f(x) = \tfrac{1}{2}\langle x, x \rangle, \tag{37}$$

then

$$\Phi'(t) = \langle x + t\eta, \eta \rangle \tag{38}$$

and

$$\Phi''(t) = \langle \eta, \eta \rangle. \tag{39}$$

It follows that $f'(x)(\eta) = \langle x, \eta \rangle$ and $f''(x)(\eta, \eta) = \langle \eta, \eta \rangle$. Observe that in this case f is infinitely Gâteaux differentiable in H.

Proposition 14. Assume that $f: H \to R$ is Gâteaux differentiable in a convex subset S of H. Then

(40) (i) f is convex in $S \Leftrightarrow f'(x)(y - x) \le f(y) - f(x), \quad \forall x, y \in S.$

(41) (ii) f is strictly convex in $S \Leftrightarrow f'(x)(y - x) < f(y) - f(x),$
$\forall x, y \in S \quad \text{and} \quad x \neq y.$

Proof [(i) $\Leftarrow$]. For $x, y \in S$ let $z = \alpha x + (1-\alpha)y$ for $0 < \alpha < 1$. By hypothesis $z \in S$ and

$$f(x) - f(z) \geq f'(z)(x - z) \tag{42}$$

and

$$f(y) - f(z) \geq f'(z)(y - z). \tag{43}$$

Therefore,

$$\begin{aligned} \alpha f(x) + (1-\alpha)f(y) - f(z) &\geq \alpha f'(z)(x - z) + (1-\alpha)f'(z)(y - z) \\ &= f'(z)[\alpha(x - z) + (1-\alpha)(y - z)] \\ &= f'(z)[\alpha x + (1-\alpha)y - z] \\ &= f'(z)(0) \\ &= 0. \end{aligned} \tag{44}$$

It follows that

$$\alpha f(x) + (1-\alpha)f(y) \geq f(\alpha x + (1-\alpha)y) \tag{45}$$

and f is convex. Exactly the same proof demonstrates [(ii) $\Leftarrow$].

[(i) $\Rightarrow$]. Suppose f is convex on S. For $x, y \in S$ and $0 < \alpha < 1$ we have

$$\alpha f(y) + (1-\alpha)f(x) \geq f(\alpha y + (1-\alpha)x). \tag{46}$$

Divide (46) by α and then let $\alpha \to 0$ to obtain

$$f(y) - f(x) \geq f'(x)(y - x). \tag{47}$$

[(ii) $\Rightarrow$]. Here the above limiting process will not suffice. Let $z = \frac{1}{2}(x + y)$, then $z - x = \frac{1}{2}(y - x)$ and by [(i) $\Rightarrow$]

$$\begin{aligned} f'(x)(y - x) &= 2f'(x)(z - x) \\ &\leq 2[f(z) - f(x)] \\ &< 2[\tfrac{1}{2}f(x) + \tfrac{1}{2}f(y) - f(x)] \\ &= f(y) - f(x). \end{aligned} \tag{48}$$

Proposition 15. Under the assumptions of Proposition 14, we have

(i) f is convex on $S \Leftrightarrow [f'(y) - f'(x)](y - x) \geq 0 \qquad \forall x, y \in S$

(ii) f is strictly convex on $S \Leftrightarrow [f'(y) - f'(x)](y - x) > 0$

$\forall x, y \in S$ and $x \neq y$.

Proof [(i) $\Rightarrow$]. Suppose f is convex on S. Then by [(i)] of Proposition 14

$$f(y) - f(x) \geq f'(x)(y - x) \tag{49}$$

and

$$f(x) - f(y) \geq f'(y)(x - y). \tag{50}$$

Now, by adding (49) to (50) we obtain

$$[f'(y) - f'(x)](y - x) \geq 0. \tag{51}$$

[(i) ⇐]. By the mean value theorem applied to $\Phi(t) = f(x + t(y - x))$ we have $\Phi(1) - \Phi(0) = \Phi'(t)$ for some $0 < t < 1$, or equivalently

$$f(y) - f(x) = f'(x + t(y - x))(y - x), \qquad 0 < t < 1. \tag{52}$$

Let $u = x + t(y - x)$. Observe that $u \in S$ and that $y - x = (u - x)/t$, so that

$$[f'(u) - f'(x)](y - x) = \frac{1}{t}[f'(u) - f'(x)](u - x) \geq 0. \tag{53}$$

Also,

$$f(y) - f(x) = [f'(u) - f'(x)](y - x) + f'(x)(y - x), \tag{54}$$

therefore,

$$f(y) - f(x) \geq f'(x)(y - x). \tag{55}$$

By [(i) ⇐] of Proposition 14 we see that (55) implies convexity. The above calculations also establish part (ii). ■

Definition 18. Consider $f: H \to R$ and a convex subset S of H. By *the cone tangent to S at x* we mean $T(x) = \{\eta \in H : \exists t > 0, \text{ such that } x + t\eta \in S\}$. Suppose f is twice Gâteaux differentiable in S. Then f'' is said to be *positive semidefinite relative to S* if for each $x \in S$ we have

$$f''(x)(\eta, \eta) \geq 0, \qquad \forall \eta \in T(x). \tag{56}$$

We say f'' is *positive definite relative to S* if strict inequality holds in (56) for $\eta \neq 0$. Finally, f'' is said to be *uniformly positive definite relative to S* if there exists $C > 0$, such that for each $x \in S$ we have

$$f''(x)(\eta, \eta) \geq C\|\eta\|^2, \qquad \forall \eta \in T(x). \tag{57}$$

Proposition 16. Assume that $f: H \to R$ is twice Gâteaux differentiable in a convex subset S of H.

Then

(i) f is convex in $S \Leftrightarrow f''$ is positive semidefinite relative to S.
(ii) f is strictly convex in $S \Leftarrow f''$ is positive definite relative to S.

Proof [(i) ⇐]. Using Taylor's theorem on $\Phi(t) = f(x + t(y - x))$, we have for $x, y \in S$

$$(58) \quad f(y) - f(x) - f'(x)(y - x) = \tfrac{1}{2} f''(x + \Theta(y - x))(y - x, y - x), \quad 0 < \Theta < 1$$

Since $x + \Theta(y - x) \in S$, it follows that $(y - x) \in T(x + \Theta(y - x))$. Consequently, Proposition 14, (58) and the fact that f'' is positive semidefinite relative to S imply f is convex.
[(ii) ⇐]. This follows exactly as in the proof of [(i) ⇐].
[(i) ⇒]. First observe that if $x \in S$ and $x + t\eta \in S$, then $x + \tau\eta \in S$ for $0 \le \tau \le t$. For $x \in S$ and $\eta \in T(x)$ we have (considering only small $t > 0$ if necessary)

$$(59) \quad f''(x)(\eta, \eta) = \lim_{t \to 0} \frac{1}{t} [f'(x + t\eta)(\eta) - f'(x)(\eta)]$$

$$= \lim_{t \to 0} \frac{1}{t^2} [f'(x + t\eta) - f'(x)](x + t\eta - x).$$

By convexity and $[(i) \Rightarrow]$ of Proposition 15 it follows that (59) implies that f'' is positive semidefinite relative to S. ■

Proposition 17. Assume that $f: H \to R$ is Gâteaux differentiable in a convex subset S of H. Then the following three statements are equivalent:

(i) f is uniformly convex in S with constant C,
(ii) $f(y) - f(x) \ge f'(x)(y - x) + C\|y - x\|^2$, $\forall x, y \in S$,
(iii) $(f'(y) - f'(x))(y - x) \ge 2C\|y - x\|^2$, $\forall x, y \in S$.

If in addition, f is twice Gâteaux differentiable in S, then the above three statements are equivalent to

(iv) f'' is uniformly positive definite relative to S with constant $2C$.

Proof [(i)] ⇒ (ii)]. We have for $0 < t < 1$ and $x, y \in S$ from (28)

$$(60) \quad tf(y) - tf(x) \ge [f(x + t(y - x)) - f(x)] + t(1 - t)C\|x - y\|^2.$$

Therefore, dividing (60) by t and then letting $t \to 0$ gives (ii).
[(ii) ⇒ (iii)]. Interchanging the roles of x and y in (ii) and adding the resulting inequality to (ii) gives (iii).
[(iii) ⇒ (ii)]. Let $t_k = \dfrac{k}{m + 1}$, $k = 0, \ldots, m + 1$ for arbitrary $m \ge 0$. The mean value theorem on $\Phi(t) = f(x + t(y - x))$ gives

$$(61) \quad \Phi(t_{k+1}) - \Phi(t_k) = \Phi'(s_k), \qquad \text{where} \quad t_k < s_k < t_{k+1}.$$

Rewriting (61) leads to

$$(62)\qquad f(x + t_{k+1}(y - x)) - f(x + t_k(y - k)) \\ = f'(x + s_k(y - x))((t_{k+1} - t_k)(y - x)).$$

Hence, from (iii) and (62) we obtain

$$(63)\quad f(y) - f(x) = \sum_{k=0}^{m} [f(x + t_{k+1}(y - x)) - f(x + t_k(y - x))]$$

$$= \sum_{k=0}^{m} [f'(x + s_k(y - x)) - f'(x)][(t_{k+1} - t_k)(y - x)]$$

$$+ f'(x)(y - x)$$

$$= \sum_{k=0}^{m} \frac{(t_{k+1} - t_k)}{s_k} [f'(x + s_k(y - x)) - f'(x)](s_k(y - x))$$

$$+ f'(x)(y - x)$$

$$\geq \sum_{k=0}^{m} \frac{2(t_{k+1} - t_k)}{s_k} s_k^2 C\|y - x\|^2 + f'(x)(y - x).$$

Now, observe that

$$(64)\qquad \sum_{k=0}^{m} (t_{k+1} - t_k)s_k \geq \sum_{k=0}^{m} (t_{k+1} - t_k)t_k = \frac{1}{(m+1)^2} \sum_{k=0}^{m} k$$

$$= \frac{1}{2} \frac{m}{m+1}.$$

Coupling (63) with (64) we arrive at

$$(65)\qquad f(y) - f(x) \geq \frac{m}{m+1} C\|y - x\|^2 + f'(x)(y - x).$$

Letting $m \to \infty$ in (65) gives (ii).

[(ii) ⇒ (i)]. Let $z = \alpha x + (1 - \alpha)y$ for x, $y \in S$ and $0 < \alpha < 1$. From (ii) we have

$$(66)\qquad f(x) - f(z) \geq f'(z)(x - z) + C\|x - z\|^2$$

and

$$(67)\qquad f(y) - f(z) \geq f'(z)(y - z) + C\|y - z\|^2.$$

Observe that

$$(68)\qquad \alpha\|x - z\|^2 + (1 - \alpha)\|y - z\|^2 = \alpha(1 - \alpha)\|x - y\|^2.$$

Now, multiplying (66) by α, (67) by $(1-\alpha)$ and then adding and recalling (68) we obtain

$$\alpha f(x) + (1-\alpha)f(y) - f(z) \geq C\alpha(1-\alpha)\|y-x\|^2. \tag{69}$$

However, (69) is merely (28) with t replaced by α.
[(iv) ⇒ (ii)]. Let $\Phi(t) = f(x + t(y-x))$. Taylor's theorem on $\Phi(t)$ gives

$$\Phi(1) - \Phi(0) - \Phi'(0) = \tfrac{1}{2}\Phi''(\theta), \qquad \text{for} \quad 0 < \theta < 1. \tag{70}$$

Coupling (70) with (iv) leads to

$$\begin{aligned} f(y) - f(x) - f'(x)(y-x) &= \tfrac{1}{2}f''(x + \theta(y-x))(y-x, y-x) \\ &\geq C\|x-y\|^2, \qquad \forall x, y \in S \end{aligned} \tag{71}$$

which is exactly (ii).
[(iii) → (iv)] For $x \in S$ and $\eta \in T(x)$ let $y = x + t\eta$ in (iii) to obtain

$$[f'(x+t\eta) - f'(x)](t\eta) \geq 2Ct^2\|\eta\|^2. \tag{72}$$

Dividing (72) by t^2 and letting $t \to 0$ gives (iv). ∎

I.4. Existence and Uniqueness of Solutions for Optimization Problems in Hilbert Space

A portion of the following material can be found in Goldstein [4]. As before, H will denote a Hilbert space. For $S \subset H$ and $f: H \to R$ we may consider the constrained optimization problem

$$\text{minimize } f(x); \quad \text{subject to } x \in S. \tag{73}$$

Solutions of problem (73) are referred to *as minimizers.*

Theorem 2 (uniqueness). If f is strictly convex and S is convex, then problem (73) has at most one minimizer.

Proof. Suppose x^* and x^{**} are both solutions of problem (73). Then, if $x^* \neq x^{**}$ we have for $0 < t < 1$ by strict convexity

$$f(tx^* + (1-t)x^{**}) < tf(x^*) + (1-t)f(x^{**}) = f(x^*). \tag{74}$$

But $x = tx^* + (1-t)x^{**} \in S$, since S is convex. This contradicts the optimality of x^*. ∎

The problem of existence of solutions to problem (73) is not as straightforward and requires the introduction of the notion of weak lower semicontinuity. Toward this end recall that the functional f is (weakly) continuous at x if given a sequence $\{x^n\} \subset H$ which converges (weakly) to $x \in H$, then given $\epsilon > 0$ there exists an integer N_ϵ, such that

$$-\epsilon < f(x^n) - f(x) < \epsilon, \qquad \forall n \geq N_\epsilon. \tag{75}$$

Definition 19. The functional f is said to be (*weakly*) *lower semicontinuous* at $x \in H$ if given a sequence $\{x^n\} \subset H$ converging (weakly) to x, then, given $\epsilon > 0$, there exists an integer N_ϵ, such that

$$-\epsilon < f(x^n) - f(x), \qquad \forall n \geq N_\epsilon. \tag{76}$$

Moreover, we say f is (*weakly*) *upper semicontinuous* at $x \in H$ if in the above we replace (76) with

$$f(x^n) - f(x) < \epsilon, \qquad \forall n \geq N_\epsilon. \tag{77}$$

Clearly, f is (weakly) continuous if and only if it is both (weakly) lower and upper semicontinuous.

Proposition 18. The functional f is (weakly) lower semicontinuous if and only if whenever $\{x^n\}$ converges (weakly) to x we have

$$f(x) \leq \underline{\lim}_n f(x^n). \tag{78}$$

Proof. The proof is a straightforward application of (76) and the definition

$$\underline{\lim}_n f(x^n) = \lim_{n \to \infty} \left(\inf_{k \geq n} f(x^k) \right). \tag{79}$$ ∎

Proposition 19. Let D be a subset of H. Then $f:D \to R$ is (weakly) lower semicontinuous in D if and only if the set

$$D_M = \{x \in D : f(x) \leq M\}$$

is (weakly) closed for all M.

Proof. Suppose that D_M is (weakly) closed $\forall M$. Consider $x \in D$ and assume that f is not (weakly) lower semicontinuous at x, then for some M and some sequence $\{x^n\}$ converging (weakly) to x we have

$$\underline{\lim}_n f(x^n) < M < f(x). \tag{80}$$

It follows that a subsequence of $\{x^n\}$ remains in D_M, and since D_M is (weakly) closed, we have $x \in D_M$. However, this contradicts (80).

Now assume that f is (weakly) lower semicontinuous and that for some M the set D_M is not (weakly) closed. Then there exists $\{x^n\} \subset D_M$, such that $\{x^n\}$ converges (weakly) to $x \notin D_M$.
This implies that $f(x) > M$. However, by Proposition 18 we have that $f(x) \leq \underline{\lim}_n f(x^n) \leq M$. This is a contradiction. ∎

Proposition 20. A continuous convex functional f defined on a closed convex subset S of H is weakly lower semicontinuous.

Proof. The set $S_M = \{x \in S : f(x) \leq M\}$ is closed, since f is continuous and S is closed. It is also convex, since S is convex and f is convex. Therefore, by Proposition 8 the set S_M is weakly closed. This means, according to Proposition 19, that f is weakly lower semicontinuous. ∎

Theorem 3 (Existence for finite dimensions.) Suppose that in problem (73) the Hilbert space H is finite dimensional, the functional f is continuous on S, and the subset S is closed and bounded. Then problem (73) has at least one minimizer.

Proof. Choose $\{x^n\} \subset S$, such that $f(x^n)$ converges to $\inf\{f(x) : x \in S\}$. By compactness of S there exists $x^* \in S$ and a subsequence $\{x^{n_i}\}$ which converges to x^*. By the continuity of f we have that $\{f(x^{n_i})\}$ converges to $f(x^*)$; hence, x^* is a minimizer to problem (73). ∎

Theorem 4 (Existence for infinite dimensions.) Suppose that in problem (73) the functional f is convex and continuous on S, and the subset S is closed, bounded, and convex. Then problem (73) has at least one minimizer.

Proof. The proof is the same as the proof of Theorem 3 if we replace continuity with the weak lower semicontinuity guaranteed by Proposition 20, compactness with the weak compactness guaranteed by Proposition 8, and then use Proposition 18. ∎

Definition 20. Consider $S \subset H$ and $f : S \to R$. We say that f has the *infinity property* in S if $\{x^n\} \subset S$ and $\|x^n\| \to \infty$ implies $f(x^n) \to +\infty$.

Theorem 5. Let S be a closed convex subset of H. If $f : S \to R$ is convex in S, continuous in S, and has the infinity property in S, then problem (73) has at least one solution.

Proof. Consider $x^0 \in S$. Let $\hat{S} = \{x \in S : f(x) \leq f(x^0)\}$. By the infinity property the set S is bounded. Moreover, by continuity $\hat{S}$ is closed. Finally, by convexity of f the set $\hat{S}$ is convex. The theorem now follows from Theorem 4, since a minimizer in $\hat{S}$ is a minimizer in S. ∎

Theorem 6. Let S be a closed and convex subset of H. If $f : S \to R$ is continuous and uniformly convex on S, then problem (73) has a unique solution.

Proof. Uniqueness follows from Theorem 2. We will show that f has the infinity property and the result will then follow from Theorem 5. There is no loss of generality in assuming $0 \in S$. For if not we could work with $S - y_0$ for some fixed $y_0 \in S$. By uniform convexity there exists $C > 0$, such that for $0 \le t \le 1$ and $x, y \in S$ we have

$$tf(x) + (1-t)f(y) - f(tx + (1-t)y) \ge Ct(1-t)\|x-y\|^2. \tag{81}$$

Choose $y = 0$ in (81). Then

$$tf(x) + (1-t)f(0) \ge f(tx) + Ct(1-t)\|x\|^2. \tag{82}$$

From (82) it follows that

$$\max(f(x), f(0)) \ge f(tx) + Ct(1-t)\|x\|^2 \tag{83}$$

Let $\hat{S} = \{x \in S : \|x\| \le 1\}$. Observe that $\hat{S}$ is closed, convex, and bounded; hence by Theorem 4 there exists α such that $f(x) \ge \alpha \ \forall x \in \hat{S}$. Given $x \in S$, such that $\|x\| \ge 1$ choose t_x, such that $\frac{1}{2} \le t_x\|x\| \le 1$. Then $0 < t_x \le 1$, and since $0 \in S$ we have $t_x x \in S$. From (83) with this x and t_x we obtain

$$\begin{aligned} \max(f(x), f(0)) &\ge \alpha + Ct_x(1-t_x)\|x\|^2 \\ &\ge \alpha + \frac{C}{2}(\|x\| - 1). \end{aligned} \tag{84}$$

Observe that (84) implies that as $\|x\| \to +\infty$ we must have $f(x) \to \infty$. ∎

The following theorem is extremely useful in applications.

Theorem 7. Let S be a closed convex subset of H. Suppose $f: H \to R$ is continuous in S, is twice Gâteaux differentiable in S, and the second Gâteaux variation is uniformly positive definite in S. Then problem (73) has a unique solution.

Proof. The theorem follows from Theorem 6 and Proposition 17. ∎

I.5. Lagrange Multiplier Necessity Conditions

Let us consider the special case of problem (73) in I.4, where

$$S = \{x \in H : g_i(x) = 0, i = 1, \dots, m\} \tag{85}$$

and g_i are functionals on H. Problem (73) can then be written as

$$\text{minimize } f(x); \quad \text{subject to } g_i(x) = 0, \qquad i = 1, \dots, m. \tag{86}$$

We assume that $f, g_i \forall i$ are continuously differentiable in H.

Definition 21. We say that $x \in H$ is a *regular point of the constraints* g_i, $i = 1, \ldots, m$ if

$$\{\nabla g_1(x), \ldots, \nabla g_m(x)\} \tag{87}$$

is a linearly independent set.

Theorem 8. Suppose that x^* solves problem (86) and also that x^* is a regular point of the constraints. Then there exist unique Lagrange multipliers $\lambda_1, \ldots, \lambda_m$, such that

$$\nabla f(x^*) + \lambda_1 \nabla g_1(x^*) + \cdots + \lambda_m \nabla g_m(x^*) = 0. \tag{88}$$

Proof. For the proof of this theorem we follow Luenberger [7]. Consider the operator $G: H \to R^{m+1}$ defined by

$$G(x) = (f(x), g_1(x), \ldots, g_m(x)). \tag{89}$$

Assume that there exists $Z \in H$ such that $\langle Z, \nabla f(x^*)\rangle \neq 0$ and $\langle Z, \nabla g_i(x^*)\rangle = 0$, $i = 1, \ldots, m$. If we write for $\eta \in H$

$$JG(x^*)(\eta) = [\langle \nabla f(x^*), \eta\rangle, \langle \nabla g_1(x^*), \eta\rangle, \ldots, \langle \nabla g_m(x^*), \eta\rangle], \tag{90}$$

then the operator $JG(x^*)$ takes H onto R^{m+1}, since $\{\nabla f(x^*), \nabla g_1(x^*), \ldots, \nabla g_m(x^*)\}$ is also a linearly independent set. By the inverse function theorem (see Luenberger [7]) for $y \in R^{m+1}$ near $G(x^*)$ there exists $x \in H$, such that $G(x) = y$. Choosing $y = (f(x^*) - \epsilon, 0, \ldots, 0)$ for any $\epsilon > 0$, we contradict the optimality of x^*. It follows that whenever $\langle Z, \nabla g_i(x^*)\rangle = 0$, $i = 1, \ldots, m$ we necessarily have that $\langle Z, \nabla f(x^*)\rangle$ is also zero. However, this implies by a well-known result (see, for example, Theorem 3.5–C of Taylor [13]) that $\nabla f(x^*)$ must be a linear combination of $\nabla g_1(x^*), \ldots, \nabla g_m(x^*)$. The coefficients in the linear combination are the Lagrange multipliers. By regularity these multipliers are unique. ∎

References

[1] Daniel, J. W. (1971). *The Approximate Minimization of Functionals.* Englewood Cliffs, New Jersey: Prentice–Hall.

[2] Dunford, H., and Schwartz, J. T. (1958). *Linear Operators, Part I.* New York: Interscience.

[3] Goffman, C., and Pedrick, G. (1965). *First Course in Functional Analysis.* Englewood Cliffs, New Jersey: Prentice–Hall.

[4] Goldstein, A. A. (1967). *Constructive Real Analysis.* New York: Harper and Row.

[5] Kantorovich, L. V., and Akilov, G. P. (1964). *Functional Analysis in Normed Spaces.* New York: Macmillan.

[6] Lions, J. L., and Magenes, E. (1972). *Non-homogeneous Boundary Value Problems and Applications.* Translated from the French by P. Kenneth. New York: Springer–Verlag.

[7] Luenberger, D. G. (1968). *Optimization by Vector Space Methods*. New York: Wiley.

[8] Ortega, J. M., and Rheinboldt, W. C. (1970). *Iterative Solution of Nonlinear Equations in Several Variables*. New York: Academic Press.

[9] Oden, J. T., and Reddy, J. N. (1976). *An Introduction to the Mathematical Theory of Finite Elements*. New York: Wiley.

[10] Rall, L. B. (1969). *Computational Solution of Nonlinear Operator Equations*. New York: Wiley.

[11] Royden, H. (1963). *Real Analysis*. New York: Macmillan.

[12] Tapia, R. A., "The differentiation and integration of nonlinear operators," in *Nonlinear Functional Analysis and Applications* (1971). Ed. L. B. Rall. New York: Academic Press.

[13] Taylor, A. E. (1958). *Introduction to Functional Analysis*. New York: Wiley.

APPENDIX

II
Numerical Solution of Constrained Optimization Problems

II.1. The Diagonalized Multiplier Method

In this section we will present a class of effective numerical algorithms for approximating solutions of constrained optimization problems called diagonalized quasi-Newton multiplier methods. These algorithms were suggested by Tapia in [2] and the reader desiring more detail is referred to that paper. The optimization problems we are concerned with in statistics require the variables to be nonnegative. In the following section of this appendix, we will describe a method for applying algorithms for problems with equality constraints to problems which also require the variables to be nonnegative. Consequently, we present the diagonalized quasi-Newton multiplier method for problems with equality constraints in this section. The reader is referred to Section II.2 for nonnegativity constraints and to [2] for the general inequality constraint.

We are interested in problem (I.86) where $H = R^n$. Specifically, consider the constrained optimization problem

(1) $$\text{minimize } f(x); \quad \text{subject to } g_i(x) = 0, \qquad i = 1, \ldots, m$$

where $f, g_i : R^n \to R$ have continuous second-order partial derivatives. In this case it is a simple matter to show that the gradient of f at x (see Definition I.17) is actually the vector of partial derivatives (some texts take this as the definition), i.e.,

(2) $$\nabla f(x) = \left(\frac{\partial f(x)}{\partial x_1}, \ldots, \frac{\partial f(x)}{\partial x_n} \right)^T.$$

We use the notation $\nabla^2 f(x)$ to denote the Hessian matrix of f at x, i.e.,

$$\nabla^2 f(x) = \left(\frac{\partial^2 f(x)}{\partial x_i \, \partial x_j}\right) \qquad i, j = 1, \ldots, n. \tag{3}$$

In (3) the notation (a_{ij}) is used to denote the matrix whose (i, j)-th element is given by a_{ij}, while in (2) the superscript T, as usual, denotes transposition. For the sake of simplicity, define $g: R^n \to R^m$ by

$$g(x) = (g_1(x), \ldots, g_m(x))^T. \tag{4}$$

It makes sense to denote the transpose of the Jacobian of g by $\nabla g(x)$, since the ith column of this matrix is merely the vector $\nabla g_i(x)$, i.e., the $n \times m$ matrix $\nabla g(x)$ can be written

$$\nabla g(x) = [\nabla g_1(x), \ldots, \nabla g_m(x)].$$

Consider the *augmented Lagrangian* for problem (1), i.e., $L: R^{n+m+1} \to R$, where

$$L(x, \lambda, C) = f(x) + \lambda^T g(x) + \frac{C}{2} g(x)^T g(x) \tag{5}$$

for $x \in R^n$, $\lambda \in R^m$, $C \geq 0$ and g given by (4). Since the augmented Lagrangian is a function of the three variables x, λ and C, we use the notation $\nabla_x L(x, \lambda, C)$ and $\nabla_x^2 L(x, \lambda, C)$ to denote differentiation with respect to x (λ and C are held fixed). In (5) λ can be thought of as an approximate Lagrange multiplier and C as a penalty constant.

Consider a quasi-Newton method for the unconstrained optimization problem

$$\underset{x}{\text{minimize}} \; L(x, \lambda, C) \tag{6}$$

with inverse Hessian update formula

$$\bar{H} = \mathscr{H}(x, \lambda, C, H), \tag{7}$$

i.e., the iteration

$$x^{k+1} = x^k - H^k \nabla_x L(x^k, \lambda, C) \tag{8}$$

and $\qquad k = 0, 1, 2, \ldots$

$$H^{k+1} = \mathscr{H}(x^k, \lambda, C, H^k),$$

where H^k is a conscious attempt to approximate, $\nabla_x^2 L(x^k, \lambda, C)^{-1}$, the inverse of the Hessian of the augmented Lagrangian. For more information on quasi-Newton methods for unconstrained optimization the reader is referred to Dennis and Moré [1].

The two main ingredients of the diagonalized quasi-Newton multiplier method are the unconstrained quasi-Newton method described in equations (6)–(8) and the *Lagrange Multiplier update formula.*

(9) $U(x, \lambda, C, H) = \lambda + (\nabla g(x)^T H \nabla g(x))^{-1}[g(x) - \nabla g(x)^T H \nabla_x L(x, \lambda, C)]$.

By a *diagonalized quasi-Newton multiplier method* for problem (1) we mean

Step 1: Determine x, λ, C and H

Step 2:

(10) $$\bar{\lambda} = U(x, \lambda, C, H)$$

(11) $$\bar{x} = x - H \nabla_x L(x, \bar{\lambda}, C)$$

Step 3:

(12) $$\bar{H} - \mathscr{H}(x, \bar{\lambda}, C, H)$$

Replace (x, λ, H) with $(\bar{x}, \bar{\lambda}, \bar{H})$ and go to Step 2.

Remark. Clearly, in implementing this algorithm a stopping criterion of the form

$$|\bar{x}_i - x_i| \le \epsilon, \qquad i = 1, \dots, n$$

and

$$|\bar{\lambda}_i - \lambda_i| \le \epsilon, \qquad i = 1, \dots, m$$

for some predetermined small number ϵ would have to be inserted between Step 2 and Step 3.

It is usual to choose C to be either zero or some small number, e.g., $C = .1$. Rules for determining optimal values of C are currently under investigation.

By the *diagonalized Newton multiplier method* we mean the choice (12) is given by

(13) $$\bar{H} = \nabla_x^2 L(\bar{x}, \bar{\lambda}, \bar{C})^{-1},$$

which corresponds to using Newton's method as the unconstrained minimization algorithm for problem (6).

Two standard choices for *diagonalized secant multiplier methods* arise by letting $\bar{H}$ in (12) be given by the following popular secant update formulas (see Dennis and Moré [1]): *Davidon–Fletcher–Powell* (DFP)

(14) $$\bar{H} = H + \frac{ss^T}{\langle s, y \rangle} - \frac{Hyy^T H}{\langle y, Hy \rangle};$$

Broyden–Fletcher–Goldfarb–Shanno (BFGS)

(15) $$\bar{H} = H + \frac{(s - Hy)s^T + s(s - Hy)^T}{\langle s, y \rangle} - \frac{\langle s - Hy, y \rangle ss^T}{\langle s, y \rangle^2},$$

with

$$s = \bar{x} - x$$

and

$$y = \nabla_x L(\bar{x}, \bar{\lambda}, C) - \nabla_x L(x, \bar{\lambda}, C).$$

Assume that $x^* \in R^n$ is a solution of problem (1), which is regular (see Definition I.20). Let $\lambda^* \in R^m$ be the Lagrange multipliers guaranteed by Theorem I.8. It can be shown that under very mild conditions there exists $\hat{C} \geq 0$, such that the Hessian matrix $\nabla_x L(x^*, \lambda^*, C)$ is positive definite for any $C \geq \hat{C}$. If we assume that $C \geq \hat{C}$, and that f and g_i in problem (1) have continuous third-order partial derivatives, then we can state the following theorem. Its proof and the above details can be found in [2].

Theorem 1. If (x, λ) is sufficiently near (x^*, λ^*) and H is sufficiently near $\nabla_x^2 L(x^*, \lambda^*, C)^{-1}$, then the diagonalized Newton multiplier method and the diagonalized BFGS and DFP secant multiplier methods are well defined and each generates a sequence (x^k, λ^k) which converges to (x^*, λ^*). Moreover, for the diagonalized Newton multiplier method the convergence is quadratic in the sense that there exists M, such that

$$\|(x^{k+1}, \lambda^{k+1}) - (x^*, \lambda^*)\| \leq M\|(x^k, \lambda^k) - (x^*, \lambda^*)\|^2, \qquad \forall k, \tag{16}$$

while for the two secant methods the convergence is superlinear in the sense that

$$\lim_{k \to \infty} \frac{\|(x^{k+1}, \lambda^{k+1}) - (x^*, \lambda^*)\|}{\|(x^k, \lambda^k) - (x^*, \lambda^*)\|} = 0. \tag{17}$$

Consider problem (1) with $n = 3$, $m = 1$

$$f(x) = (x_1 - 1)^2 + (x_1 - x_2)^2 + (x_2 - x_3)^4$$

and

$$g(x) = x_1(1 + (x_2)^2) + (x_3)^4 - 4 - 3\sqrt{2}.$$

This problem has a relative minimizer at

$$\lambda^* = (1.1049, 1.1967, 1.5353),$$

with associated Lagrange multiplier $\lambda^* = -.0107$. Initialize as follows:

$$\begin{aligned} x^0 &= (1.5, 1.5, 1.5) \\ \lambda^0 &= [\nabla g(x^0)^T \nabla g(x^0))^{-1}[g(x^0) - \nabla g(x^0)^t \nabla f(x^0)] \\ C^0 &= 0 \\ H^0 &= \nabla_x^2 L(x^0, \lambda^0, 0)^{-1}. \end{aligned} \tag{18}$$

Table II.1. Values of $\|x^k - x^*\|$

k	Newton	BFGS
1	$.31 \times 10^{-0}$	$.31 \times 10^{-0}$
2	$.14 \times 10^{-0}$	$.22 \times 10^{-0}$
3	$.35 \times 10^{-1}$	$.12 \times 10^{-0}$
4	$.29 \times 10^{-2}$	$.38 \times 10^{-1}$
5	$.19 \times 10^{-4}$	$.11 \times 10^{-1}$
6	$.90 \times 10^{-9}$	$.88 \times 10^{-3}$
7		$.29 \times 10^{-4}$
8		$.19 \times 10^{-5}$
9		$.44 \times 10^{-8}$

Note: The results of Table II.1 clearly indicate the quadratic and superlinear convergence phenomena.

Table II.1 compares the diagonalized Newton multiplier method with the diagonalized BFGS secant multiplier method for the initialization values given by (18).

II. 2. Optimization Problems with Nonnegativity Constraints

Let us consider problem (1) with addition of nonnegativity constraints. Specifically, we are interested in the constrained optimization problem

(19) minimize $f(x)$; subject to

$$g_i(x) = 0, \qquad i = 1, \ldots, m$$

and

$$x_i \geq 0, \qquad i = 1, \ldots, n,$$

where $f, g_i : R^n \to R$. Define the functions $\hat{f}, \hat{g}_i : R^n \to R$ by

$$\hat{f}(x_1, \ldots, x_n) = f(x_1^2, \ldots, x_n^2) \tag{20}$$

and

$$\hat{g}_i(x_1, \ldots, x_n) = g_i(x_1^2, \ldots, x_n^2), \tag{21}$$

and consider the equality constrained optimization problem

$$\text{minimize } \hat{f}(x); \quad \text{subject to } \hat{g}_i(x) = 0, \qquad i = 1, \ldots, m. \tag{22}$$

Proposition 1. If $x = (x_1, \ldots, x_n)$ solves problem (19), then $\sqrt{x} = (\sqrt{x_1}, \ldots, \sqrt{x_n})$ solves problem (22). Conversely, if $x = (x_1, \ldots, x_n)$ solves problem (22), then $x_1^2 = (x_1^2, \ldots, x_n^2)$ solves problem (19).

Proof. The proof follows almost by definition. ■

Remark. Many authors feel that solving problem (19) by first solving problem (22) is undesirable and not the optimal approach. In principle, we certainly agree; however, we maintain that whether or not this approach should be used depends on the algorithm employed, as well as the particular problem. Moreover, we have found this approach to be very satisfactory for the problems described in Chapter 5 when using the diagonalized multiplier methods of Section II.1.

References

[1] Dennis, J. E., Jr., and Moré, J. J. (1977). "Quasi-Newton methods: motivation and theory." *SIAM Review* 10: 46–89.

[2] Tapia, R. A. (1977). "Diagonalized multiplier methods and quasi-Newton methods for constrained optimization." *Journal of Optimization theory and Applications* 22: 135–94.

APPENDIX

III

Optimization Algorithms for Noisy Problems

III.1. The Nelder-Mead Approach

The algorithms which work so beautifully for deterministic problems frequently are disastrous for those involving noisy or simulated data. It is difficult to carry out decent estimation of derivatives if noise is present. If we use a standard unmodified Newton procedure, a "great leap forward" Newton step is very likely to put us in a spot where the function to be minimized, say, is woefully large.

One way to avoid this is to employ the robust optimization technique of Nelder and Mead [3]. It is significant that this algorithm was developed by statisticians who saw the necessity of creating an approach applicable in the real world rather than in an idealized one. Instead of moving directly to a supposed minimum, the Nelder-Mead uses the combination of a zig-zag stepping and an envelopment approach.

Let us suppose we are using the SIMEST approach to estimate $\chi^2(\Theta)$, where $\Theta = (\theta_1, \theta_2, \ldots, \theta_p)$. We would like to find the value of Θ which approximately minimizes $\chi^2(\Theta)$. To do this, we start out with $p+1$ trial values of $\Theta = (\theta_1, \theta_2, \ldots, \theta_p)$ of rank p. We evaluate $\chi^2(\Theta)$ at each of these values, identifying the three values where $\chi^2(\Theta)$ is the minimum (B), the maximum (W), and the second largest ($2W$). Averaging all the $\Theta's$ except W, we obtain the centrum C. In Figure III.1, we sketch the algorithm for the case where $p = 2$. The flowchart in Figure III.2, however, works for any $p \geq 2$.

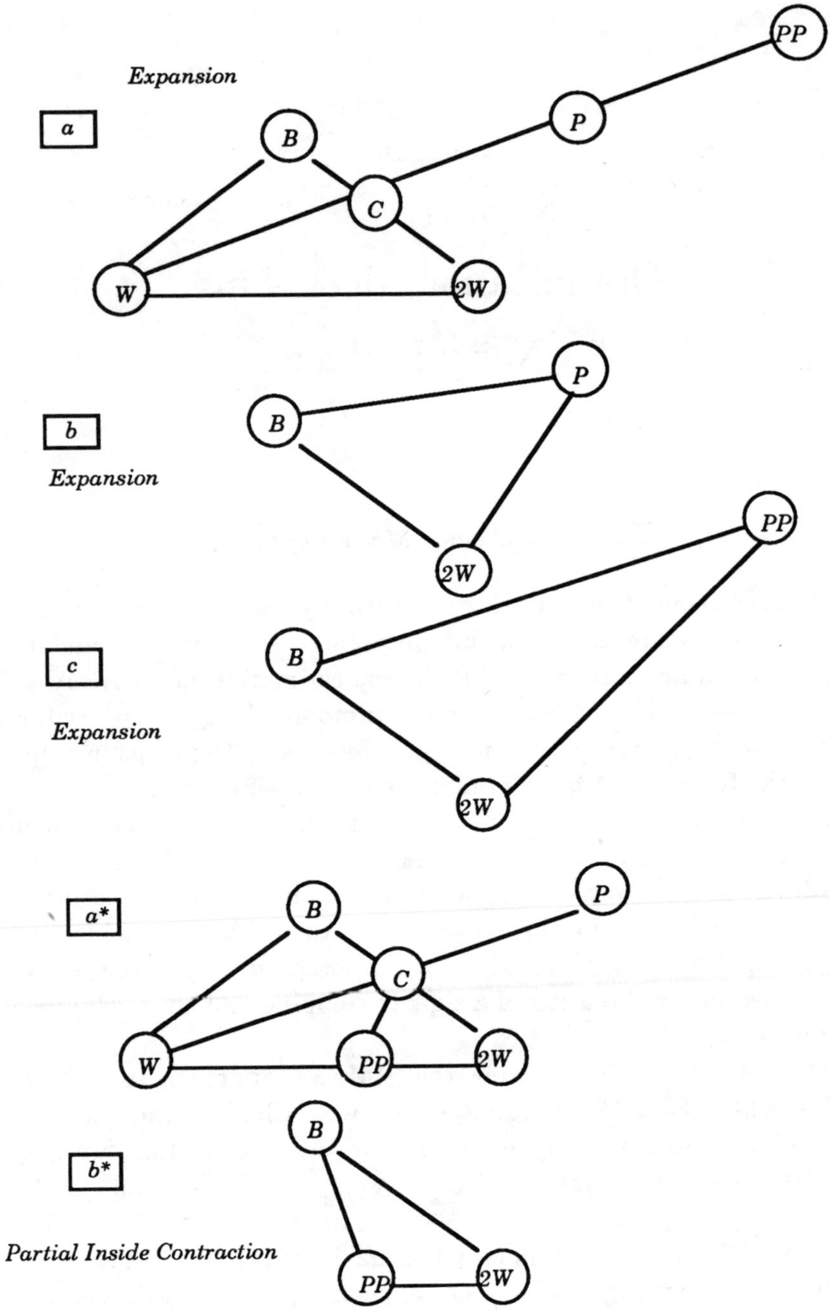
Expansion
a
PP
P
B
C
W
2W
b
Expansion
B
P
2W
c
Expansion
B
PP
2W
a*
B
P
C
W
PP
2W
b*
Partial Inside Contraction
B
PP
2W

Figure III.1.

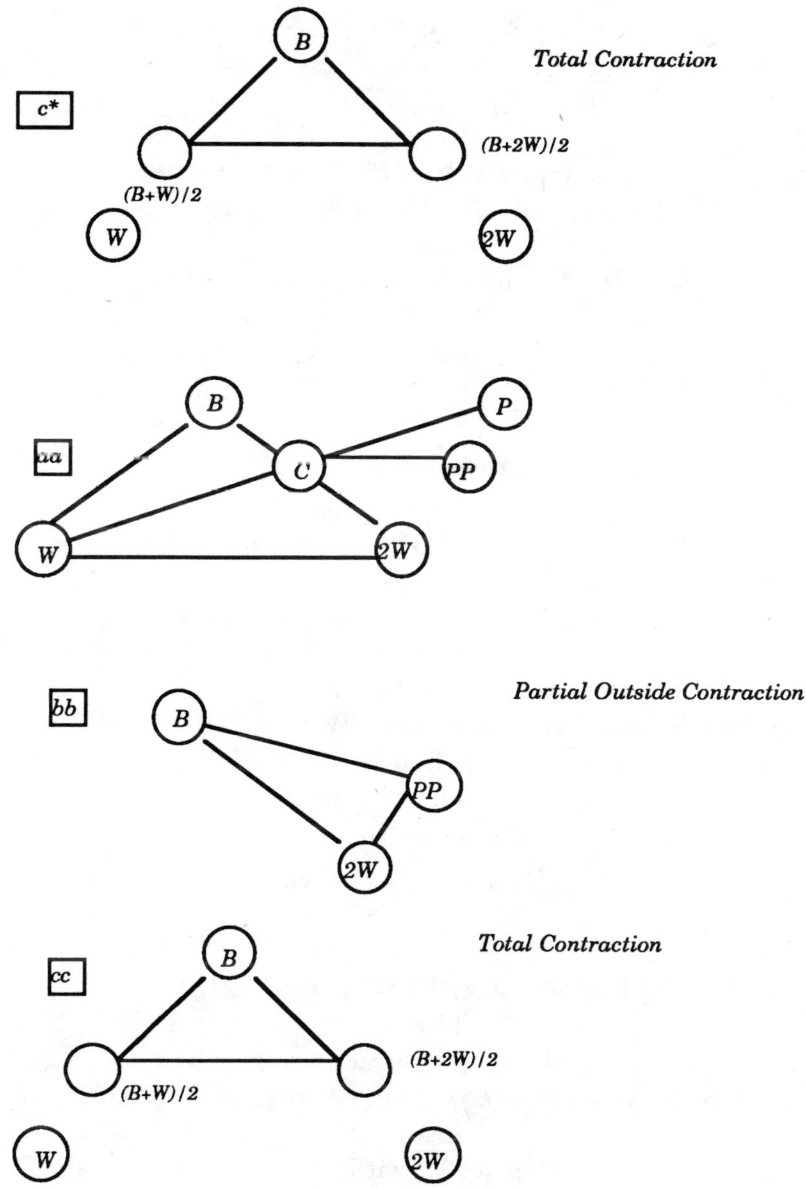

Figure III.1 (continued).

Expansion

$P = C + \gamma_R(C - W)$ where typically $\gamma_R = \gamma_E = 1$
If $\chi^2(P) < \chi^2(B)$, then
[a] $PP = P + \gamma_E(C - W)$
If $\chi^2(PP) < \chi^2(P)$, then
[c] replace W with PP as new vertex
Else
[b] Accept P as new vertex
End If
Else
[b] If $\chi^2(P) < \chi^2(2W)$, then accept P as new vertex
Else

Contraction

If $\chi^2(W) < \chi^2(P)$, then
[a*] $PP = C + \gamma_C(W - B)$ (typically $\gamma_C = 1/2$)
If $\chi^2(PP) < \chi^2(W)$, then
[b*] Replace W with PP as new vertex
Else
[c*] (total contraction)
Replace W with $(W + B)/2$ and $2W$ with $(2W + B)/2$
[c*] End If
Else

Contraction

If $\chi^2(2W) < \chi^2(P)$ then
[aa] $PP = C + \gamma_C(P - B)$
If $\chi^2(PP) < \chi^2(P)$, then
[bb] Replace W with PP as new vertex
Else
[cc] (total contraction)
Replace W with $(W + B)/2$ and $2W$ with $(2W + B)/2$
Else
Replace W with P
End If
End If

Figure III.2.

The advantages of the Nelder-Mead algorithm include its simplicity. Once learned, it is hard to forget, and can generally be programmed in a few lines. It is robust to errors and noise in the data. Constraints are easily added to the program. Disadvantages include relative slowness and a difficulty in efficient parallelization (due to the fact that it changes one point in the simplex at a time). This latter difficulty has been addressed by the multidirectional modification of Torczon [4].

III.2. An Algorithm Based on the Rotatable Designs of Box and Hunter

There is an easy way to modify the evaluation of the criterion function at various points in the parameter space so that the standard procedures of optimization theory can be utilized. To carry out this approach, we evaluate the criterion function at several points in the parameter space and fit a smooth parametric function in such a way as to minimize, say, the least squares fit of the parametric function to the pointwise evaluations of the criterion function. For example, we can approximate the goodness of fit between the data and the simulated data via the local quadratic model:

$$J_3(\Theta) = \beta_0 + \sum_{i=1}^{p} \beta_i \Theta_i + \sum_{i=1}^{p} \sum_{j=i}^{p} \beta_{ij} \Theta_i \Theta_j + \epsilon. \tag{1}$$

We shall assume that we are standing at the current best guess for that value of Θ, say Θ_n which gives the minimum value for the goodness of fit statistic. We have carried out several numerical experiments to evaluate χ^2 for Θ values around this current best guess. Having fit the coefficients to the data via least squares, i.e.,

$$\beta_{n+1} = (\Theta_n^T \Theta_n)^{-1} \Theta_n^T \chi_n^2, \tag{2}$$

we can move to our next best guess by taking the partial derivatives of $J_3(\Theta_n)$ and setting them equal to zero.

A means of selecting the experimental Θ points around the last best guess was suggested, in the context of chemical engineering experiments, by Box and Hunter [2]. To oversimplify the Box-Hunter technique, we will first take our last fit and linearly transform the $\Theta's$ to obtain

$$J_3(\Theta) = \beta_0 + \sum_{i=1}^{p} \tilde{\beta}_i \tilde{\Theta_i}^2. \tag{3}$$

Then, we pick the design unit τ depending on

$$(4) \qquad \|\Theta_{\text{opt},n+1} - \Theta_{\text{opt},n}\| = \sqrt{\frac{1}{p}\sum_{i=1}^{p}(\theta_{\text{opt},i,n+1} - \theta_{\text{opt},i,n})^2}.$$

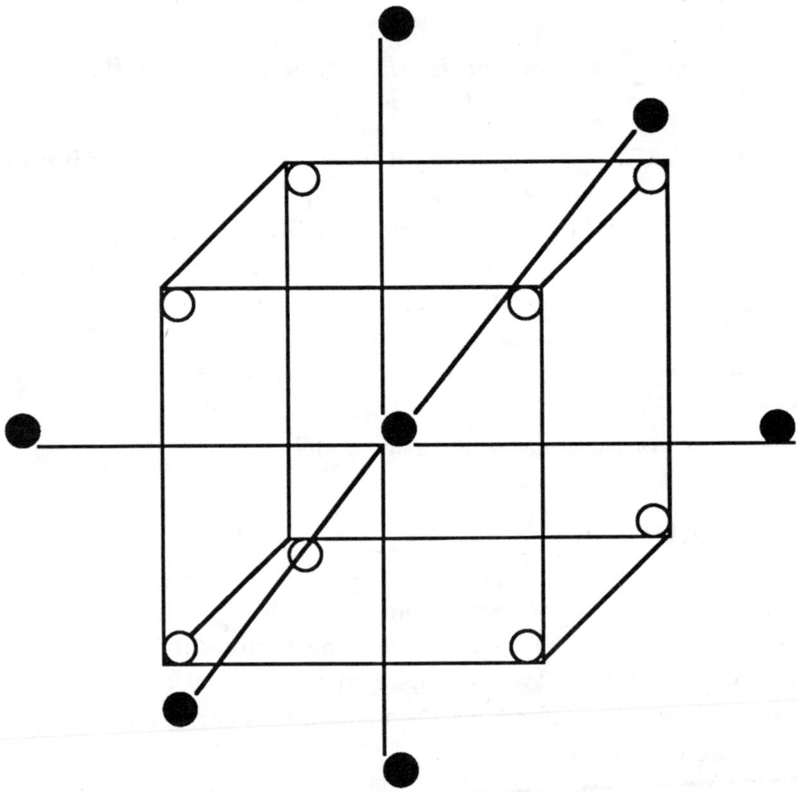

Figure III.3 (continued).

The basic idea of Box and Hunter was to select design points in such a way that the variance would be constant on hyperspheres going out from the current best guess for the optimum. They put points on the trivial hypersphere at the origin, at the vertices of a hypercube inscribed by a larger hypersphere, and points at the face centers of a hypercube inscribed by a still larger hypersphere. The Box-Hunter rotatable designs were created in the light of designs where each experiment might cost thousands of dollars. In the case of simulation based estimation, our experimental (i.e., simulation) costs are relatively trivial. Consequently, we may find it useful to come up with different design strategies. Nevertheless, the Box-Hunter formulation has the advantage of easy parallelization. We discuss their strategy below.

Having picked a unit as a function of τ, in Θ space, we perform experiments at the 2^p vertices of the inner hypercube, i.e., at points of the type $(\pm1, \pm1, \ldots, \pm1)$. Also, we perform experiments at the $2p$ "star" points at the centers of faces of a larger hypercube, namely points of sort $(\pm\alpha, 0, \ldots, 0), (0, \pm\alpha, 0, \ldots, 0), \ldots, (0, 0, \ldots, \pm\alpha)$. And, finally, we perform experiments at our current estimate for Θ_{optimum}. To illustrate this design more clearly, we show it in Figure III.3 for the case where $p = 3$.

For dimensions through $p = 7$, reasonable results can be obtained by taking experiments at roughly equal ratios at the hypercube vertices and the star points, with twice as many points at the center of the design as at any star or hypercube vertex. In such a case, α is computed as follows:

$$\alpha = (2k)^{1/4}. \tag{5}$$

In the case of a three dimensional Θ, if we have a parallel processor with 16 cpu's, we put eight of these to work on the vertices, six on the star points, and two at the center of the design. Since we are carrying out simulations, essentially no "handshaking" between the cpu's is required. We simply dump our simulation results to the central processor for evaluation of the least squares estimation when any one of them has reached an appropriate number of simulations, say 1,000. For dimension $p = 5$, we have 32 vertices, plus 10 star points, plus 2 center points with which to contend. Few parallel processors will have precisely 44 cpu's. A certain amount of swapping in using the cpu's is appropriate. For example, if we have a 16 cpu unit, we can assign 16 of the design points to the cpu's and then as the requisite number of simulations is reached on a given cpu, we can shift it to another design point, etc.

As has been mentioned, the Box-Hunter design strategy should probably be modified in the case of simulation. One such strategy would be, standing at the current best guess for the minimizer of the goodness of fit, Θ_n, to sample from a normal distribution $N(\Theta_n, f(\tau)I)$, where $f(.)$ is an appropriate expansion-contraction operator. Since we realize that our quadratic fit works only locally, we may wish to discard design points too far from the current best guess of the optimum, say for $||\Theta - \Theta_n|| \geq 3\tau$. One advantage of replacing the classical design of Box and Hunter by one more suitable to cheap experiments is that it enables us to use any number of cpu's in a parallel processor efficiently. We simply put the same sampling rule on each cpu and pool our results when a lower bound threshold for the number of simulations has been crossed on one of them.

The disadvantages of the modified Box-Hunter approach when contrasted to the Nelder-Mead algorithm is the relative robustness of the latter. The major advantage is that, in smoothing away simulation noise, it puts us back in a situation where we can achieve the greater efficiencies of Newtonian algorithms.

References

[1] Box,G.E.P. and Draper, N.R. (1987). *Empirical Model-Building and Response Surfaces.* New York: John Wiley and Sons.

[2] Box, G.E.P. and Hunter, J.S. (1957)."Multifactor experimental designs for exploring response surfaces." *Annals of Mathematical Statistics* 28: 195-241.

[3] Nelder, J.A. and Mead, R. (1965). "A simplex method for function minimization." *Computational Journal* 7: 308-313.

[4] Torczon, V.J. (1989). *Multi-Directional Search: A Direct Search Algorithm for Parallel Machines.* Doctoral Dissertation, Rice University.

APPENDIX

IV
A Brief Primer in Simulation

IV.1. Introduction

There are many views as to what constitutes simulation. To the statistician, simulation generally involves randomness as a key component. Engineers, on the other hand, tend to consider simulation as a deterministic process. If, for example, an engineer wishes to simulate tertiary recovery from an oil field, he will probably program a finite element approximation to a system of partial differential equations. If a statistician attacked the same problem, he might use a random walk algorithm for approximating the pointwise solution to the system of differential equations.

In the broadest sense, we may regard the simulation of a process as the examination of any process simpler than that under consideration. The examination will frequently involve a "mathematical model," i.e., an oversimplified mathematical analogue of the real world situation of interest. The related simpler process might be very close to the more complex process of interest. For example, we might simulate the success of a proposed chain of fifty grocery stores by actually building a single store and seeing how it progressed. At a far different level of abstraction, we might attempt to describe the functioning of the chain by writing down a series of equations to approximate the functioning of each store, together with other equations to approximate the local economies and so on. Clearly, it is this second level of abstraction which will be of more interest to us.

It is to be noted that the major component of simulation is neither stochasticity nor determinism, but rather analogy. Needless to say, our visions of reality are always other than reality itself. When we see a forest, it is really a biochemical reaction in our minds which produces something to which we relate the notion of "forest." When we talk of the "real world," we really talk of perceptions of that world which are clearly other than that world but are (hopefully) in strong correlation with it. So, in a very real sense, analogy is "hardwired" into the human cognitive system. But to carry analogy beyond that which it is instinctive to do involves a learning process more associated with some cultures than with others. And it is the ability of the human intellect to construct analogies which makes modern science and technology a possibility. Interestingly, like so many other important advances in human thought, the flowering of reasoning by analogy started with Socrates, Plato and Aristotle. Analogy is so much a part of Western thinking that we tend to take it for granted. In simulation we attempt to enhance our abilities to analogize to a level consistent with the tools at our disposal.

The modern digital computer, at least at the present time, is not particularly apt at analogue formulations. However, the rapid digital computing power of the computer has enormous power as a device complementary to the human ability to reason by analogy. Naturally, during the majority of the scientific epoch, the digital computer simply did not exist. Accordingly, it is not surprising that the mindset of most scientists is still oriented to methodologies in which the computer does not play an intimate part.

The inventor of the digital computer, Johann von Neumann, created the device in order to perform something like random quadrature rather than to change fundamentally the precomputer methodology of modeling and analogy. And indeed the utilization of the computer by von Neumann was oriented as a fast calculator with a large memory. This kind of mindset, which is a carryover of modeling techniques in the precomputer age, led to something rather different from what I call simulation, namely the "Monte Carlo method." According to this methodology we essentially start to work on the abstraction of a process (through differential equations and the like) as though we had no computer. Then, when we find difficulties in obtaining a "closed form solution," we use the computer as a means of facilitating pointwise function evaluation. There is no clear demarcation between the Monte Carlo method on the one hand and simulation on the other. However, as we argue in Chapter 8, a more full utilization of the computer frequently enables us to dispense altogether with abstraction strategies suitable to a precomputer age.

As an example of the fundamental change that the modern digital computer makes in the modeling process, let us consider a situation where we wish to examine particles emanating from a source in the interior of an irregular and heterogeneous medium. The particles interact with the medium by collisions with it and travel in an essentially random fashion. The classical approach for a regular and symmetrically heterogeneous medium would be to model the aggregate process by looking at differential equations which track the average behavior of the micro-axioms governing the progress of the particles. For an irregular and nonsymmetrically heterogeneous medium, the Monte Carlo investigator would attempt to use random walk simulations of these differential equations with pointwise change effects for the medium. In other words, the Monte Carlo approach would be to start with a precomputer age methodology and use the computer as a means for random walk implementations of that methodology.

If we wish to use the power of the digital computer more fully, then we can go immediately from the micro-axioms to random tracking of a large number of the particles. It is true that even with current computer technology, we will still not be in a position to deal with 10^{16} particles. However, a simulation dealing with 10^5 particles is both manageable and probably sufficient to make very reasonable conjectures about the aggregate of the 10^{16} particle system. In distinguishing between "simulation" and the Monte Carlo method, we will, in the former, be attempting the modeling in the light of the existence of a computer which may take us very close indeed to a precise emulation of the process under consideration. Clearly, then, "simulation" is a moving target. The faster the computer and the larger the storage, the closer we can come to a true "simulation."

IV.2. Random Number Generation

Many simulations will involve some aspect of randomness. Theorem 1 shows that, at least at the one dimensional level, randomness can be dealt with if only we can find a random number generator from the uniform distribution on the unit interval $U(0,1)$.

Theorem 1. Let X be a continuous random variable with distribution function $F(.)$, i.e., let $F(x) = P(X \leq x)$. Consider the random variable $Y = F(x)$. Let the distribution function of Y be given by $G(y) = P(Y \leq y)$. Then Y is distributed as $U(0,1)$.
Proof.

$$G(y) = P(Y \leq y) = P(F(x) \leq y) = P(x \leq F^{-1}(y)) \tag{1}$$

but $P(x \leq F^{-1}(y))$ is simply the probability that X is less than or equal to that value of X than which X is less y of the time. Now this is precisely the distribution function of the uniform distribution on the unit interval. This proves the theorem. ■

For the simulator, Theorem 1 has importance rivaling that of the Central Limit Theorem, for it says that all that is required to obtain a satisfactory random number generator for any continuous one dimensional random variable, for which we have a means of inverting the distribution function, is a good $U(0,1)$ generator. This is conceptually quite easy. For example, we might have an electrical oscillator in which a wave front travels at essentially the speed of light in a linear medium callibrated from 0 to 1 in increments of, say, 10^{-10}. Then, we simply sense the generator at random times which an observer will pick. Aside from the obvious fact that such a procedure would be prohibitively costly, there seems to be a real problem with paying the observer and then reading the numbers into the computer. Of course, once it were done, we could use table look-up forever, being sure never to repeat a sequence of numbers once used.

Realizing the necessity for a generator which might be employed by the computer itself, without the necessity of human intervention except, perhaps, at the start of the generation process, von Neumann developed such a scheme. He dubbed the generator he developed the "midsquare" method. To carry out such a procedure, we take a number of, say, four digits, square it, and then take the middle four digits, which are used for the generator for the next step. If using base ten numbering, we then simply put a decimal before the first of the four digits. Let us show how this works with the simple example below.

We will start with $X_0 = 3333$. Squaring this, we obtain $11\underline{1088}89$. Taking the middle four digits, we have $X_1 = 1088$. Squaring 1088 we have 1183744. This gives $X_2 = 8374$, and so on. If we are using base 10, then this gives us the string of supposed $U(0,1)$ random variates.

The midsquare method is highly dependent on the starting value. Depending on the seed X_0 the generator may be terrible or satisfactory. Once we obtain a small value such as 0002, we will be stuck in a rut of small values until we climb out of the well. Moreover, as soon as we obtain 0, we have to obtain a new starter, since 0 is not changed by the midsquare operation.

Examinations of the midsquare method may be rather complicated mathematically if we are to determine, for example, the *cycle length*, the length of the string at which it starts to repeat itself. Some have opined that since this generator was used in rather crucial computations concerning nuclear reactions, civilization is fortunate that no catastrophe was encountered as a result of its use. As a matter of fact, for reasonable selections of "seeds," i.e., starting values, the procedure can be quite satisfactory for most applications. It is, however, the specificity of behavior based on starting values which makes the method rather unpopular.

The midsquare method embodies more generally many of the attributes of "random number generators" on the digital computer. First of all, it is to be noted that such generators are not really random, since when we see part of the string, given the particular algorithm for a generator, we can produce the rest of the string. We might decide to introduce a kind of randomness by, say, by using the time on a computer clock as a seed value. However, it is fairly clear that we need to obtain generators realizing that the very nature of realistic generation of "random numbers" on the digital computer will produce problems. Attempting to wish these problems away by introducing factors which are "random" simply because we do not know what they are is probably a bad idea. Knuth [2] gives an example of an extremely complex generator of this sort which would appear to give very random-looking strings of random numbers, but in fact easily gets into the rut of reproducing the seed value forever.

The maxim of dealing with the devil we know dominates the practical creation of random number generators. We need to obtain algorithms which are conceptually simple so that we can readily discern their shortcomings. The most widely used of the current random number generation algorithms is the *congruential random number generator*. The apparent inventor of the congruential random number generation scheme is D.H. Lehmer who introduced the algorithm in 1948 [3].

First of all, we note how incredibly simple the scheme is. Starting with a "seed" X_0 we build up the succession of "pseudo-random" numbers via the rule

$$X_{n+1} = aX_n + b(\text{mod } m). \tag{2}$$

One of the considerations given with such a scheme is the length of the string length of pseudo-random numbers before we have the first repeat. Clearly, by the very nature of the algorithm, with its one step memory, once we have a repeat, the new sequence will exactly repeat itself. If our only concern is the length of the cycle before a repeat, a very easy fix is available. Choosing $a = b = 1$ and $X_0 = 0$ we have for any m,

$$\begin{aligned} X_1 &= 1 \\ X_2 &= 2 \\ X_3 &= 3 \\ \ldots & \quad \ldots \\ X_m &= m \\ X_{m+1} &= 1. \end{aligned} \tag{3}$$

Seemingly, then, we have achieved something really spectacular, for we have a string which does not repeat itself until we get to an arbitrary length of m. Of course, the string bears little resemblance to a random string, since it marches straight up to m and then collapses back to 1. We have to come up with a generator such that any substring of any length appears to be random. Let us consider what happens when we choose $m = 90$, $a = 5$, and $b = 0$. Then, if we start with $X_0 = 7$, we have

$$\begin{aligned} X_1 &= 35 \\ X_2 &= 85 \\ X_3 &= 65 \\ X_4 &= 55 \end{aligned} \tag{4}$$

$$
\begin{aligned}
X_5 &= 5 \\
X_6 &= 25 \\
X_7 &= 35.
\end{aligned}
$$

It could be argued that this string appears random, but clearly its cycle length of six is far too short for most purposes. Our task shall be to find a long cycle length which also appears random. The rather simple proof of the following theorem is due to Morgan [3].

Theorem 2. Let $X_{n+1} = aX_n + b(\text{mod } m)$.
Let $m = 2^k$, $a = 4c + 1$, b be odd.
Then the string of pseudo-random numbers so generated has cycle length $m = 2^k$.
Proof.
Let $Y_{n+1} = aY_n + b$.
Without loss of generality, we can take $X_0 = Y_0 = 0$.
Then

$$
\begin{aligned}
Y_1 &= aY_0 + b = b \\
Y_2 &= ab + b = b(1 + a) \\
Y_3 &= aY_2 + b = ab(a + a) + b = b(a + a + a^2) \\
&\dots \\
Y_n &= b(1 + a + a^2 + a^3 + \ldots + a^{n-1}).
\end{aligned}
\tag{5}
$$

We observe that $X_i = Y_i - h_1 2^k$.
Now suppose that $X_i = X_j$ for $i > j$.
We wish to show that $i - j \geq 2^k$.
Now, $Y_i - Y_j = b(a^j + a^{j+1} + \ldots + a^{i-1})$.
If $X_i = X_j$, then
$ba^j W_{i-j} = ba^j(1 + a + a^2 + \ldots + a^{i-j-1}) = h_2 2^k$
where $W_n = 1 + a + a^2 + \ldots + a^{n-1}$ for $n \geq 1$.
In order to prove the theorem, we must show that W_{i-j} cannot equal an integer multiple of 2^k if $i - j < 2^k$, i.e.,
$W_{i-j} \neq h_3 2^k$ for $i - j < 2^k$.
We shall suppose first of all that <u>*i - j is odd.*</u>
Then $i - j = 2t + 1$ for $t \geq 0$.
(This is the place we use the fact that $a = 4c + 1$.)

$W_{2t} = \frac{1-a^{2t}}{1-a} = \frac{(1+4c)^{2t}-1}{4c} = [(1+4c)^t - 1][(1+4c)^t + 1]/(4c)$
$= [(1+4c)^t + 1][\sum_{i=1}^{t}(4c)^{i-1}\begin{pmatrix} t \\ i \end{pmatrix}]$

But $1 + (1+4c)^t = 2 + 4c\sum_{i=1}^{t}(4c)^{i-1}\begin{pmatrix} t \\ i \end{pmatrix}$

So, $W_{2t+1} = W_{2t} + a^{2t+1}$ is odd, since a is odd.
Hence, if $i - j$ is odd, then $W_{i-j} \neq h_3 2^k$, since W_{i-j} is odd.
Next we wish to consider the case where *i - j is even.*
If $i - j$ is even, then there exists an s such that $i - j = \alpha 2^s$, for some odd integer α.

$$
\begin{aligned}
(6) \quad W_{i-j} &= W_{\alpha 2^s} \\
&= 1 + a + \ldots + a^{\alpha 2^{s-1}-1} + a^{\alpha 2^{s-1}} + \ldots + a^{\alpha 2^s - 1} \\
&= W_{\alpha 2^{s-1}} + a^{\alpha 2^{s-1}}(1 + a + a^2 + \ldots + a^{\alpha 2^{s-1}-1}) \\
&= W_{\alpha 2^{s-1}} + a^{\alpha 2^{s-1}} W_{\alpha 2^{s-1}} \\
&= W_{\alpha 2^{s-1}} + a^{\alpha 2^{s-1}} W_{\alpha 2^{s-1}} \\
&= W_{\alpha 2^{s-1}}[1 + a^{\alpha 2^{s-1}}].
\end{aligned}
$$

Similarly, we have

$$(7) \qquad W_{\alpha 2^{s-1}} = W_{\alpha 2^{s-2}}[1 + a^{\alpha 2^{s-2}}].$$

Continuing the decomposition,

$$
\begin{aligned}
(8) \qquad W_{i-j} &= W_{\alpha 2^{s-2}}[1 + a^{\alpha 2^{s-2}}][1 + a^{\alpha 2^{s-1}}] \\
&= W_{\alpha}[1 + a^{\alpha}][1 + a^{\alpha 2}] \ldots [1 + a^{\alpha 2^{s-1}}].
\end{aligned}
$$

Recalling that

$$(9) \qquad 1 + (4c+1)^k = 2 + 4c\sum_{i=1}^{k}\begin{pmatrix} k \\ i \end{pmatrix}(4c)^i$$

we see that $W_{i-j} = W_\alpha \gamma 2^s$.
Note that we have shown in the first part of the proof that for α odd, W_α is also odd. Furthermore, we note that the product of terms like $1 + even$ is also odd, so γ is odd. Thus if $W_{i-j} = h_2 2^k$, we must have $s = k$. ■

The following more general theorem is stated without proof.

Theorem 3. Let $X_{n+1} = aX_n + b \pmod{m}$. Then the cycle of the generator is m if and only if
(i) b and m have no common factor other than 1,
(ii) $(a-1)$ is a multiple of every prime number that divides m,
(iii) $(a-1)$ is a multiple of 4 if m is a multiple of 4.

So far, we have seen how to construct a sequence of arbitrarily long cycle length. Obviously, if we wish our numbers to lie on the unit interval, we will simply divide the sequence members by m. Thus, the j'th random number would be X_j/m. It would appear that there remains the design problem of selecting a and b to give the generator seemingly random results. In order to do this, we need to examine congruential generators in the light of appropriate tests of randomness. We shall address this issue in the next section.

IV.3. Testing for "Randomness"

In a sense, it is bizarre that we ask whether a clearly deterministic sequence, such as one generated by a "congruential random number generator" is random. Many investigators have expressed amazement at early "primitive" schemes, such as those of von Neumann, devised for random number generation. As a matter of fact, I am personally unaware of any tragedy or near tragedy caused by any of these primitive schemes (although I have experienced catastrophes myself when I inadvertently did something putatively wrong, such as using, repetitively, the same seed to start runs of, supposedly different, congruential random numbers). The fact is, of course, that the issue is always decided on the basis of what are the minimal requirements for a sequence of "random numbers."

For many applications, it will simply be sufficient if we can show that for any small volume, say ϵ of a hypercube of, say, unit volume, for a large number of generated random numbers, say N, the number of these falling in the volume will be approximately ϵN. So a goodness of fit statistic procedure may be a sufficient test. On the other hand, for some applications, we need to look at any glaringly nonrandom behavior of the generator. Such behavior is observed for the once popular RANDU generator of IBM.

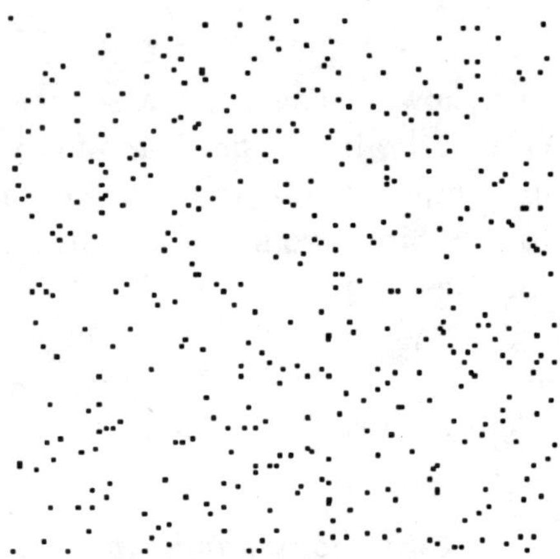

Figure IV.1.

$$x_{i+1} = (2^{16} + 3)x_i \bmod(2^{31}). \tag{10}$$

A little work reveals the following relation [3]:

$$x_{i+1} = (6x_i - 9x_{i-1}) \bmod(2^{31}). \tag{11}$$

Such a relationship between successive triples is probably not disastrous for most applications, but it looks very bad when graphed from the proper view. Using D^2 Software's MacSpin, we observe two views of this generator, one, in Figure IV.1 seemingly random, the other in Figure IV.2 very much not.

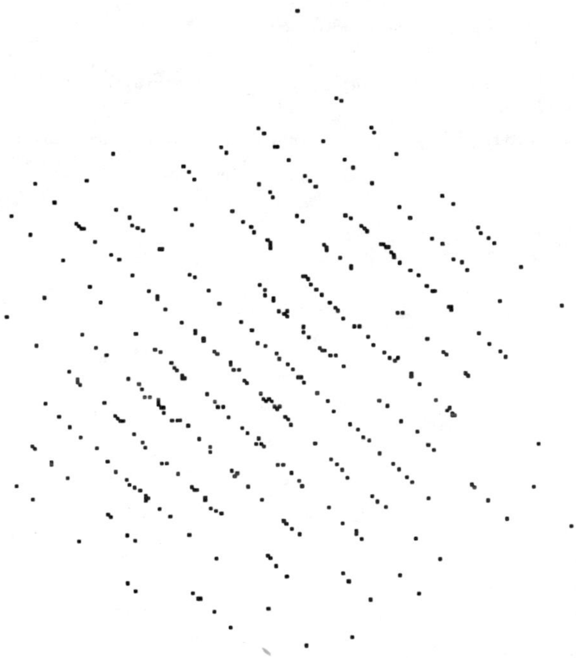

Figure IV.2.

Knuth [2] suggests as a means of minimizing the difficulties associated with such latticing that one choose values of a, b, and m to give large values of

$$C_n = \frac{\pi^{n/2}\nu_n^n}{(n/2)!m}, \tag{12}$$

where the wave number ν_n is given as the solution to

$$\nu_n = \min \sqrt{s_1^2 + s_2^2 + \ldots + s_n^2} \tag{13}$$

subject to

$$s_1 + s_2a + s_3a^2 + \ldots + s_na^{n-1} = 0(\bmod m). \tag{14}$$

Knuth suggests that we restrict ourselves to generators with $C_n \geq 1$ for $n = 2$, 3 and 4. One such random generator, where $a = 5^{15}$, $m = 2^{35}$, $b = 0$, has $C_2 = 2.02$, $C_3 = 4.12$, and $C_4 = 4.0$.

References

[1] Kennedy, W.J. and Gentle, J.E. (1980). *Statistical Computing.* New York: Marcel Dekker.

[2] Knuth, D.E. (1981). *The Art of Computing, v.2, Seminumerical Algorithms.* Reading: Addison-Wesley.

[3] Morgan, B.J.T. (1984). *Elements of Simulation*, London: Chapman and Hall.

Index